Published for
OXFORD INTERNATIONAL AQA EXAMINATIONS

International A2 Level
MATHEMATICS
Pure and Statistics

Sue Chandler
Janet Crawshaw
Joan Chambers

T0177905

Great Clarendon Street, Oxford, OX2 6DP, United Kingdom

Oxford University Press is a department of the University of Oxford. It furthers the University's objective of excellence in research, scholarship, and education by publishing worldwide. Oxford is a registered trade mark of Oxford University Press in the UK and in certain other countries

British Library Cataloguing in Publication Data
Data available

978-0-19-837597-5

13

Paper used in the production of this book is a natural, recyclable product made from wood grown in sustainable forests. The manufacturing process conforms to the environmental regulations of the country of origin.

Printed and bound by CPI Group (UK) Ltd, Croydon, CR0 4YY

Acknowledgements
The publishers would like to thank the following for permissions to use their photographs:

Cover: Colin Anderson/Getty Images.

Header: Shutterstock.

Although we have made every effort to trace and contact all copyright holders before publication this has not been possible in all cases. If notified, the publisher will rectify any errors or omissions at the earliest opportunity.

Links to third party websites are provided by Oxford in good faith and for information only. Oxford disclaims any responsibility for the materials contained in any third party website referenced in this work.

AQA material is reproduced by permission of AQA.

Contents

iv

About this book

This book has been specially created for the Oxford AQA International A2 Level Mathematics examination (9660).

It has been written by an experienced team of teachers, consultants and examiners and is designed to help you obtain the best possible grade in your maths qualification.

In each chapter the lessons are organised in a logical order to help you to progress through each topic. At the start of each chapter you can see an Introduction, to show you how you will use the knowledge in this chapter, a Recap of what prior knowledge you will need to recall and a clear list of the Objectives that you will fulfil by the end of the chapter.

The Note boxes give you help and support as you work through the examples and exercises.

Clear, worked examples show you how to tackle each question and the steps needed to reach the answer.

Key points are in bold and the chapter colour to make it clear that this information is important.

Exercises allow you to apply the skills that you have learned, and give the opportunity to practise your reasoning and problem solving abilities.

At the end of a chapter you will find a summary of what you have learned, together with a review section that allows you to test your fluency in the basic skills. Finally there is an Assessment section where you can practise exam-style questions.

At the end of the book you will find a comprehensive glossary of key phrases and terms and a full set of answers to all of the exercises.

We wish you well with your studies and hope that you enjoy this course and achieve exam success.

1 Functions

Introduction
This chapter extends the work on functions introduced at AS-level and gives various methods for expressing algebraic fractions in simpler forms. These methods are needed later in the course for integrating and differentiating fractions.

Recap
You will need to remember...
- The properties and the shapes of the graphs of linear, quadratic, exponential and trigonometric functions.
- The effect of simple transformations on a graph, including translations, one-way stretches and reflections in the x- and y-axes.
- The Cartesian equation of a curve gives the relationship between the x- and y-coordinates of points on the curve.
- How to complete the square for a quadratic function.
- How to factorise quadratic expressions.
- The remainder theorem.

Objectives
By the end of this chapter, you should know how to...
- Define a function, range of a function and domain of a function.
- Introduce inverse functions, composite functions and modulus functions.
- Use combinations of transformations to help to sketch graphs.
- Simplify an algebraic fraction by dividing by common factors.
- Decompose algebraic fractions into simpler fractions.

1.1 Functions

When you substitute any number for x in the expression $x^2 - 2x$, you get a single answer.

For example when $x = 3$, $x^2 - 2x = 3$.

However, when you substitute a positive number for x in the expression $\pm\sqrt{x}$, you have two possible answers.

For example when $x = 4$, $\pm\sqrt{x} = -2$ or 2.

> A *function* of one variable is such that when a number is substituted for the variable, there is only one answer.

Therefore $x^2 - 2x$ is an example of a function f and can be written as $f(x) = x^2 - 2x$.

However, $\pm\sqrt{x}$ is not a function of x because any positive value of x gives two answers.

Domain and range

> The set of values which the variable in a function can take is called the *domain* of the function.

The domain does not have to contain all possible values of the variable; it can be as wide, or as restricted, as needed. Therefore to define a function fully, the domain must be stated.

If the domain is not stated, assume that it is the set of all **real numbers** (the set of real numbers is denoted by $\mathbb{R}$).

> For each domain, there is a corresponding set of values of $f(x)$. These are values which the function can take for values of x in that particular domain. This set is called the *range* of the function.

Look at the expression $x^2 + 3$.

A function f for this expression can be defined over any domain. Some examples, with their graphs are given.

1. $f(x) = x^2 + 3$ for $x \in \mathbb{R}$
   (the symbol $\in$ means 'is a member of').
   The range is $f(x) \geq 3$.

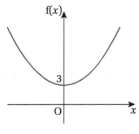

2. $f(x) = x^2 + 3$ for $x \geq 0$.
   The range is also $f(x) \geq 3$.

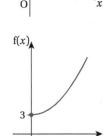

> **Note**
>
> The point on the curve where $x = 0$ is included and this is denoted this by a solid dot. If the domain were $x > 0$, then the point would not be part of the curve and this is indicated by a hollow dot.

3. $f(x) = x^2 + 3$ for $x = 1, 2, 3, 4, 5$.
   The range is the set of numbers 4, 7, 12, 19, 28.

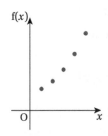

> **Note**
>
> This time the graphical representation consists of just five separate points.

Example 1

Question

The function, f, is defined by $f(x) = x^2$ for $x \leq 0$

and $f(x) = x$ for $x > 0$.

a Find $f(4)$ and $f(-4)$.

b Sketch the graph of $f(x)$.

c Give the range of f.

Answer

a For $x > 0$, $f(x) = x$,

therefore $f(4) = 4$.

For $x \leq 0$, $f(x) = x^2$,

therefore $f(-4) = (-4)^2 = 16$.

(continued)

(continued)

b To sketch the graph of a function, use what you know about lines and curves in the xy-plane.

So $f(x) = x$ for $x > 0$ is the part of the line $y = x$ which corresponds to positive values of x, and $f(x) = x^2$ for $x \le 0$ is the part of the parabola $y = x^2$ that corresponds to negative values of x.

c The range of f is $f(x) \ge 0$.

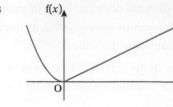

Exercise 1

1 Find the range of f in each of the following cases.

    a $f(x) = 2x - 3$ for $x \ge 0$

    b $f(x) = x^2 - 5$ for $x \le 0$

    c $f(x) = 1 - x$ for $x \le 1$

    d $f(x) = \dfrac{1}{x}$ for $x \ge 2$

2 Sketch the graph of each function given in question 1.

3 The function f is such that $f(x) = -x$ for $x < 0$

                     and $f(x) = x$ for $x \ge 0$.

    a Find the value of f(5), f(−4), f(−2) and f(0).

    b Sketch the graph of the function.

4 The function f is such that $f(x) = x$ for $0 \le x \le 5$

                     and $f(x) = 5$ for $x > 5$.

    a Find the value of f(0), f(2), f(4), f(5) and f(7).

    b Sketch the graph of the function.

    c Give the range of the function.

1.2 Composite functions

Look at the two functions f and g given by $f(x) = x^2$ and $g(x) = \dfrac{1}{x}$ for $x \ne 0$.

When $g(x)$ replaces x in $f(x)$ this gives the **composite function**

$$f[g(x)] = f\left(\frac{1}{x}\right) = \frac{1}{x^2} \text{ for } x \ne 0$$

A composite function formed this way is also called a **function of a function** and it is denoted by fg.

For example, if $f(x) = 3^x$ and $g(x) = 1 - x$ then gf(x) means the function g of f(x).

$\Rightarrow$ $gf(x) = g(3^x) = 1 - 3^x$

Also $fg(x) = f(1 - x) = 3^{(1-x)}$

This example shows that gf(x) is *not* always the same as fg(x).

Exercise 2

1. The functions f, g and h are defined by $f(x) = x^2$, $g(x) = \dfrac{1}{x}$ for $x \neq 0$ and $h(x) = 1 - x$.
   Find

   a $fg(x)$ **b** $fh(x)$ **c** $hg(x)$ **d** $hf(x)$ **e** $gf(x)$

2. When $f(x) = 2x - 1$ and $g(x) = x^3$ find the value of

   a $gf(3)$ **b** $fg(2)$ **c** $fg(0)$ **d** $gf(0)$

3. Given that $f(x) = 2x$, $g(x) = 1 + x$ and $h(x) = x^2$, find

   a $hg(x)$ **b** $gh(x)$ **c** $gf(x)$

4. When $f(x) = \sin x$ and $g(x) = 3x - 4$ find

   a $fg(x)$ **b** $gf(x)$

1.3 Inverse functions

Look at the function f where $f(x) = 2x$ for $x = 2, 3, 4$.

The domain of f is $\{2, 3, 4\}$ and the range of f is $\{4, 6, 8\}$. The relationship between the domain and range is shown in the arrow diagram.

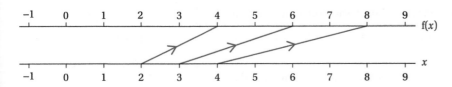

It is possible to reverse this process, so that each member of the range can be mapped back to the corresponding member of the domain by halving each member of the range.

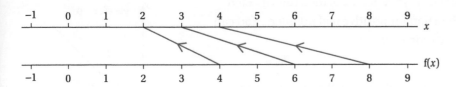

This process can be expressed algebraically.

When $x = 4, 6, 8$, then $x \to \dfrac{1}{2}x$ maps 4 to 2, 6 to 3 and 8 to 4.

This reverse mapping is a function in its own right and it is called the **inverse function** of f where $f(x) = 2x$.

Denoting this inverse function by f^{-1} we can write $f^{-1}(x) = \dfrac{1}{2}x$ for $x = 4, 6, 8$.

The function $f(x) = 2x$ for $x \in \mathbb{R}$ also has an inverse function, given by $f^{-1}(x) = \dfrac{1}{2}x$ which also has domain $x \in \mathbb{R}$.

> If a function g exists that maps the range of f back to its domain,
> then g is called the inverse of f and it is denoted by f^{-1}.

The graph of a function and its inverse

Consider the curve g(x) that is obtained by reflecting $y = f(x)$ in the line $y = x$ (see graph).

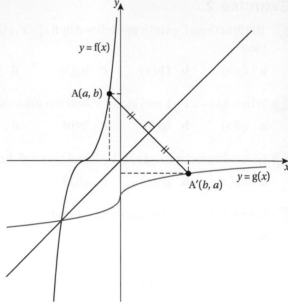

A point A(a, b) on the curve $y = f(x)$ is reflected onto a point A′ on the curve $y = g(x)$, whose coordinates are (b, a). Hence, interchanging the x- and y-coordinates of A gives the coordinates of A′.

> The equation of $y = g(x)$ is found by interchanging x and y in the equation $y = f(x)$.

The coordinates of A on $y = f(x)$ are [a, f(a)].

Therefore the coordinates of A′ on $y = g(x)$ are [f(a), a].

So the range of $y = f(x)$ becomes the domain of $y = g(x)$.

> When the equation of the reflected curve is $y = g(x)$ then g is the inverse of f, so $g = f^{-1}$.

Any curve whose equation can be written in the form $y = f(x)$ can be reflected in the line $y = x$. However this reflected curve may not have an equation that can be written in the form $y = f^{-1}(x)$.

For example, look at the curve $y = x^2$ and its reflection in the line $y = x$ (see graph).

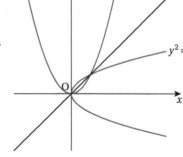

The equation of the reflected curve is $x = y^2$, giving $y = \pm\sqrt{x}$ and $\pm\sqrt{x}$ is not a function.

Therefore the function f where $f(x) = x^2$ does not have an inverse. You can also see this from the diagram, because on the reflected curve, one value of x maps to two values of y. So in this case y cannot be written as a function of x.

> Not every function has an inverse.

However, by changing the definition of f to $f(x) = x^2$ for $x \geq 0$, then the reflected curve is $y = \sqrt{x}$ for $x \geq 0$, and $\sqrt{x}$ is a function for positive real numbers. You can see this in the graph. Therefore $f^{-1}(x) = \sqrt{x}$ for $x \geq 0$.

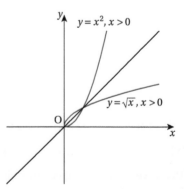

To summarise:

▶ The inverse of a function undoes the function, i.e. it maps the range of a function to its domain.

▶ The inverse of the function f is written f^{-1}.

▶ Not all functions have an inverse.

▶ When the curve whose equation is $y = f(x)$ is reflected in the line $y = x$, the equation of the reflected curve is $x = f(y)$.

▶ If this equation can be written in the form $y = g(x)$ then g is the inverse of f, so $g(x) = f^{-1}(x)$, and the domain of g is the range of f.

Example 2

Question

Determine whether there is an inverse of the function f given by $f(x) = 2 + \dfrac{1}{x}$, $x \neq 0$

If f^{-1} exists, express it as a function of x and give its domain.

Answer

The sketch of $f(x) = 2 + \dfrac{1}{x}$ shows that one value of $f(x)$ maps to one value of x, therefore the reverse mapping is a function.

The equation of the reflection of $y = 2 + \dfrac{1}{x}$ can be written as $x = 2 + \dfrac{1}{y}$ $\Rightarrow$ $y = \dfrac{1}{x-2}$

Therefore when $f(x) = 2 + \dfrac{1}{x}$, $f^{-1}(x) = \dfrac{1}{x-2}$ for $x \in \mathbb{R}$, provided that $x \neq 2$.

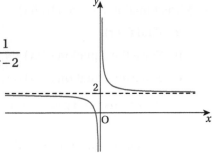

Example 3

Question

The function f is given by $f(x) = 5x - 1$.

a Find $f^{-1}(4)$.

b Solve the equation $f^{-1}(x) = x$.

Answer

a Let $y = f(x)$, that is $y = 5x - 1$.

The equation of the reflected line is

$$x = 5y - 1 \quad \Rightarrow \quad y = \frac{1}{5}(x+1)$$

So $f^{-1}(x) = \dfrac{1}{5}(x+1)$

Therefore $f^{-1}(4) = \dfrac{1}{5}(4+1) = 1$

b $f^{-1}(x) = x \quad \Rightarrow \quad \dfrac{1}{5}(x+1) = x$

$\Rightarrow \quad x + 1 = 5x$

Therefore $x = \dfrac{1}{4}$.

Exercise 3

1 Sketch the graphs of $y = f(x)$ and $y = f^{-1}(x)$ on the same axes.

   a $f(x) = 3x - 1$

   b $f(x) = (x-1)^3$

   c $f(x) = 2 - x$

   d $f(x) = \dfrac{1}{x-3}$

   e $f(x) = \dfrac{1}{x}$

2 Determine whether f has an inverse function and, if it does, find it.

   a $f(x) = x + 1$

   b $f(x) = x^2 + 1$

   c $f(x) = x^3 + 1$

5 Describe a sequence of transformations that maps the graph of $y = 2^x$ to the graph of $y = 3 + 2^{-x}$.

6 Simplify

a $\dfrac{x^2 - 9}{2x - 6}$

b $\dfrac{4x^2 - 25}{4x^2 + 20x + 25}$

7 Express $\dfrac{x - 3}{x + 6}$ as a number plus a proper fraction.

8 Express $\dfrac{3x^2 - 5x + 1}{x + 3}$ as a linear polynomial plus a proper fraction.

9 Express $\dfrac{x^3 - 4x^2 + 5}{x - 1}$ as a quadratic polynomial plus a proper fraction.

10 Express in partial fractions.

a $\dfrac{4}{(2x + 1)(x - 3)}$

b $\dfrac{(3x - 2)}{(x + 1)(4x - 3)}$

c $\dfrac{2t}{(t^2 - 1)}$

11 Express in partial fractions.

a $\dfrac{x + 4}{(x + 3)(x - 5)}$

b $\dfrac{(2x - 3)}{(x - 2)(4x - 3)}$

c $\dfrac{4x^2}{4x^2 - 9}$

12 Express in partial fractions.

a $\dfrac{3x}{2x^2 - 2x - 4}$

b $\dfrac{3x - 1}{x^2(x - 3)}$

Assessment

1 The function f is defined by $f(x) = \sqrt{x - 1}$ for $x \geq 1$.

  a State the domain and range of f and find $f^{-1}(x)$.

  b Solve the equation $f^{-1}(x) = 2x$.

2 **a** Express $\dfrac{x^2}{x^2 - 4}$ as a linear function plus a proper fraction.

  b Hence express $\dfrac{x^2}{x^2 - 4}$ in partial fractions.

3 **a** Describe a sequence of two transformations that maps the graph of $y = |x + 1|$ to the graph of $y = 1 - |1 + x|$.

  b Sketch the graph of $y = 1 - |1 + x|$.

  c Find the coordinates of the points of intersection of the graphs of $y = |x + 1|$ and $y = 1 - |1 + x|$.

  d Hence find the possible values of x for which $|x + 1| > 1 - |1 + x|$.

4 Express each rational function in partial fractions.

a $\dfrac{4}{x^2 - 7x - 8}$

b $\dfrac{2x - 1}{(2x + 1)(x - 2)^2}$

c $\dfrac{3}{x(2x + 1)}$

5 **a** Sketch the graph of $f(x) = \cos x$ for the domain $0 \leq x \leq 2\pi$.

  b State the range of f.

  c Given that $g(x) = 1 - |\cos x|$, find $fg(x)$.

  d Find the value of $fg\left(\dfrac{\pi}{2}\right)$.

6 The curve with equation $y = \dfrac{63}{4x-1}$ is sketched below for $1 \le x \le 16$.

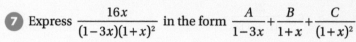

The function f is defined by $f(x) = \dfrac{63}{4x-1}$ for $1 \le x \le 16$.

a Find the range of f.

b The inverse of f is f^{-1}.

   i Find.

   ii Solve the equation $f^{-1}(x) = 1$.

c The function g is defined by $g(x) = x^2$ for $-4 \le x \le -1$

   i Write down an expression for $fg(x)$.

   ii Solve the equation $fg(x) = 1$.

AQA MPC3 January 2012

7 Express $\dfrac{16x}{(1-3x)(1+x)^2}$ in the form $\dfrac{A}{1-3x} + \dfrac{B}{1+x} + \dfrac{C}{(1+x)^2}$

AQA MPC4 June 2014 (part question)

8 a Sketch the curve with equation $y = 4 - |2x+1|$, indicating the coordinates where the curve crosses the axes.

b Solve the equation $x = 4 - |2x+1|$.

c Solve the inequality $x < 4 - |2x+1|$.

d Describe a sequence of two geometrical transformations that maps the graph of $y = |2x+1|$ onto the graph of $y = 4 - |2x+1|$.

AQA MPC3 June 2015

2 Binomial Series

Introduction

Many functions can be represented by an infinite polynomial expression. These infinite expressions can be used to find polynomial approximations for the given function, and hence approximate values for such functions at given values of x.

Objectives

By the end of this chapter, you should know how to ...
- Give the expansion of a binomial expression to a power that is not a positive integer.
- Use this expansion to find approximations.

Recap

You will need to remember...
- The expansion of $(a + b)^n$ where n is a positive integer.
- The meaning of a convergent series.
- The sum to infinity of a geometric series.
- How to express a rational function in partial fractions.

2.1 The binomial series for any value of n

The expansion of $(1 + x)^n$ as a series when n is not a positive integer is very similar to the expansion of $(1 + x)^n$ when n is a positive integer. However, there are two important differences:

- the series does not terminate but carries on to infinity,
- the series converges only for values of x in the range $-1 < x < 1$,

so $(1+x)^n = 1 + nx + \dfrac{n(n-1)}{2!}x^2 + \dfrac{n(n-1)(n-2)}{3!}x^3 + \cdots$

for *any* value of n provided that $|x| < 1$.

This series converges only when x has a value within the stated range, therefore this range *must be stated* for every expansion.

There are two particular expansions that illustrate this.

Using the expansion of $(1 + x)^n$ with $n = -1$ gives

$(1+x)^{-1} = 1 + (-1)x + \dfrac{(-1)(-2)}{2!}x^2 + \dfrac{(-1)(-2)(-3)}{3!}x^3 + \dfrac{(-1)(-2)(-3)(-4)}{4!}x^4 + \cdots$

$= 1 - x + x^2 - x^3 + x^4 - \cdots$

The right-hand side is a geometric series with common ratio $-x$, and so has a sum to infinity of $\dfrac{1}{1-(-x)} = \dfrac{1}{1+x} = (1+x)^{-1}$ provided that $|x| < 1$, so

$(1+x)^{-1} = 1 - x + x^2 - x^3 + x^4 - \ldots$ provided that $|x| < 1$.

Replacing x with $-x$ gives

$(1-x)^{-1} = 1 + x + x^2 + x^3 + x^4 + \cdots$ provided that $|x| < 1$.

Example 1

Question

Expand each of the following expressions as a series in ascending powers of x up to and including the term in x^3. State the values of x for which each expansion is valid.

a $(1+x)^{\frac{1}{2}}$ **b** $(1-2x)^{-3}$ **c** $(2-x)^{-2}$

Answer

For $|x| < 1$, $(1+x)^n = 1 + nx + \dfrac{n(n-1)}{2!}x^2 + \dfrac{n(n-1)(n-2)}{3!}x^3 + \ldots$ [1]

a Replacing n by $\dfrac{1}{2}$ in [1] gives

$$(1+x)^{\frac{1}{2}} = 1 + \frac{1}{2}x + \frac{\frac{1}{2}\left(\frac{1}{2}-1\right)}{2!}x^2 + \frac{\frac{1}{2}\left(\frac{1}{2}-1\right)\left(\frac{1}{2}-2\right)}{3!}x^3 + \ldots$$

$$= 1 + \frac{1}{2}x + \frac{\frac{1}{2}\left(-\frac{1}{2}\right)}{2!}x^2 + \frac{\frac{1}{2}\left(-\frac{1}{2}\right)\left(-\frac{3}{2}\right)}{3!}x^3 + \ldots$$

$$= 1 + \frac{x}{2} - \frac{x^2}{8} + \frac{x^3}{16} - \ldots \text{ for } |x| < 1$$

b Replacing n by -3 and x by $-2x$ in [1] gives

$$(1-2x)^{-3} = 1 + (-3)(-2x) + \frac{(-3)(-4)}{2!}(-2x)^2 + \frac{(-3)(-4)(-5)}{3!}(-2x)^3 + \ldots$$

$$= 1 + 6x + 24x^2 + 80x^3 + \ldots$$

provided that $|2x| < 1 \implies -\dfrac{1}{2} < x < \dfrac{1}{2}$.

c $(2-x)^{-2} = 2^{-2}\left(1 - \dfrac{1}{2}x\right)^{-2}$

Replacing n by -2 and x by $-\dfrac{1}{2}x$ in [1] gives

$$(2-x)^{-2} = \frac{1}{4}\left[1 + (-2)\left(-\frac{1}{2}x\right) + \frac{(-2)(-3)}{2!}\left(-\frac{1}{2}x\right)^2 + \frac{(-2)(-3)(-4)}{3!}\left(-\frac{1}{2}x\right)^3 + \ldots\right]$$

$$= \frac{1}{4}\left(1 + x + \frac{3}{4}x^2 + \frac{1}{2}x^3 + \ldots\right)$$

$$= \frac{1}{4} + \frac{1}{4}x + \frac{3}{16}x^2 + \frac{1}{8}x^3 + \ldots$$

The expansion of $\left(1 - \dfrac{1}{2}x\right)^{-2}$ is valid for $\left|\dfrac{1}{2}x\right| < 1 \implies -2 < x < 2$.

Therefore the expansion $(2-x)^{-\frac{1}{2}}$ also is valid for $-2 < x < 2$.

Exercise 1

Expand each the following expressions as a series in ascending powers of x up to and including the term in x^3. In each case give the range of values of x for which the expansion converges.

1 $(1-2x)^{\frac{1}{2}}$

2 $(1+5x)^{-2}$

3 $\left(1-\frac{1}{2}x\right)^{-3}$

4 $(1+x)^{\frac{3}{2}}$

5 $(3+x)^{-1}$

6 $(4-x)^{\frac{1}{2}}$

7 $\left(1+\frac{x}{2}\right)^{-\frac{1}{2}}$

8 $\dfrac{1}{(1-x)^2}$

9 $\dfrac{1}{(2+x)^3}$

10 $\sqrt{\dfrac{1}{1+x}}$

11 $\left(1+\frac{x^2}{9}\right)^{-1}$

12 $(4-3x)^{\frac{1}{2}}$

Series expansion of rational functions

When a rational function such as $\dfrac{5}{(1+3x)(1-2x)}$ is expressed in partial fractions, each fraction can be expanded as a series. These two series can then be added to give a single series as the binomial expansion of $\dfrac{5}{(1+3x)(1-2x)}$. This is shown in Example 2.

Example 2

Question

Express $\dfrac{5}{(1+3x)(1-2x)}$ in partial fractions.

Hence expand $\dfrac{5}{(1+3x)(1-2x)}$ as a series in ascending powers of x up to and including the term in x^3.

State the range of values of x for which the expansion is valid.

Answer

Expressing $\dfrac{5}{(1+3x)(1-2x)}$ in partial fractions gives

$$\frac{5}{(1+3x)(1-2x)}=\frac{3}{(1+3x)}+\frac{2}{(1-2x)}=3(1+3x)^{-1}+2(1-2x)^{-1}$$

Using $(1+x)^{-1}=1-x+x^2-x^3+\dots$ for $-1<x<1$ and replacing x by $3x$ gives

$$(1+3x)^{-1}=1-3x+(3x)^2-(3x)^3+\dots$$
$$=1-3x+9x^2-27x^3+\dots \quad \text{for} \quad -1<3x<1$$

Also using $(1-x)^{-1}=1+x+x^2+x^3+\dots$ for $-1<-x<1$ and replacing x by $2x$ gives

$$(1-2x)^{-1}=1+(2x)+(2x)^2+(2x)^3+\dots$$
$$=1+2x+4x^2+8x^3+\dots \quad \text{for} \quad -1<-2x<1$$

Hence $\dfrac{5}{(1+3x)(1-2x)}=3(1+3x)^{-1}+2(1-2x)^{-1}$

$$=(3+2)+(-9+4)x+(27+8)x^2+(-81+16)x^3+\dots$$
$$=5-5x+35x^2-65x^3+\dots$$

provided that $-\dfrac{1}{3}<x<\dfrac{1}{3}$ and $-\dfrac{1}{2}<x<\dfrac{1}{2}$.

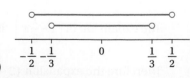

(continued)

(continued)

Therefore the first four terms of the series are $5 - 5x + 35x^2 - 65x^3$.

The expansion is valid for the range of values of x satisfying both

$-\dfrac{1}{3} < x < \dfrac{1}{3}$ and $-\dfrac{1}{2} < x < \dfrac{1}{2}$. As you can see from the diagram,

the expansion is therefore valid when $-\dfrac{1}{3} < x < \dfrac{1}{3}$.

Exercise 2

In questions 1 to 8, express the fraction in partial fractions and hence expand the expression as a series in ascending powers of x up to and including the term in x^3. In each case give the range of values of x for which the expansion converges.

1 $\dfrac{1}{(1-x)(1+x)}$

2 $\dfrac{1+3}{x-1}$

3 $\dfrac{2}{x(1-x)}$

4 $\dfrac{1}{(2-x)(1+2x)}$

5 $\dfrac{2}{9-x^2}$

6 $\dfrac{x}{(1+x)(1-2x)}$

7 $\dfrac{3x}{2x^2-2x-4}$

8 $\dfrac{-x^2}{4-x^2}$

In questions 9 to 11, express the fraction in partial fractions and hence expand the expression as a series in ascending powers of x, up to and including the term in x^2. Give the range of values of x for which the expansion converges.

9 $\dfrac{3x-1}{x(3-x)^2}$

10 $\dfrac{1-x-x^2}{(1-2x)(1-x)^2}$

11 $\dfrac{1}{(1-3x)(1-x)^2}$

2.2 Approximations

A binomial expansion of a function can be used to find a finite polynomial that is approximately equal to the function, for values of x which lie within the interval for which the binomial series converges.

A series can also be used to find an approximate value for an irrational number to a given number of decimal places.

Example 3

Show that for small values of x, $(1-2x)^{\frac{1}{2}} \approx 1 - x - \dfrac{1}{2}x^2$

$$(1-2x)^{\frac{1}{2}} = 1 + \frac{1}{2}(-2x) + \frac{\left(\dfrac{1}{2}\right)\left(-\dfrac{1}{2}\right)}{2!}(-2x)^2 + \ldots$$

$$\approx 1 - x - \frac{1}{2}x^2$$

Example 4

Use the expansion of $(1-x)^{\frac{1}{2}}$ with $x = 0.02$ to find the decimal value of $\sqrt{2}$ correct to nine decimal places.

$$(1-x)^{\frac{1}{2}} = 1 - \frac{1}{2}x + \frac{\left(\frac{1}{2}\right)\left(-\frac{1}{2}\right)}{2!}(-x)^2 + \frac{\left(\frac{1}{2}\right)\left(-\frac{1}{2}\right)\left(-\frac{3}{2}\right)}{3!}(-x)^3$$

$$+ \frac{\left(\frac{1}{2}\right)\left(-\frac{1}{2}\right)\left(-\frac{3}{2}\right)\left(-\frac{5}{2}\right)}{4!}(-x)^4 + \frac{\left(\frac{1}{2}\right)\left(-\frac{1}{2}\right)\left(-\frac{3}{2}\right)\left(-\frac{5}{2}\right)\left(-\frac{7}{2}\right)}{5!}(-x)^5 - \dots$$

$$= 1 - \frac{1}{2}x - \frac{1}{8}x^2 - \frac{1}{16}x^3 - \frac{5}{128}x^4 - \frac{7}{256}x^5 - \dots$$

This series converges when $-1 < x < 1$ and so converges when $x = 0.02$.

Replacing x by 0.02 gives

$(0.98)^{\frac{1}{2}} = 1 - 0.01 - 0.000\,05 - 0.000\,000\,5 - 0.000\,000\,006\,25 - 0.000\,000\,000\,087\,5 - \dots$

The next term in the series is 1.3125×10^{-12} and since this does not contribute to the first ten decimal places we do not need it, or any further terms.

So $\sqrt{\dfrac{98}{100}} = 0.989\,949\,493\,7$ to 10 decimal places

$\Rightarrow \dfrac{7}{10}\sqrt{2} = 0.989\,949\,493\,7$ to 10 decimal places

Therefore $\sqrt{2} = 1.414\,213\,562$ correct to 9 decimal places.

Exercise 3

1. Show that, when x is so small that x^2 and higher powers of x may be neglected, $\sqrt{1-x} \approx 1 - \dfrac{1}{2}x$.

2. a Express $\dfrac{1}{(1-x)(2+x)}$ in partial fractions.

   b Hence show that if x is so small that terms in x^2 and higher powers of x may be neglected, then $\dfrac{1}{(1-x)(2+x)} \approx \dfrac{1}{2} + \dfrac{1}{4}x$.

3. a Use partial fractions and the binomial series to find a linear expression that is an approximation for $\dfrac{3}{(1-2x)(2-x)}$.

   b Give the range of values of x for which the approximation is valid.

4. Find a quadratic expression that approximates to $f(x) = \dfrac{1}{\sqrt[3]{(1-3x)^2}}$

   and give the range of values of x for which the approximation is valid.

5 a Show that $\sqrt{121-2x} \approx 11 - \dfrac{x}{11} - \dfrac{x^2}{2662}$ and give the range of values of x for which the approximation is valid.

b Hence find $\sqrt{119}$ giving your answer correct to 5 decimal places.

6 a Show that $(125-x)^{\frac{1}{3}} \approx 5 - \dfrac{x}{75} - \dfrac{x^2}{28125}$ and give the range of values of x for which the approximation is valid.

b Hence find $\sqrt[3]{124}$ giving your answer correct to 5 decimal places.

7 a Show that $(625-4x)^{\frac{1}{4}} \approx 5 - \dfrac{1}{125}x - \dfrac{3}{156250}x^2$ and give the range of values of x for which the approximation is valid.

b Hence find $\sqrt[4]{621}$ giving your answer to 5 decimal places.

8 a Find the binomial expansion of $(169-2x)^{\frac{1}{2}}$ up to and including the term in x^2. Give the range of values of x for which the expansion is valid.

b Hence find $\sqrt{167}$ giving your answer to 5 decimal places.

9 a Find a quadratic expression that approximates to $(27-2x)^{\frac{1}{3}}$ and give the range of values of x for which the approximation is valid.

b Hence find $\sqrt[3]{25}$ giving your answer to 3 decimal places.

10 By substituting 0.08 for x in $(1+x)^{\frac{1}{2}}$ and its expansion, find a value for $\sqrt{3}$ correct to four significant figures.

11 By substituting $\dfrac{1}{10}$ for x in $(1-x)^{-\frac{1}{2}}$ and its expansion, find a value for $\sqrt{10}$ correct to six significant figures.

Summary

► When n is not a positive integer

$$(1+x)^n = 1 + nx + \frac{n(n-1)}{2!}x^2 + \frac{n(n-1)(n-2)}{3!}x^3 + \dots$$

for *any* value of n provided that $|x| < 1$.

► In particular $(1+x)^{-1} = 1 - x + x^2 - x^3 + x^4 - \dots$
and $(1-x)^{-1} = 1 + x + x^2 + x^3 + x^4 - \dots$ provided that $|x| < 1$.

► When a rational function is expressed in partial fractions, each fraction can be expanded as a binomial series. These two series can then be added to give a single series.

Review

1 Expand $\dfrac{1}{1+2x}$ as a series in ascending powers of x, giving the first three terms.

2 Expand $(1-3x)^{-3}$ as a series in ascending powers of x up to and including the term in x^2.

3 **a** Express $\dfrac{1}{(1-x)(1-2x)}$ in partial fractions.

b Hence expand $\dfrac{1}{(1-x)(1-2x)}$ as a series in ascending powers of x up to and including the term in x^3, and state the range of values of x for which the series converges.

4 Find a linear approximation to the curve $y = \dfrac{1}{(1-2x)^2}$ for small values of x.

5 **a** Show that, when x is small $\dfrac{1}{\sqrt[3]{27-x}} \approx \dfrac{1}{3} + \dfrac{1}{243}x + \dfrac{1}{36561}x^2$.

b Use a binomial expansion with $x=1$ to show that $\dfrac{1}{\sqrt[3]{26}} \approx 0.337$.

Assessment

1 **a** Show that $\sqrt[3]{1-2x} \approx 1 - \dfrac{2x}{3} - \dfrac{4x^2}{9} - \dfrac{40x^3}{81}$ and give the values of x for which this approximation is valid.

b Show that when $x = \dfrac{1}{10}$, $\sqrt[3]{1-2x} \approx \dfrac{2}{\sqrt[3]{10}}$.

c Hence find an approximate value for $\sqrt[3]{10}$, giving your answer in the form $\dfrac{a}{b}$ where a and b are integers.

2 **a** Show that when $x = \dfrac{1}{3}$ then $\dfrac{1}{\sqrt{1-x^2}} = \dfrac{3\sqrt{2}}{4}$.

b Show that $\dfrac{1}{\sqrt{1-x^2}} \approx 1 + \dfrac{x^2}{2} + \dfrac{ax^4}{8} + \dfrac{bx^6}{16}$ and state the values a and b.

c Hence find the value of $\sqrt{2}$ giving your answer to 3 decimal places.

3 **a** Express $\dfrac{2}{(3-2x)(1-2x)}$ in partial fractions.

b Expand $\dfrac{2}{(3-2x)(1-2x)}$ as a series in ascending powers of x up to and including the term in x^2. State the range of values of x for which the series converges.

4 It is given that $f(x) = \dfrac{7x-1}{(1+3x)(3-x)}$.

a Express $f(x)$ in the form $\dfrac{A}{3-x} + \dfrac{B}{1+3x}$, where A and B are integers.

b i Find the first three terms of the binomial expansion of $f(x)$ in the form $a + bx + cx^2$, where a, b and c are rational numbers.

  ii State why the binomial expansion cannot be expected to give a good approximation to $f(x)$ at $x = 0.4$.

AQA MPC4 January 2013

5 a Find the binomial expansion of $(1+6x)^{-\frac{1}{3}}$ up to and including the term in x^2.

b i Find the binomial expansion of $(27+6x)^{-\frac{1}{3}}$ up to and including the term in x^2, simplifying the coefficients.

  ii Given that $\sqrt[3]{\dfrac{2}{7}} = \dfrac{2}{\sqrt[3]{28}}$, use your binomial expansion from part (b)(i) to obtain an approximation to $\sqrt[3]{\dfrac{2}{7}}$ giving your answer to six decimal places.

AQA MPC4 June 13

3 Trigonometric Functions and Formulae

Introduction

This chapter extends the work on trigonometric functions started at AS level. It also introduces more trigonometric formulae that can be used to derive further formulae, solve equations and eliminate parameters.

Recap

You will need to remember...

▶ The definition of a function and an inverse function.
▶ The properties and the graphs of the functions $f(x) = \sin x$, $f(x) = \cos x$ and $f(x) = \tan x$.
▶ The exact values of the sine, cosine and tangent of $\dfrac{\pi}{6}, \dfrac{\pi}{4}, \dfrac{\pi}{3}$.
▶ How to describe a combination of transformations.
▶ The cosine formula and Pythagoras' theorem.
▶ The formulae $\tan\theta = \dfrac{\sin\theta}{\cos\theta}$ and $\cos^2\theta + \sin^2\theta = 1$.

Objectives

By the end of this chapter, you should know how to...

▶ Define the inverse trigonometric functions.
▶ Define the reciprocal trigonometric functions and associated formulae.
▶ Derive and use the compound angle formulae.
▶ Express $a\cos\theta + b\sin\theta$ in the form $r\sin(\theta \pm \alpha)$ or $r\cos(\theta \pm \alpha)$ and use this form to solve equations.
▶ Define and use the double angle formulae.

3.1 The inverse trigonometric functions

Look at the function given by $f(x) = \sin x$ for $x \in \mathbb{R}$.

The inverse mapping is given by $\sin x \to x$, but this mapping is not a function because one value of $\sin x$ maps to many values of x. As such, $f(x) = \sin x$ does not have an inverse function for the domain $x \in \mathbb{R}$.

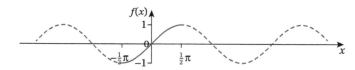

However, when the function $f(x) = \sin x$ is restricted to the domain $-\dfrac{1}{2}\pi \leq x \leq \dfrac{1}{2}\pi$, one value of $\sin x$ maps to only one value of x. Therefore $f(x) = \sin x$ for the domain $-\dfrac{1}{2}\pi \leq x \leq \dfrac{1}{2}\pi$ does have an inverse, so $f^{-1}(x)$ exists. The curve $y = f^{-1}(x)$ is found by reflecting $y = \sin x$ in the line $y = x$. Interchanging x and y in the equation $y = \sin x$ gives the equation of the curve $y = f^{-1}(x)$, which in this case is $\sin y = x$, so y is the angle between $-\dfrac{1}{2}\pi$ and $\dfrac{1}{2}\pi$ whose sine is x.

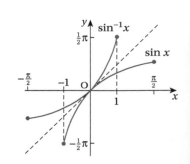

Using $y = \sin^{-1} x$ to mean 'y is the angle between $-\frac{1}{2}\pi$ and $\frac{1}{2}\pi$ whose sine is x', where the range of $y = \sin x$ is the domain of the function $y = \sin^{-1} x$, we get the result

Note

To remind yourself about inverse functions, look back at Chapter 1, Section 1.3.

when $f(x) = \sin x$ for $-\frac{1}{2}\pi \leq x \leq \frac{1}{2}\pi$

then $f^{-1}(x) = \sin^{-1} x$ for $-1 \leq x \leq 1$

Remember that $\sin^{-1} x$ is an angle, and that angle is between $-\frac{1}{2}\pi$ and $\frac{1}{2}\pi$.

For example, $\sin^{-1} 0.5$ is the angle between $-\frac{1}{2}\pi$ and $\frac{1}{2}\pi$ whose sine is 0.5.

Therefore $\sin^{-1} 0.5 = \frac{1}{6}\pi$.

Now look at the function $f(x) = \cos x$ for $0 \leq x \leq \pi$.

The diagram shows that f^{-1} exists and it is denoted by $\cos^{-1}$ where $\cos^{-1} x$ means 'the angle between 0 and π whose cosine is x'.

The range of $\cos x$ is $-1 \leq \cos x \leq 1$, and therefore

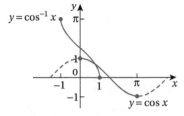

when $f(x) = \cos x$ for $0 \leq x \leq \pi$

then $f^{-1}(x) = \cos^{-1} x$ for $-1 \leq x \leq 1$

For example, $\cos^{-1} (-0.5)$ is the angle between 0 and π whose cosine is -0.5.

Therefore, $\cos^{-1} (-0.5) = \frac{2}{3}\pi$.

Similarly, when $f(x) = \tan x$ for $-\frac{1}{2}\pi < x < \frac{1}{2}\pi$, then f^{-1} exists and is written $\tan^{-1}$ where $\tan^{-1} x$ means 'the angle between $-\frac{1}{2}\pi$ and $\frac{1}{2}\pi$ whose tangent is x'.

The range of $\tan x$ is $\tan x \in \mathbb{R}$, and therefore

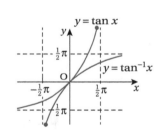

when $f(x) = \tan x$ for $\frac{1}{2}\pi < x < \frac{1}{2}\pi$

then $f^{-1}(x) = \tan^{-1} x$ for $x \in \mathbb{R}$.

Note that the domain of $\tan^{-1} x$ is all values of x.

Example 1

Question

Find in terms of π the exact value of   **a** $\tan^{-1} 1$ **b** $\sin^{-1}\left(-\frac{\sqrt{2}}{2}\right)$

Answer

a $\tan^{-1} 1$ means the angle whose tangent is 1, so if this angle is α then $\tan \alpha = 1$. Hence, $\tan^{-1} 1 = \frac{\pi}{4}$.

b $\sin^{-1}\left(-\frac{\sqrt{2}}{2}\right)$ means the angle whose sine is $-\frac{\sqrt{2}}{2}$,

so if this angle is β then $\sin \beta = -\frac{\sqrt{2}}{2}$. Hence $\sin^{-1}\left(-\frac{\sqrt{2}}{2}\right) = -\frac{\pi}{4}$.

Example 2

a Sketch the graph of $y = \cos^{-1}(2x - 1)$.

b Solve the equation $\cos^{-1}(2x - 1) = \dfrac{\pi}{2}$

a The curve $y = \cos^{-1}(2x - 1)$ is a one-way stretch of $y = \cos^{-1} x$ by a factor $\dfrac{1}{2}$ in the direction of the x-axis followed by the translation $\begin{bmatrix} 1 \\ 0 \end{bmatrix}$.

b $\cos^{-1}(2x - 1) = \dfrac{\pi}{2}$

$2x - 1 = \cos\left(\dfrac{\pi}{2}\right)$

$2x - 1 = 0$

$x = \dfrac{1}{2}$

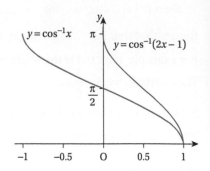

Exercise 1

Find, in terms of π, the value of

1 $\tan^{-1}\sqrt{3}$

2 $\sin^{-1}(-1)$

3 $\cos^{-1} 0$

4 $\sin^{-1}\left(-\dfrac{\sqrt{3}}{2}\right)$

5 $\cos^{-1}\left(-\dfrac{1}{2}\right)$

6 $\tan^{-1}(-1)$.

Solve these equations.

7 $\cos^{-1}(2x - 1) = \dfrac{\pi}{3}$

8 $\sin^{-1}(4x) = \dfrac{\pi}{6}$

9 $\tan^{-1}(x - 2) = \dfrac{\pi}{6}$

10 $\tan^{-1}(2 - x) = \dfrac{\pi}{4}$

11 $2\cos^{-1} x = \pi$

12 $2\sin^{-1} x = \dfrac{\pi}{2}$

3.2 The reciprocal trigonometric functions

The reciprocals of the three main trigonometric functions have their own names.

$$\frac{1}{\sin\theta} = \operatorname{cosec}\theta, \quad \frac{1}{\cos\theta} = \sec\theta, \quad \frac{1}{\tan\theta} = \cot\theta$$

where cosec, sec and cot are abbreviations of **cosecant**, **secant** and **cotangent** respectively.

The graph of $f(\theta) = \operatorname{cosec}\theta$ is shown here.

This graph shows that the cosec function is not continuous as it is undefined when θ is any integer multiple of π (because these are values of θ where $\sin\theta = 0$, and at these values $\operatorname{cosec}\theta = \dfrac{1}{0}$ which is undefined).

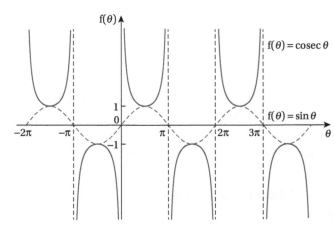

The graph of $f(\theta) = \sec \theta$ is similar to the graph of $y = \operatorname{cosec} \theta$.

However, $\sec(\theta)$ is not defined when the value of θ is $(2n+1)\dfrac{\pi}{2}$, where n is any integer.

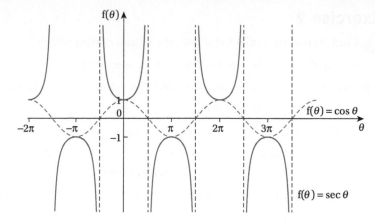

The graph of $f(\theta) = \cot \theta$ is given here. In a similar way to the cosec function, $\cot \theta$ is undefined when θ is any integer multiple of π.

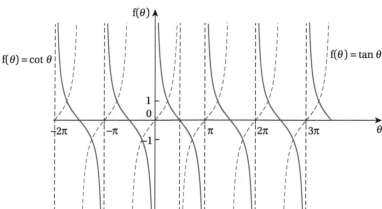

In the right-angled triangle shown in the margin,
$$\tan \alpha = \frac{a}{b} \text{ and } \cot \beta = \frac{a}{b}\left(\cot \beta = \frac{1}{\tan \beta}\right)$$

Now $\alpha + \beta = 90°$, so the cotangent of an angle is equal to the tangent of its complement.

This is true for all values of θ as can be seen from the graph.

Reflecting the curve $y = \tan \theta$ in the vertical axis gives $y = \tan(-\theta)$.

Then translating this curve $\dfrac{1}{2}\pi$ to the left gives $y = \tan\left(\dfrac{1}{2}\pi - \theta\right)$ which is the curve $y = \cot \theta$.

For *any* angle θ, $\cot \theta = \tan\left(\dfrac{1}{2}\pi - \theta\right)$

Example 3

Find the values of θ for which $\operatorname{cosec} \theta = -8$ for $0 \le \theta \le 360°$

$$\sin \theta = \frac{1}{\operatorname{cosec} \theta} = -\frac{1}{8} = -0.125$$

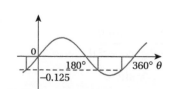

From a calculator $\theta = -7.2°$

From the sketch, the required values of θ are $187.2°$ and $352.8°$.

Exercise 2

1 Find, in the interval $0 \le \theta \le 360°$, the values of θ for which

    a $\sec \theta = 2$ **b** $\cot \theta = 0.6$ **c** $\operatorname{cosec} \theta = 1.5$.

2 Find, in the interval $-180° \le \theta \le 180°$, the values of θ for which

    a $\cot \theta = 1.2$ **b** $\sec \theta = -1.5$ **c** $\operatorname{cosec} \theta = -2$.

3 **a** Use $\tan \theta = \dfrac{\sin \theta}{\cos \theta}$ to write $\cot \theta$ in terms of $\sin \theta$ and $\cos \theta$.

    b Hence show that $\cot \theta - \cos \theta = 0$ can be written in the form
        $\cos \theta (1 - \sin \theta) = 0$, provided that $\sin \theta \ne 0$.

    c Find the values in the interval $-\pi \le \theta \le \pi$ for which $\cot \theta - \cos \theta = 0$.

4 Find the values of

    a $\cot \dfrac{1}{4}\pi$ **b** $\sec \dfrac{5}{4}\pi$ **c** $\operatorname{cosec} \dfrac{11}{6}\pi$.

5 Sketch the graph of $f(\theta) = \sec\left(\theta - \dfrac{1}{4}\pi\right)$ for $0 \le \theta \le 2\pi$ and give the values
of θ for which $f(\theta) = 1$.

6 Sketch the graph of $f(\theta) = \cot\left(\theta + \dfrac{1}{3}\pi\right)$ for $-\pi \le \theta \le \pi$. Hence give the values
of θ in this interval for which $f(\theta) = 1$.

3.3 Trigonometric formulae

When the formula $\cos^2 \theta + \sin^2 \theta = 1$ is divided by $\cos^2 \theta$ this gives

$1 + \dfrac{\sin^2 \theta}{\cos^2 \theta} \equiv \dfrac{1}{\cos^2 \theta}$. Hence

    $1 + \tan^2 \theta = \sec^2 \theta$

When $\cos^2 \theta + \sin^2 \theta = 1$ is divided by $\sin^2 \theta$ this gives $\dfrac{\cos^2 \theta}{\sin^2 \theta} + 1 = \dfrac{1}{\sin^2 \theta}$. Hence

    $\cot^2 \theta + 1 = \operatorname{cosec}^2 \theta$

These formulae can be used to:

▶ simplify trigonometrical expressions

▶ eliminate trigonometrical terms from pairs of equations

▶ derive a variety of further trigonometrical relationships

▶ solve equations.

Example 4

Question

Simplify $\dfrac{\sin \theta}{1 + \cot^2 \theta}$.

Answer

$\dfrac{\sin \theta}{1 + \cot^2 \theta} = \dfrac{\sin \theta}{\operatorname{cosec}^2 \theta}$

$= \sin^3 \theta$

> **Note**
>
> Using $1 + \cot^2 \theta = \operatorname{cosec}^2 \theta$
> and $\operatorname{cosec} \theta = \dfrac{1}{\sin \theta}$

Example 5

Question

Prove that $(1 - \cos A)(1 + \sec A) = \sin A \tan A$.

Answer

Do not start by assuming that the relationship is correct. The left and right hand sides must be isolated throughout the proof, by working on only one of these sides at a time. It often helps to express all ratios in terms of sine and/or cosine as, in general, these are easier to work with.

LHS $= (1 - \cos A)(1 + \sec A) = 1 + \sec A - \cos A - \cos A \sec A$

$$= 1 + \sec A - \cos A - \cos A \left(\frac{1}{\cos A} \right)$$

$$= \sec A - \cos A = \frac{1}{\cos A} - \cos A$$

$$= \frac{1 - \cos^2 A}{\cos A} = \frac{\sin^2 A}{\cos A}$$

$$= \sin A \left[\frac{\sin A}{\cos A} \right] = \sin A \tan A = \text{RHS}$$

Note

$\cos^2 A + \sin^2 A = 1$

Example 6

Question

Find the solution of the equation $\cot\left(\frac{1}{3}\theta - 90° \right) = 1$, for which $0 \le \theta \le 540°$.

Answer

Using $\frac{1}{3}\theta - 90° = \phi$ gives $\cot\left(\frac{1}{3}\theta - 90° \right) = \cot\phi$

The solution of the equation $\cot\phi = 1$ is $\phi = 45°$

But $\phi = \frac{1}{3}\theta - 90°$, so $\frac{1}{3}\theta - 90° = 45°$

Therefore $\frac{1}{3}\theta = 135°$

$\Rightarrow \qquad \theta = 405°$

Note

As you require θ in the range $0 \le \theta \le 540°$, find ϕ in the range $\frac{1}{3}(0) - 90° \le \phi \le \frac{1}{3}(540°) - 90°$ i.e. $-90° \le \phi \le 90°$.

Exercise 3

Simplify these expressions given that θ is an acute angle.

1. $\dfrac{1 - \sec^2 A}{1 - \operatorname{cosec}^2 A}$

2. $\dfrac{\sin\theta}{\sqrt{(1 - \cos^2\theta)}}$

3. $\dfrac{\sin\theta}{\cos\theta} + \dfrac{\cos\theta}{\sin\theta}$

4. $\dfrac{\sqrt{(1 + \tan^2\theta)}}{\sqrt{(1 - \sin^2\theta)}}$

5. $\dfrac{1}{\cos\theta\sqrt{(1 + \cot^2\theta)}}$

6. $\dfrac{\sin\theta}{1 + \cot^2\theta}$

Eliminate θ from these pairs of equations.

7 $x = 4\sec\theta$
$y = 4\tan\theta$

8 $x = a\operatorname{cosec}\theta$
$y = b\cot\theta$

9 $x = 2\tan\theta$
$y = 3\cos\theta$

10 $x = 2 + \tan\theta$
$y = 2\cos\theta$

11 $x = a\sec\theta$
$y = b\sin\theta$

Prove these formulae.

12 $\cot\theta + \tan\theta = \sec\theta\operatorname{cosec}\theta$

13 $\dfrac{\cos A}{1 - \tan A} + \dfrac{\sin A}{1 - \cot A} = \sin A + \cos A$

14 $\tan^2\theta + \cot^2\theta = \sec^2\theta + \operatorname{cosec}^2\theta - 2$

15 $\dfrac{\sin A}{1 + \cos A} = \dfrac{1 - \cos A}{\sin A}$ $\longleftarrow$

> **Note**
>
> *Hint for question 15:* Multiply top and bottom of LHS by $(1 - \cos A)$.

16 $\dfrac{\sin A}{1 + \cos A} + \dfrac{1 + \cos A}{\sin A} = \dfrac{2}{\sin A}$

Solve the equations for values of θ in the interval $0 \le \theta \le 360°$, giving your answers correct to 1 decimal place.

17 $\sec^2\theta + \tan^2\theta = 6$

18 $\cot^2\theta = \operatorname{cosec}\theta$

19 $\tan\theta + \cot\theta = 2\sec\theta$

20 $\tan\theta + 3\cot\theta = 5\sec\theta$

Solve the equations for angles in the interval $-\pi \le \theta \le \pi$, giving your answers correct to 3 significant figures.

21 $4\cot^2\theta + 12\operatorname{cosec}\theta + 1 = 0$

22 $4\sec^2\theta - 3\tan\theta = 5$

Solve the equations for values of θ in the interval $0 \le \theta \le \pi$.

23 $\sec 5\theta = 2$

24 $\cot\dfrac{1}{2}\theta = -1$

25 $\cot\left(\theta + \dfrac{1}{4}\pi\right) = 1$

26 $\sec\left(2\theta - \dfrac{1}{3}\pi\right) = -2$

3.4 Compound angle formulae

It is tempting to think that $\cos(A - B)$ can be written as $\cos A - \cos B$ but $\cos(A - B)$ is NOT equal to $\cos A - \cos B$.

The correct formula for $\cos(A - B)$ is found using the diagram below, which shows a circle of radius 1 unit centre O.

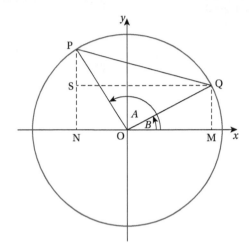

Using the cosine formula in triangle OPQ gives

$$PQ^2 = 1^2 + 1^2 - 2\cos(A - B) = 2 - 2\cos(A - B)$$

$$OM = \cos B \text{ and } ON = -\cos A, \quad \text{so} \quad QS = (-\cos A + \cos B)$$

$$QM = \sin B \text{ and } PN = \sin A \quad \text{so} \quad PS = (\sin A - \sin B)$$

Using Pythagoras' theorem in triangle PQS gives

$$PQ^2 = (-\cos A + \cos B)^2 + (\sin A - \sin B)^2$$

$$= \cos^2 A + \cos^2 B - 2\cos A \cos B + \sin^2 A + \sin^2 B - 2\sin A \sin B$$

$$= 2 - 2(\cos A \cos B + \sin A \sin B)$$

Equating the two expressions for PQ^2 gives

$$\cos(A - B) = \cos A \cos B + \sin A \sin A$$

This formula is true for all angles.

The formula derived above is one of the compound angle formulae and we can use it to derive others.

$$\cos(A - B) = \cos A \cos B + \sin A \sin A \qquad [1]$$

Replacing B by $-B$ in [1],
$$\cos(A + B) = \cos A \cos(-B) + \sin A \sin(-B)$$
$$= \cos A \cos B - \sin A \sin B \qquad [2]$$

Replacing A by $\frac{1}{2}\pi - A$ in [1],
$$\cos\left(\left(\frac{1}{2}\pi - A\right) + B\right) = \cos\left(\frac{1}{2}\pi - A\right)\cos B + \sin\left(\frac{1}{2}\pi - A\right)\sin B$$
$$\sin(A + B) = \sin A \cos B + \cos A \sin B \qquad [3]$$

Replacing B by $-B$ in [3],
$$\sin(A - B) = \sin A \cos(-B) + \cos A \sin(-B)$$
$$= \sin A \cos B - \cos A \sin B \qquad [4]$$

Dividing [1] by [3] gives $\dfrac{\sin(A+B)}{\cos(A+B)} = \dfrac{\sin A \cos B + \cos A \sin B}{\cos A \cos B - \sin A \sin B}$

$$\tan(A+B) = \dfrac{\dfrac{\sin A \cos B}{\cos A \cos B} + \dfrac{\cos A \sin B}{\cos A \cos B}}{\dfrac{\cos A \cos B}{\cos A \cos B} - \dfrac{\sin A \sin B}{\cos A \cos B}} = \dfrac{\tan A + \tan B}{1 - \tan A \tan B} \quad [5]$$

Replacing B by $-B$ in [5] gives

$$\tan(A-B) = \dfrac{\tan A - \tan B}{1 + \tan A \tan B} \qquad [6]$$

Collecting these formulae together gives

$$\sin(A+B) = \sin A \cos B + \cos A \sin B$$

$$\sin(A-B) = \sin A \cos B - \cos A \sin B$$

$$\cos(A+B) = \cos A \cos B - \sin A \sin B$$

$$\cos(A-B) = \cos A \cos B + \sin A \sin B$$

$$\tan(A+B) = \dfrac{\tan A + \tan B}{1 - \tan A \tan B}$$

$$\tan(A-B) = \dfrac{\tan A - \tan B}{1 + \tan A \tan B}$$

You do not need to learn these formulae, but you need to be able to recognise the right hand side of these formulae as equivalents to the left hand sides.

You can use these formulae to prove further trigonometrical formulae, as you will see in Examples 7, 8 and 9.

Example 7

Find exact values for

a $\sin 75°$

b $\cos 105°$

To find exact values, express the given angle in terms of angles whose trig ratios are known exact values, that is 30°, 60°, 45°, 90°.

a $\sin 75° = \sin(45° + 30°) = \sin 45° \cos 30° + \cos 45° \sin 30°$

$$= \left(\dfrac{\sqrt{2}}{2}\right)\left(\dfrac{\sqrt{3}}{2}\right) + \left(\dfrac{\sqrt{2}}{2}\right)\left(\dfrac{1}{2}\right) = \dfrac{\sqrt{2}}{4}(\sqrt{3}+1)$$

b $\cos 105° = \cos(60° + 45°) = \cos 60° \cos 45° - \sin 60° \sin 45°$

$$= \left(\dfrac{1}{2}\right)\left(\dfrac{\sqrt{2}}{2}\right) - \left(\dfrac{\sqrt{3}}{2}\right)\left(\dfrac{\sqrt{2}}{2}\right) = \dfrac{\sqrt{2}}{4}(1-\sqrt{3})$$

Example 8

Simplify $\sin \theta \cos \frac{1}{3}\pi - \cos \theta \sin \frac{1}{3}\pi$ and hence find the smallest positive value of θ for which the expression has a minimum value.

Answer

$$f(\theta) = \sin\theta\cos\frac{1}{3}\pi - \cos\theta\sin\frac{1}{3}\pi = \sin\left(\theta - \frac{1}{3}\pi\right)$$

The graph of $f(\theta) = \sin\left(\theta - \frac{1}{3}\pi\right)$ is a sine wave, but translated $\frac{1}{3}\pi$ in the direction of the positive θ-axis.

Therefore $f(\theta)$ has a minimum value of -1 and the smallest positive value of θ at which this occurs is $\frac{3}{2}\pi + \frac{1}{3}\pi = \frac{11}{6}\pi$.

> **Note**
>
> $\sin\theta\cos\frac{1}{3}\pi - \cos\theta\sin\frac{1}{3}\pi$
>
> is the expansion of $\sin(A - B)$
>
> with $A = \theta$ and $B = \frac{1}{3}\pi$.

Example 9

Question

Prove that $\dfrac{\sin(A-B)}{\cos A \cos B} = \tan A - \tan B$.

Answer

Expanding the numerator, the LHS becomes

$$\frac{\sin A\cos B - \cos A\sin B}{\cos A\cos B} = \frac{\sin A\cos B}{\cos A\cos B} - \frac{\cos A\sin B}{\cos A\cos B}$$

$$= \tan A - \tan B$$

Example 10

Question

Find all the solutions of the equation $2\cos\theta = \sin\left(\theta + \frac{1}{6}\pi\right)$ in the interval $0 \le \theta \le 2\pi$.

Answer

$$2\cos\theta = \sin\left(\theta + \frac{1}{6}\pi\right) = \sin\theta\cos\frac{1}{6}\pi + \cos\theta\sin\frac{1}{6}\pi = \frac{\sqrt{3}}{2}\sin\theta + \frac{1}{2}\cos\theta$$

Therefore $\dfrac{3}{2}\cos\theta = \dfrac{\sqrt{3}}{2}\sin\theta$

$\Rightarrow \qquad \dfrac{3}{\sqrt{3}} = \dfrac{\sin\theta}{\cos\theta}$

$\Rightarrow \qquad \tan\theta = \sqrt{3}$

Now $\tan\frac{1}{3}\pi = \sqrt{3}$, so the solution is $\theta = \frac{1}{3}\pi, \frac{4}{3}\pi$.

Exercise 4

Find the exact value of each expression, leaving your answer in surd form where necessary.

1. $\cos 40° \cos 50° - \sin 40° \sin 50°$

2. $\sin 37° \cos 7° - \cos 37° \sin 7°$

3. $\cos 75°$

4. $\tan 105°$

5. $\sin 165°$

6. $\cos 15°$

Simplify each expression.

7. $\sin\theta\cos 2\theta + \cos\theta\sin 2\theta$

8. $\cos\alpha\cos(90° - \alpha) - \sin\alpha\sin(90° - \alpha)$

9. $\dfrac{\tan A + \tan 2A}{1 - \tan A\tan 2A}$

10. $\dfrac{\tan 3\beta - \tan 2\beta}{1 + \tan 3\beta\tan 2\beta}$

11 Find the greatest value of each expression and the value of θ between 0 and 360° at which it occurs.

    a $\sin \theta \cos 25° - \cos \theta \sin 25°$ **b** $\sin \theta \sin 30° + \cos \theta \cos 30°$

    c $\cos \theta \cos 50° - \sin \theta \sin 50°$ **d** $\sin 60° \cos \theta - \cos 60° \sin \theta$

Prove the formulae.

12 $\cot (A+B) = \dfrac{\cot A \cot B - 1}{\cot A + \cot B}$ **13** $\sin (A+B) + \sin (A-B) = 2 \sin A \cos B$

14 $\cos (A+B) + \cos (A-B) = 2 \cos A \cos B$ **15** $\dfrac{\sin (A+B)}{\cos A \cos B} = \tan A + \tan B$

Solve the equations for values of θ in the interval $0 \le \theta \le 360°$.

16 $\cos (45° - \theta) = \sin \theta$ **17** $3 \sin \theta = \cos (\theta + 60°)$

18 $\tan (A - \theta) = \dfrac{2}{3}$ and $\tan A = 3$ **19** $\sin (\theta + 60°) = \cos \theta$

3.5 Expressions of the form $f(\theta) = a \cos \theta + b \sin \theta$

The expression $a \cos \theta + b \sin \theta$ can be expressed in the form $r \sin (\theta + \alpha)$ where $r > 0$.

Starting with $r \sin (\theta + \alpha) = a \cos \theta + b \sin \theta$, expanding the LHS using the compound angle formulae gives

$$r \underline{\sin} \theta \cos \alpha + r \underline{\cos} \theta \sin \alpha = a \underline{\cos} \theta + b \underline{\sin} \theta$$

Comparing coefficients of $\cos \theta$ and of $\sin \theta$ gives

$$r \sin \alpha = a \qquad\qquad\qquad\qquad\qquad [1]$$

and $\qquad r \cos \alpha = b \qquad\qquad\qquad\qquad\qquad [2]$

Squaring and adding equations [1] and [2] gives

$$r^2(\sin^2 \alpha + \cos^2 \alpha) = a^2 + b^2 \quad \Rightarrow \quad r = \sqrt{a^2 + b^2}$$

Dividing equation [1] by equation [2] gives

$$\frac{r \sin \alpha}{r \cos \alpha} = \frac{a}{b} \quad \Rightarrow \quad \tan \alpha = \frac{a}{b}$$

Therefore $r \sin (\theta + \alpha) = a \cos \theta + b \sin \theta$

where $r = \sqrt{a^2 + b^2}$ and $\tan \alpha = \dfrac{a}{b}$

Using a similar method it is also possible to express $a \cos \theta + b \sin \theta$ as $r \sin (\theta - \alpha)$ or $r \cos(\theta \pm \alpha)$.

Example 11

Express $3 \sin \theta - 2 \cos \theta$ as $r \sin (\theta - \alpha)$. Give α to the nearest 0.1°.

Question

$3 \sin \theta - 2 \cos \theta = r \sin (\theta - \alpha) \implies 3 \underline{\sin} \theta - 2 \underline{\cos} \theta = r \underline{\sin} \theta \cos \alpha - r \underline{\cos} \theta \sin \alpha$

Comparing coefficients of $\sin \theta$ and of $\cos \theta$ gives

$$\left. \begin{array}{l} 3 = r \cos \alpha \\ 2 = r \sin \alpha \end{array} \right\} \implies \begin{cases} 13 = r^2 \implies r = \sqrt{13} \\ \tan \alpha = \dfrac{2}{3} \implies \alpha = 33.7° \end{cases}$$

Therefore $3 \sin \theta - 2 \cos \theta = \sqrt{13} \, \sin (\theta - 33.7°)$.

Example 12

Find the maximum value of $f(x) = 3 \cos x + 4 \sin x$ and the smallest positive value of x at which the maximum occurs. Give your answer to the nearest $0.1°$.

Expressing $f(x)$ in the form $r \sin (x \pm \alpha)$ or $r \cos (x \pm \alpha)$ means that its maximum value, and the values of x at which this maximum occurs, can be seen from a sketch. In this question you can choose the form in which to express $f(x)$. The solution given here uses $r \cos (x - \alpha)$ because of the plus sign in $f(x)$ it fits better than $r \sin (x + \alpha)$.

$3 \underline{\cos} x + 4 \underline{\sin} x = r \cos (x - \alpha) = r \underline{\cos} x \cos \alpha + r \underline{\sin} x \sin \alpha$

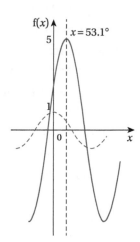

Hence $\left. \begin{array}{l} r \cos \alpha = 3 \\ r \sin \alpha = 4 \end{array} \right\} \implies \begin{cases} r^2 = 25 \implies r = 5 \\ \tan \alpha = \dfrac{4}{3} \implies \alpha = 53.1° \end{cases}$

Therefore $f(x) = 5 \cos (x - 53.1°)$

The graph of $f(x)$ is a cosine wave stretched by a factor of 5 parallel to the y-axis and translated by $53.1°$ along the x-axis.

Therefore $f(x)$ has a maximum value of 5 and the smallest positive value of x at which it occurs is $53.1°$.

Example 13

Find, for $-\pi \leq x \leq \pi$, the solution of the equation $\sqrt{3} \cos x + \sin x = 1$.

First express $\sqrt{3} \cos x + \sin x$ in the form $r \cos (x - \alpha)$.

$\sqrt{3} \underline{\cos} x + \underline{\sin} x = r \cos (x - \alpha)$

$\qquad\qquad = r \underline{\cos} x \cos \alpha + r \underline{\sin} x \sin \alpha$

Comparing coefficients gives $\begin{cases} r \cos \alpha = \sqrt{3} \\ r \sin \alpha = 1 \end{cases} \implies \begin{cases} r^2 = 4 \implies r = 2 \\ \tan \alpha = \dfrac{1}{\sqrt{3}} \implies \alpha = \dfrac{1}{6}\pi \end{cases}$

Therefore $\sqrt{3} \cos x + \sin x = 2 \cos (x - \dfrac{1}{6}\pi)$

so $\sqrt{3} \cos x + \sin x = 1 \implies 2 \cos (x - \dfrac{1}{6}\pi) = 1$

$$\Rightarrow \qquad \cos\left(x - \frac{1}{6}\pi\right) = \frac{1}{2}$$

$$\Rightarrow \qquad x - \frac{1}{6}\pi = \pm\frac{1}{3}\pi$$

Therefore $\qquad x = \frac{1}{2}\pi, -\frac{1}{6}\pi.$

Exercise 5

1 Find the values of r and α for which

   a $\sqrt{3}\cos\theta - \sin\theta = r\cos(\theta + \alpha)$

   b $\cos\theta + 3\sin\theta = r\cos(\theta - \alpha)$

   c $4\sin\theta - 3\cos\theta = r\sin(\theta - \alpha)$.

   Give angles in degrees to the nearest $0.1°$

2 Express $\cos 2\theta - \sin 2\theta$ in the form $r\cos(2\theta + \alpha)$. Give α in radians to three significant figures.

3 Express $2\cos 3\theta + 5\sin 3\theta$ in the form $r\sin(3\theta + \alpha)$. Give α in degrees to the nearest $0.1°$.

4 **a** Express $\cos\theta - \sqrt{3}\sin\theta$ in the form $r\sin(\theta - \alpha)$. Give α in degrees to the nearest $0.1°$. Hence sketch the graph of $f(\theta) = \cos\theta - \sqrt{3}\sin\theta$.

   b Give the maximum and minimum values of $f(\theta)$ and the values of θ between 0 and $360°$ at which they occur.

5 **a** Express $7\cos\theta - 24\sin\theta$ in the form $r\cos(\theta + \alpha)$. Give α in degrees to the nearest $0.1°$ Hence sketch the graph of $f(\theta) = 7\cos\theta - 24\sin\theta + 3$.

   b Give the maximum and minimum values of $f(\theta)$ and the values of θ between 0 and $360°$ at which they occur.

6 Find the solution of these equations in the interval $0 \le x \le 360°$.

   a $\cos x + \sin x = \sqrt{2}$ **b** $7\cos x + 6\sin x = 2$

   c $\cos x - 3\sin x = 1$ **d** $2\cos x - \sin x = 2$

3.6 Double angle formulae

The compound angle formulae are true for any two angles A and B, and can therefore be used for two equal angles, that is, when $B = A$.

Replacing B by A in the trigonometric formulae for $(A + B)$ gives you the set of double angle formulae.

$$\sin 2A = 2\sin A\cos A$$

$$\cos 2A = \cos^2 A - \sin^2 A$$

$$\tan 2A = \frac{2\tan A}{1 - \tan^2 A}$$

Using $\cos^2 A + \sin^2 A = 1$, you can change the right-hand side of the formula for $\cos 2A$ into an expression involving *only* $\sin^2 A$ or $\cos^2 A$.

$$\cos^2 A - \sin^2 A = \begin{cases} (1 - \sin^2 A) - \sin^2 A = 1 - 2\sin^2 A \\ \cos^2 A - (1 - \cos^2 A) = 2\cos^2 A - 1 \end{cases}$$

$$\cos 2A = \begin{cases} \cos^2 A - \sin^2 A \\ 1 - 2\sin^2 A \\ 2\cos^2 A - 1 \end{cases}$$

The last two forms of the cosine double angle formulae can be rearranged to give $\sin^2 A$ or $\cos^2 A$ in terms of $\cos 2A$. These are useful when you need change either $\sin^2 A$ or $\cos^2 A$ into an expression involving $\cos 2A$.

Starting with $\cos 2A = 2\cos^2 A - 1$, and rearranging gives

$$\cos^2 A = \frac{1}{2}(1 + \cos 2A)$$

and starting with $\cos 2A = 1 - 2\sin^2 A$, and rearranging gives

$$\sin^2 A = \frac{1}{2}(1 - \cos 2A).$$

> **Note**
>
> You need to learn all the compound angle and double angle formulae.

Example 14

When $\tan \theta = \frac{3}{4}$, show that $\tan 2\theta = \frac{24}{7}$. Hence find the value of $\tan 4\theta$.

Using $\tan 2A = \frac{2\tan A}{1 - \tan^2 A}$ with $A = \theta$ and $\tan \theta = \frac{3}{4}$ gives

$$\tan 2\theta = \frac{2\left(\frac{3}{4}\right)}{1 - \left(\frac{3}{4}\right)^2} = \frac{24}{7}$$

Using the formula for $\tan 2A$ again, but this time with $A = 2\theta$, gives

$$\tan 4\theta = \frac{2\tan 2\theta}{1 - \tan^2 2\theta} = \frac{2\left(\frac{24}{7}\right)}{1 - \left(\frac{24}{7}\right)^2} = -\frac{336}{527}$$

Example 15

Eliminate θ from the equations $x = \cos 2\theta$, $y = \sec \theta$.

Using $\cos 2\theta = 2\cos^2 \theta - 1$ gives

$$x = 2\cos^2 \theta - 1 \text{ and } y = \frac{1}{\cos \theta}$$

therefore $x = 2\left(\frac{1}{y}\right)^2 - 1 \implies (x+1)y^2 = 2$

Example 16

Prove that $\sin 3A = 3\sin A - 4\sin^3 A$

$$\sin 3A = \sin (2A + A)$$
$$= \sin 2A \cos A + \cos 2A \sin A$$
$$= (2 \sin A \cos A) \cos A + (1 - 2 \sin^2 A) \sin A$$
$$= 2 \sin A \cos^2 A + \sin A - 2 \sin^3 A$$
$$= 2 \sin A (1 - \sin^2 A) + \sin A - 2 \sin^3 A$$
$$= 3 \sin A - 4 \sin^3 A$$

Example 17

Solve the equation $\cos 2x + 3 \sin x = 2$, giving values of x in the interval $-\pi \le x \le \pi$.

When an equation involves a trigonometric functions of different multiples of an angle, it is sensible to express the equation in a form where all the trigonometric functions are of the same angle and, when possible, involving only one trigonometric function.

Using $\cos 2x = 1 - 2 \sin^2 x$ gives

$$1 - 2 \sin^2 x + 3 \sin x = 2$$
$$\Rightarrow \qquad 2 \sin^2 x - 3 \sin x + 1 = 0$$
$$\Rightarrow \qquad (2 \sin x - 1)(\sin x - 1) = 0$$

therefore $\quad \sin x = \dfrac{1}{2} \quad$ or $\quad \sin x = 1$

When $\sin x = \dfrac{1}{2}$, $x = \dfrac{1}{6}\pi, \dfrac{5}{6}\pi \quad$ and $\quad$ when $\sin x = 1$, $x = \dfrac{1}{2}\pi$

Therefore the solution is $x = \dfrac{1}{6}\pi, \dfrac{5}{6}\pi, \dfrac{1}{2}\pi$.

Example 18

Express $4 \cos^2 x + 1$ in terms of the angle $2x$.

Using $\cos^2 x = \dfrac{1}{2}(1 + \cos 2x)$ gives

$$4 \cos^2 x + 1 = 4 \times \dfrac{1}{2}(1 + \cos 2x) + 1$$
$$= 2(1 + \cos 2x) + 1$$
$$= 3 + 2 \cos 2x$$

Exercise 6

For questions 1 to 8, simplify the expressions, giving an exact value where this is possible.

1 $2 \sin 15° \cos 15°$

2 $\cos^2 \dfrac{1}{8}\pi - \sin^2 \dfrac{1}{8}\pi$

3 $\sin \theta \cos \theta$

4 $1 - 2 \sin^2 4\theta$

5 $\dfrac{2 \tan 75°}{1 - \tan^2 75°}$

6 $\dfrac{2 \tan 3\theta}{1 - \tan^2 3\theta}$

7 $2\cos^2\dfrac{3}{8}\pi - 1$

8 $1 - 2\sin^2\dfrac{1}{8}\pi$

9 Find the value of $\cos 2\theta$ and $\sin 2\theta$ for an acute angle θ.

    a $\cos\theta = \dfrac{3}{5}$
    b $\sin\theta = \dfrac{7}{25}$
    c $\tan\theta = \dfrac{12}{5}$

10 Given that $\tan\theta = -\dfrac{7}{24}$ and θ is obtuse, find

    a $\tan 2\theta$
    b $\cos 2\theta$
    c $\sin 2\theta$
    d $\cos 4\theta$.

11 Eliminate θ from the pairs of equations.

    a $x = \tan 2\theta,\ y = \tan\theta$
    b $x = \cos 2\theta,\ y = \cos\theta$

    c $x = \cos 2\theta,\ y = \operatorname{cosec}\theta$
    d $x = \sin 2\theta,\ y = \sec 4\theta$

12 Express in terms of $\cos 2x$.

    a $2\sin^2 x - 1$
    b $4 - 2\cos^2 x$
    c $2\cos^2 x + \sin^2 x$

    d $2\cos^2 x(1 + \cos^2 x)$
    e $\cos^4 x$ (Hint: $\cos^4 x = (\cos^2 x)^2$)
    f $\sin^4 x$

13 Prove that these identities are correct.

    a $\dfrac{1 - \cos 2A}{\sin 2A} = \tan A$
    b $\sec 2A + \tan 2A = \dfrac{\cos A + \sin A}{\cos A - \sin A}$

    c $\cos 4A = 8\cos^4 A - 8\cos^2 A + 1$
    d $\sin 2\theta = \dfrac{2\tan\theta}{1 + \tan^2\theta}$

    e $\cos 2\theta = \dfrac{1 - \tan^2\theta}{1 + \tan^2\theta}$

14 Find solutions of the equations for angles from 0 to 2π.

    a $\cos 2x = \sin x$
    b $\sin 2x + \cos x = 0$
    c $\cos 2x = \cos x$

    d $\sin 2x = \cos x$
    e $4 - 5\cos\theta = 2\sin^2\theta$
    f $\sin 2\theta - 1 = \cos 2\theta$

    g $\cos^2\theta = \sin\theta - 1$
    h $\cos 2\theta = 1 + \sin\theta$

Summary

Inverse trigonometric functions

$\sin^{-1} x$ means the angle whose sine is x.

$f(x) = \sin^{-1} x$ has domain $-1 \le x \le 1$ and range $-\dfrac{1}{2}\pi \le f(x) \le \dfrac{1}{2}\pi$.

$\cos^{-1} x$ means the angle whose cosine is x.

$f(x) = \cos^{-1} x$ has domain $-1 \le x \le 1$ and range $0 \le f(x) \le \pi$.

$\tan^{-1} x$ means the angle whose tangent is x.

$f(x) = \tan^{-1} x$ has domain $x \in \mathbb{R}$ and range $-\dfrac{1}{2}\pi < f(x) < \dfrac{1}{2}\pi$.

Reciprocal trigonometric functions

$$\operatorname{cosec}\theta = \frac{1}{\sin\theta},\ \sec\theta = \frac{1}{\cos\theta},\ \cot\theta = \frac{1}{\tan\theta}$$

Trigonometric formulae

$$1 + \tan^2\theta = \sec^2\theta$$

$$\cot^2\theta + 1 = \operatorname{cosec}^2\theta$$

Compound angle formulae

$\sin(A + B) = \sin A \cos B + \cos A \sin B$

$\sin(A - B) = \sin A \cos B - \cos A \sin B$

$\cos(A + B) = \cos A \cos B - \sin A \sin B$

$\cos(A - B) = \cos A \cos B + \sin A \sin B$

$\tan(A + B) = \dfrac{\tan A + \tan B}{1 - \tan A \tan B}$

$\tan(A - B) = \dfrac{\tan A - \tan B}{1 + \tan A \tan B}$

$a \cos \theta \pm b \sin \theta$ can be expressed as $r \sin(\theta \pm \alpha)$ or $r \cos(\theta \pm \alpha)$

where $r = \sqrt{a^2 + b^2}$ and $\tan \alpha = \dfrac{a}{b}$ or $\tan \alpha = \dfrac{b}{a}$

Double angle formulae

$\sin 2A = 2 \sin A \cos A$

$\tan 2A = \dfrac{2 \tan A}{1 - \tan^2 A}$ $\qquad$ $\cos^2 A = \dfrac{1}{2}(1 + \cos 2A)$

$\cos 2A = \begin{cases} \cos^2 A - \sin^2 A \\ 1 - 2\sin^2 A \\ 2\cos^2 A - 1 \end{cases}$ $\qquad$ $\sin^2 A = \dfrac{1}{2}(1 - \cos 2A)$

Review

1 Eliminate α from the equations $x = \cos \alpha$, $y = \operatorname{cosec} \alpha$.

2 Find the solution of the equation $\sec \theta + \tan^2 \theta = 5$ for values of θ in the interval $0 \le \theta \le 360°$.

3 Prove that $\left(\cot \theta + \operatorname{cosec} \theta\right)^2 \equiv \dfrac{1 + \cos \theta}{1 - \cos \theta}$.

4 Simplify $\sec^4 \theta - \sec^2 \theta$.

5 Eliminate θ from the equations $x = \sec \theta - 3$, $y = 2 - \tan \theta$.

6 State the value of $\cos^{-1} \dfrac{\sqrt{3}}{2}$.

7 Solve the equation $\sin^{-1}(3x - 1) = \dfrac{\pi}{2}$.

8 Eliminate θ from the equations $x = \sin \theta$ and $y = \cos 2\theta$.

9 Prove the formula $\dfrac{\sin 2\theta}{1 + \cos 2\theta} = \tan \theta$.

10 Prove that $\tan\left(\theta + \dfrac{1}{4}\pi\right) \tan\left(\dfrac{1}{4}\pi - \theta\right) = 1$.

11 When $\cos A = \dfrac{4}{5}$ and $\cos B = \dfrac{5}{13}$ find the possible values of $\cos(A + B)$.

12 Eliminate θ from the equations $x = \cos 2\theta$ and $y = \cos^2 \theta$.

13 Solve the equation $8 \sin \theta \cos \theta = 3$ for values of θ from $-180°$ to $180°$.

14 Find the solution of the equation $\cos^2 \theta - \sin^2 \theta = 1$ for $-\pi \le \theta \le \pi$.

15 Show that $\cos^4 \theta - \sin^4 \theta = \cos 2\theta$.

16 Simplify the expression $\dfrac{1+\cos 2x}{1-\cos 2x}$.

17 Find the values of A between 0 and 360° for which
$\sin(60° - A) + \sin(120° - A) = 0$.

18 **a** Express $2\sin^2\theta + 1$ in terms of $\cos 2\theta$.

b Express $4\cos^2 2A$ in terms of $\cos 4A$ (Hint: Use $2A = x$).

19 Express $4\sin\theta - 3\cos\theta$ in the form $r\sin(\theta - \alpha)$. Hence find the local maximum and local minimum values of $\dfrac{7}{4\sin\theta - 3\cos\theta + 2}$.

20 Express $\cos x + \sin x$ in the form $r\cos(x - \alpha)$. Hence find the smallest positive value of x for which $\dfrac{1}{(\cos x + \sin x)}$ has a minimum value.

Assessment

1 **a** Express $3\cos x - 4\sin x$ in the form $r\cos(x + \alpha)$.

b Hence express $4 + \dfrac{10}{3\cos x - 4\sin x}$ in the form $4 + k\sec(x + \alpha)$.

c Sketch the graph of $y = 4 + \dfrac{10}{3\cos x - 4\sin x}$.

2 **a** Express $\sin 2\theta - \cos 2\theta$ in the form $r\sin(2\theta - \alpha)$.

b Hence find the smallest positive value of θ for which $\sin 2\theta - \cos 2\theta$ has a maximum value.

3 **a** Express $4\sin x + 3\cos x$ in the form $r\sin(x + \alpha)$.

b Find the local maximum value of $\dfrac{3}{4\sin x + 3\cos x}$.

4 **a** Find all the values of x between 0 and 360° for which $\cos x - 2\sin x = 1$.

b Solve the equation $3\cos x - 2\sin x = 1$ for values of x in the interval $0 \le x \le 360°$.

5 **a** Find the values of x in the range $0 \le x \le 360°$ that satisfy the equation $3\sec^2 x + 5\tan x = 5$.

b Eliminate t from the equations $x = 4\cos 2t$ and $y = 3\sin t$.

6 Sketch the curve with equation $y = \sin^{-1} 3x$, where y is in radians.

State the exact values of the coordinates of the end points of the graph.

AQA MPC3 June 2015 part question

7 **a** Express $2\cos x - 5\sin x$ in the form $R\cos(x + \alpha)$, where $R > 0$ and $0 < \alpha < \dfrac{\pi}{2}$, giving your value of α, in radians, to three significant figures.

b **i** Hence find the value of x in the interval $0 < x < 2\pi$ for which $2\cos x - 5\sin x$ has its maximum value. Give your value of x to three significant figures.

ii Use your answer to part (a) to solve the equation $2\cos x - 5\sin x + 1 = 0$ in the interval $0 < x < 2\pi$, giving your solutions to three significant figures.

AQA MPC4 June 2015

4 Exponential and Logarithmic Functions

Introduction

In this chapter you will use what you learnt about logarithms at AS level to calculate values associated with exponential growth and decay values. This chapter also introduces the number e, which is one of the most important numbers in mathematics, together with the associated exponential and logarithmic functions.

Recap

You will need to remember...

▶ The properties of $f(x) = a^x$ and the shape of its graph, where a is a positive number.
▶ The meaning of $\log_a b$, and that $\log_a b = c \iff b = a^c$.
▶ The laws of logarithms.
▶ How to solve equations of the form $a^x = b$.
▶ How to solve inequalities.
▶ The meaning of an inverse function.
▶ How to use a combination of transformations to sketch curves.

Objectives

By the end of this chapter, you should know how to...

▶ Use logarithms to solve problems involving growth or decay by a constant factor over equal time intervals.
▶ Define the irrational number e.
▶ Define the exponential and logarithmic functions in terms of e.

Applications

Many quantities are said to increase exponentially, such as the infection rate of a disease, the rate of inflation or the growth of capital in a savings account. All these can be calculated when the rate of increase is known.

4.1 Exponential growth and decay

There are many examples where a quantity grows or decays by a constant factor over equal time intervals. This is called **exponential growth** or **exponential decay.**

Exponential growth

Suppose a debt of £100 has 2% interest added each month then, if no repayment is made,

the debt grows to $£100 \times \left(1 + \dfrac{2}{100}\right) = £100 \times 1.02$ after one month

to $£100 \times 1.02 \times 1.02$, that is $£100 \times (1.02)^2$, after two months

and so on, to $£100 \times (1.02)^n$ after n months.

1.02 is called the **growth factor** per month.

Therefore when £y is the debt after x months, $y = 100(1.02)^x$.

This is an **exponential function** that *increases* in value as the exponent x increases, so y grows exponentially.

Exponential decay

Some business assets (such as vehicles) depreciate over time. Suppose that at the end of each year, the value of a lorry is half its value at the start of the year. In this case, a lorry which was initially worth £A is worth £$A \times (1 - 0.5) = £A \times (0.5)$ after one year, is worth £$A \times (0.5)^2$ after two years, and so on.

If £y is the lorry's value after n years, then $y = A \times (0.5)^x = A\left(\dfrac{1}{2}\right)^n = A \times 2^{-n}$.

The function 2^{-n} is an exponential function that *decreases* in value as n increases, so y decays exponentially.

In each example above, the mathematical expression for the relationship is obtained by making certain assumptions and it is not necessarily valid at all times. The first example assumes that interest rates remain constant, and this is not usually the case. Therefore this relationship is valid only for the time during which the interest rate is 2%.

The assumption in the second case is that the rule for writing down the value of assets never changes, but this is not normally true in practice. After a few years the vehicle will have a value small enough to be written off completely.

Example 1

Ashis invests £1000 in a savings account at a fixed interest rate of 1.5% per year.

a Find, to the nearest pound, the value of the investment after 8 years.

b The money is kept invested for N whole years. Find the value of N for which the value of the investment first exceeds £2000.

c Rajiv invests £800 in a savings account with a different bank at a fixed interest rate of 2% per year. Ashis and Rajiv invest their money at the same time. Find the number of complete years after which Rajiv's investment first exceeds Ashis's investment.

a The value of Ashis's investment after n years is £$1000(1.015)^n$.

When $n = 8$, $1000(1.015)^8 = 1126.4...$

Therefore the value of the investment after 8 years is £1126 correct to the nearest pound.

b After N years, the value of the investment is $1000(1.015)^N$.

$$1000(1.015)^N > 2000 \quad \Rightarrow \quad (1.015)^N > 2$$

Therefore $N \log(1.015) > \log 2$

$\Rightarrow \qquad N > \dfrac{\log 2}{\log(1.015)} = 46.5...$

Therefore the value of N for which the value of the investment first exceeds £2000 is 47 (since N is a whole number).

> **Note**
>
> You can use $\log_{10}$ on your calculator to do this calculation

(continued)

c The value of Rajiv's investment after n years is £800$(1.02)^n$.

The value of Rajiv's investment is greater than the value of Ashis's investment when £800$(1.02)^n$ > £1000$(1.015)^n$.

$$800(1.02)^n > 1000(1.015)^n \quad \Rightarrow \quad \left(\frac{1.02}{1.015}\right)^n > 1.25$$

Therefore $n\log\left(\dfrac{1.02}{1.015}\right) > \log 1.25 \quad \Rightarrow \quad n > \dfrac{\log(1.25)}{\log\left(\dfrac{1.02}{1.015}\right)} = 45.4...$

Therefore Rajiv's investment first exceeds Ashis's investment after 46 complete years.

Exercise 1

1 When a toll bridge opens, the cost for a car to cross the bridge is £3. At the end of the first year, and after each subsequent year, the cost will increase by a fixed rate of 3%. Find the cost for a car to cross the bridge after 5 years.

2 The cost of a new industrial machine is £10 000. Its value depreciates at the fixed rate of 10% per year.

   a Find the value of the machine after 4 years.

   b Find the number of whole years after which the value of the machine is less than £5000.

3 Mr Brown buys a house for £250 000. The value increases by a constant 5% each year.

   Find the number of whole years after which the value of the house is more than £300 000.

4 The cost of a new car is £20 000. The value of the car depreciates by a constant 10% each year.

   Find the number of whole years after which the car has lost half its original value.

5 Mr Said takes out a loan. The cost of the loan, including interest, is £5000.

   Mr Said pays back a fixed amount of 10% of the outstanding loan each year.

   a Find the amount that Mr Said will owe 4 years after he took the loan.

   b Find the number of complete years after which the loan is less than £500.

6 Mr Mendes invests £1000 at Bank Avro at a fixed interest rate of 2% per year.

   a Find the number of complete years after which the value of the investment first exceeds £1500.

   b Mr Nero invests £900 at Bank Bifra at a fixed rate of interest of 2.5% per year.

      Find the number of complete years after which the value of Mr Nero's investment is greater than the value of Mr Mendes' investment.

7 Mr Ahmed pays $200 000 for a house. The value of the house increases at the fixed rate of 4% per year.

   Mr Daud pays $250 000 for a yacht. The value of the yacht decreases at the fixed rate of 5% per year.

   Find the number of complete years after which the value of the yacht is less than the value of the house.

Answer

8 Mr Carlos buys a painting for $100 000. The value of the painting is expected to increase by a constant 6% per year.

At the same time Mr Edwards buys a gold goblet for $50 000. The value of the goblet is expected to increase by a constant 10% per year.

Find the number of complete years after which the value of the goblet is expected to be greater than the value of the painting.

9 A culture dish was seeded with 4 mm² of mould. The table shows the area of the mould at six-hourly intervals.

Time, x hours	0	6	12	18	24	30	36
Area, y mm²	4	8.1	15.9	33	68	118	190

a Show that, for a time, the growth factor in the area for each interval of 6 hours is roughly 2 (that is, that the area approximately doubles every 6 hours). Hence show that it is reasonable to use an exponential function to model the relationship between time and area, and suggest a possible function.

b Give reasons why

   i   the model is approximate

   ii the model ceases to be reasonable after some time.

10 The equation $y = A(1.3)^{-t}$, where A is constant, is used to predict the value $y of a company vehicle when it is t years old.

a What is the meaning of the constant A?

b A new car is valued at $15 000. Use the model to predict the value of this car when it is two years old. Give one reason why this value should only be considered as approximate.

4.2 The exponential function

The number e is an irrational number (like π) so it cannot be expressed exactly as a decimal.

The value of e can be calculated, to as many decimal places as needed, from the infinite series

$$1 + \frac{1}{1!} + \frac{1}{2!} + \frac{1}{3!} + \frac{1}{4!} + \frac{1}{5!} + \cdots \text{ and is equal to 2.71823 correct to 5 decimal places.}$$

The exponential function (as opposed to *an* exponential function) is defined as $f(x) = e^x$ for all real values of x.

For any value of a ($a > 0$), a^x is *an* exponential function, however for the base e (e ≈ 2.718), e^x is *the* exponential function.

The diagrams show sketches of $y = e^x$ and some simple variations of this function.

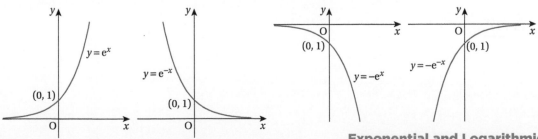

Exercise 2

1 Find the value, correct to 3 significant figures, of

   a e^2 **b** e^{-1} **c** $e^{1.5}$ **d** $e^{-0.3}$.

2 Sketch the curve whose equation is

   a $y = 1 - e^x$ **b** $y = e^x + 1$ **c** $y = e^{(x-1)}$

   d $y = 2 - e^x$ **e** $y = 1 + e^{-x}$ **f** $y = e^{2x}$

   g $y = 2 - e^{2x}$.

4.3 Natural logarithms

Logarithms to the base e are called **natural** (or **Naperian**) **logarithms**. The natural logarithm of a, that is $\log_e a$ is written as $\ln a$, so

$$\ln a = b \iff a = e^b$$

Logarithms to the base 10 are called **common logarithms** and are written as log.

The laws used for working with logarithms to a general base also apply to natural logarithms, so

$$\ln a + \ln b = \ln ab$$
$$\ln a - \ln b = \ln \frac{a}{b}$$
$$\ln a^n = n \ln a$$

Example 2

Question

Separate $\ln (\tan x)$ into two terms.

Answer

$$\ln (\tan x) = \ln \left(\frac{\sin x}{\cos x} \right)$$
$$= \ln (\sin x) - \ln (\cos x)$$

Example 3

Question

Express $4\ln(x+1) - \frac{1}{2}\ln x$ as a single logarithm.

Answer

$$4\ln(x+1) - \frac{1}{2}\ln x = \ln(x+1)^4 - \ln\sqrt{x}$$

$$= \ln \frac{(x+1)^4}{\sqrt{x}}$$

4.4 The logarithmic function

Look at the curve with equation $y = f(x)$ where $f(x) = \ln x$.

When $y = \ln x$ then $x = e^y$, so

the logarithmic function is the inverse of the exponential function.

It follows that the curve $y = \ln x$ is the reflection of the curve $y = e^x$ in the line $y = x$.

There is no part of the curve $y = \ln x$ for which x is negative.

This is because, when $x = e^y$ (that is when $y = \ln x$), x is positive for all real values of y.

Therefore

$\ln x$ **does not exist for values of** x **where** $x \leq 0$.

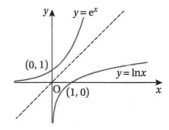

Example 4

Question

Sketch the curve whose equation is $y = \ln(x - 3)$.

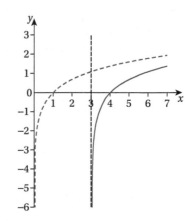

Answer

The curve $y = \ln(x - 3)$ is a translation of the curve $y = \ln x$ by the vector $\begin{bmatrix} 3 \\ 0 \end{bmatrix}$.

The curve $y = \ln x$ crosses the x-axis where $x = 1$, and the y-axis is an asymptote.

Therefore the curve $y = \ln(x - 3)$ crosses the x-axis where $x = 4$, and the line $x = 3$ is an asymptote.

Exercise 3

1 Write these in logarithmic form.

a $e^x = 4$ **b** $e^2 = y$ **c** $e^{2x} = 3$

d $e^{(x-1)} = 5$ **e** $e = x$

2 Write these in index form.

a $\ln x = 4$ **b** $\ln 0.5 = x$ **c** $\ln x = y$

d $2 \ln x = 3$ **e** $2 \ln(1 - x) = 1.5$

3 Find the value of

a $\ln 48$ **b** $\ln e$ **c** $\ln 1$.

4 Express each as a sum or difference of logarithms.

a $\ln \dfrac{x^2}{(x+1)}$ **b** $\ln(a^2 - b^2)$

c $\ln(\cot x)$ **d** $\ln(\sin^2 x)$

5 Express each as a single logarithm.

a $\ln(\cos x) - \ln(\sin x)$ **b** $1 + \ln x$ **c** $\dfrac{2}{3}\ln(x-1)$

6 Solve the equations for x.

   a $e^x = 8.2$ **b** $e^{2x} + e^x - 2 = 0$ ◄

   c $e^{2x-1} = 3$ **d** $e^{4x} - e^x = 0$

> **Note**
>
> Hint: Use $e^{2x} = (e^x)^2$

7 Sketch each curve and mark the vertical asymptote on your sketch.

   a $y = -\ln x$ **b** $y = \ln(-x)$ **c** $y = 2 + \ln x$

   d $y = \ln(x^2)$ **e** $y = \ln(3 - x)$ **f** $y = 3 - \ln x$

Summary

▶ Exponential growth or decay is where a quantity grows or decays by a constant factor over equal time intervals.

▶ The number $e \approx 2.718...$ is an irrational number.

▶ The function f where $f(x) = e^x$ is the exponential function, and is defined for all real numbers x.

▶ The logarithmic function f where $f(x) = \ln x$, is the inverse of the exponential function.

▶ The function f where $f(x) = \ln x$ only exists for positive values of x.

Review

1 Find the value of

   a e^4 **b** $e^{-1.5}$ **c** $e^{\frac{1}{2}}$.

2 Sketch the curve whose equation is

   a $y = e^{3x}$ **b** $y = e^x - 2$.

3 Write in logarithmic form.

   a $e^x = 2$ **b** $e^3 = y$.

4 Write in index form.

   a $\ln x = 2$ **b** $\ln 0.4 = x$ **c** $\ln(x - 1) = y$

5 Find the value of $\ln 4$.

6 Express as a sum or difference of logarithms.

   a $\ln \dfrac{x}{x^2 + 1}$ **b** $\ln(x^2 - 2x + 1)$

7 Express as a single logarithm.

   a $\ln(x - 1) - \ln x$ **b** $\ln(\sin x) - \ln(\cos x)$

8 Solve the equation $e^x = 10$.

9 Sketch the curve whose equation is $y = 2 - \ln(x + 1)$.

Assessment

1 Juan buys a vintage car for £2000. The car is expected to increase in value at the fixed rate of 2.5% per year.

   a   Find the value of the car after 5 years correct to the nearest pound.

   b   The car is kept for N whole years. Find the value of N for which the value of the car first exceeds £3000.

c Mujad invests £1500 in a savings account at a fixed interest rate of 3% per year.

Mujad invests his money at the same time at which Juan buys the car. Find the number of complete years after which Mujad's investment first exceeds the value of Juan's car.

2 a Find the value of $e^{-\frac{1}{3}}$.

b Sketch the curve whose equation is $y = 1 - e^{2x}$.

3 a Write $e^{(x-1)} = 3$ in logarithmic form.

b Solve the equation $e^{x-1} = 1.5$.

4 a Express $2\ln(x+1) - \ln(x-1)$ as a single logarithm.

b Show the equation $2\ln(x+1) - \ln(x-1) = 0$ has no real roots.

5 The equation $y = A(1.02)^t$, where A is constant, is used to predict the value \$$y$ of a company share t years after it is purchased.

a What is the meaning of constant A?

b Aisha bought 100 of these shares when they cost \$1.20 each. Find the cost of these shares.

c Use the model to predict the value of these shares five years later. Give one reason why this value should only be considered as approximate.

6 The functions f is defined by $f(x) = 5 - e^{3x}$ for all real values of x

a Find the range of f.

b The inverse of f is f^{-1}.

   i Find $f^{-1}(x)$.   **ii** Solve the equation $f^{-1}(x) = 0$.

AQA MPC3 June 2015 (part question)

7 The curve with equation $y = f(x)$, where $f(x) = \ln(2x - 3)$, $x > \dfrac{3}{2}$, is sketched below.

a The inverse of f is f^{-1}.

   i Find $f^{-1}(x)$.

   ii State the range of f^{-1}.

   iii Sketch, the curve with equation $y = f^{-1}(x)$, indicating the value of the y-coordinate of the point where the curve intersects the y-axis.

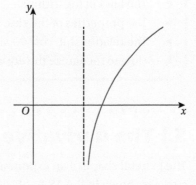

b The function g is defined by $g(x) = e^{2x} - 4$, for all real values of x

   i Find $gf(x)$, giving your answer in the form $(ax - b)^2 - c$, where a, b and c are integers.

   ii Write down an expression for $fg(x)$, and hence find the exact solution of the equation $fg(x) = \ln 5$.

AQA MPC3 June 2013

5 Differentiation

Introduction

This chapter extends differentiation to exponential, logarithmic and trigonometric functions and to combinations of these functions.

Sometimes the equation of a curve giving the direct relationship between x and y is awkward to work with. In this case it helps to express x and y each in terms of a third variable, called a parameter, which makes the equation much easier to work with.

All the differentiation techniques you have seen so far have been used on equations that can be expressed in the form $y = f(x)$. However some curves have equations that cannot easily be written in this way. This chapter shows how gradients of such curves can be found.

Recap

You will need to remember...

▶ The meaning of the functions e^x and $\ln x$ and the trigonometric functions.
▶ How to differentiate polynomials.
▶ How to find stationary values on a curve.
▶ How to find the equation of a tangent and a normal to a curve at a given point on the curve.
▶ The laws of logarithms.
▶ The properties of the sine and cosine functions.
▶ The relationship $\cos^2\theta + \sin^2\theta = 1$.
▶ How to recognize the equation of a circle.

Objectives

By the end of this chapter, you should know how to...

▶ Find and use the derivative of the functions e^x, $\ln x$, $\sin x$, and $\cos x$.
▶ Find and use formulae for differentiating products and quotients of function for differentiating $\tan x$, and for differentiating a composite function.
▶ Explain the meaning of an implicit function and how to differentiate such a function with respect to x.
▶ Understand parameters.
▶ Convert a parametric equation of a curve into the Cartesian equivalent.
▶ Find $\dfrac{dy}{dx}$ from the parametric equation of a curve and the equation of a tangent or a normal to the curve, at a general point on the curve.

5.1 The derivative of e^x

The general shape of an exponential curve is shown in the AS Book.

This diagram shows three exponential curves.

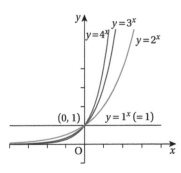

They all pass through the point (0, 1) because, when $y = a^x$ for any positive value of a, then $a^0 = 1$.

Each exponential function has a unique property: when the gradient at any point on the curve $y = a^x$ is divided by the value of y at that point, the result is always the same number.

The table shows the value of $\frac{dy}{dx} \div y$ for some exponential functions.

$y =$	2^x	3^x	4^x
$\frac{dy}{dx} \div y \approx$	0.7	1.1	1.4

The table shows that there is a number, somewhere between 2 and 3, for which

$$\frac{dy}{dx} \div y = 1, \quad \text{that is} \quad \frac{dy}{dx} = y.$$

This number is e, which you met in Chapter 4.

When $y = e^x$ **then** $\frac{dy}{dx} = e^x$.

The function f where $f(x) = e^x$ is the only function which is unchanged when differentiated.

Example 1

a Find the gradient at the point on the curve $y = 2e^x - x$ where $x = 2$.

b Find the exact value of x at which $f(x) = 2e^x - x$ has a stationary value.

a The derivative of $cf(x) =$ is $cf'(x)$ therefore $\frac{d}{dx}(2e^x) = 2\frac{d}{dx}(e^x)$

$$y = 2e^x - x \implies \frac{dy}{dx} = 2e^x - 1$$

When $x = 2$, $\frac{dy}{dx} = 2e^2 - 1$

Note

$f'(x)$ means the derivative of $f(x)$ with respect to x.

b $f(x)$ has a stationary value where $2e^x - 1 = 0$

$$\implies \quad e^x = \frac{1}{2} \quad \implies \quad x \ln e = \ln\left(\frac{1}{2}\right)$$

$\ln e = 1$ therefore $x = \ln\left(\frac{1}{2}\right) = -\ln 2$

$f(x)$ has a stationary value where $x = -\ln 2$.

Exercise 1

1 Write down the derivative of

    a $2e^x$ **b** $x^2 - e^x$ **c** e^x.

In questions 2 to 4, find the gradient of each curve at the specified value of x.

2 $y = e^x - 2x$ where $x = 2$

3 $y = x^2 + 2e^x$ where $x = 1$

4 $y = e^x - 3x^3$ where $x = 0$

5 Find the value of x at which $f(x) = e^x - x$ has a stationary value.

6 Find the value of x at which $f(x) = 4x - e^x$ has a stationary value.

7 Explain why $f(x) = e^x + x$ does not have a stationary value.

5.2 The derivative of ln x

To find the derivative of ln x, a relationship is needed between $\dfrac{dy}{dx}$ and $\dfrac{dx}{dy}$.

The equation $y = f(x)$, where $f(x)$ is any function of x, is such that

$$\frac{dy}{dx} = \lim_{\delta x \to 0} \frac{\delta y}{\delta x} = \lim_{\delta x \to 0} \left(\frac{1}{\dfrac{\delta x}{\delta y}} \right)$$

Now $\delta y \to 0$ as $\delta x \to 0$, $\displaystyle\lim_{\delta y \to 0} \left(\frac{1}{\dfrac{\delta x}{\delta y}} \right)$, so

$$\frac{dy}{dx} = \frac{1}{\dfrac{dx}{dy}}$$

This relationship can be used to find the derivative of *any* function if the derivative of its inverse is known.

To differentiate $y = \ln x$, remember that

$$y = \ln x \iff x = e^y$$

Differentiating e^y with respect to y gives

$$\frac{dx}{dy} = e^y = x$$

Therefore $\quad \dfrac{dy}{dx} = \dfrac{1}{\left(\dfrac{dx}{dy} \right)} = \dfrac{1}{x} \quad$ so

$$\frac{d}{dx} \ln x = \frac{1}{x}.$$

This result can be used to differentiate many logarithmic functions if they are first simplified by applying the laws of logarithms.

Example 2

Find the derivative, with respect to x, of ln $(2x)$.

$f(x) = \ln (2x) = \ln 2 + \ln x$

$\dfrac{d}{dx}\{f(x)\} = 0 + \dfrac{1}{x} \quad$ (as ln 2 is a number)

The derivative of ln $(2x)$ is $\dfrac{1}{x}$.

Example 3

Find the derivative, with respect to x, of

a $\quad \ln\left(\dfrac{1}{x^3}\right)$

b $\quad \ln\left(4\sqrt{x}\right)$.

Answer

a $f(x) = \ln\left(\dfrac{1}{x^3}\right) = \ln(x^{-3}) = -3\ln x$

$\dfrac{d}{dx}\{f(x)\} = \dfrac{d}{dx}\{-3\ln x\} = \dfrac{-3}{x}$

b $f(x) = \ln\left(4\sqrt{x}\right) = \ln 4 + \ln\left(\sqrt{x}\right) = \ln 4 + \dfrac{1}{2}\ln x$

$\dfrac{d}{dx}\{f(x)\} = \dfrac{d}{dx}(\ln 4) + \dfrac{d}{dx}\left(\dfrac{1}{2}\ln x\right) = 0 + \dfrac{\frac{1}{2}}{x} = \dfrac{1}{2x}$

Exercise 2

1 Write down the derivative, with respect to x, of each of these expressions.

   a $\ln x^3$ **b** $\ln(3x)$ **c** $\ln(x^{-2})$

   d $\ln\left(\dfrac{3}{\sqrt{x}}\right)$ **e** $\ln\left(\dfrac{1}{x^5}\right)$ **f** $\ln\left(2x^{\frac{1}{2}}\right)$

   g $\ln\left(x^{-\frac{3}{2}}\right)$ **h** $\ln\left(\dfrac{x^3}{\sqrt{x}}\right)$

2 Find the stationary points on each curve.

   a $y = \ln x - x$ **b** $y = x^3 - 2\ln x^3$ **c** $y = \ln x - \sqrt{x}$

Using $\dfrac{dy}{dx} = \dfrac{1}{\dfrac{dx}{dy}}$ to differentiate an equation of the form $x = f(y)$

When, for example, the equation of a curve is $x = y^3$,

you can find $\dfrac{dy}{dx}$ by first finding $\dfrac{dx}{dy}$.

In this case $\dfrac{dx}{dy} = 3y^2$.

Then using $\dfrac{dy}{dx} = \dfrac{1}{\dfrac{dx}{dy}}$, you can see that $\dfrac{dy}{dx} = \dfrac{1}{3y^2}$.

Example 4

Question

A curve has equation $x = y + \ln y^2$. Find the value of $\dfrac{dy}{dx}$ when $y = 2$.

Answer

$x = y + \ln y^2 \implies x = y + 2\ln y$

$\dfrac{dx}{dy} = 1 + \dfrac{2}{y} = \dfrac{y+2}{y}$ so $\dfrac{dy}{dx} = \dfrac{1}{\dfrac{(y+2)}{y}} = \dfrac{y}{y+2}$

When $y = 2$, $\dfrac{dy}{dx} = \dfrac{2}{4} = \dfrac{1}{2}$

Example 5

Find the equation of the tangent to the curve whose equation is $x = 3y^2 - 2y$ at the point where $y = 3$.

To find the equation of the tangent at the point where $y = 3$, you need to know the coordinates of the point where $y = 3$, and the gradient of the curve at that point.

When $y = 3$, $x = 21$.

$$x = 3y^2 - 2y \quad \Rightarrow \quad \frac{dx}{dy} = 6y - 2$$

Therefore $\frac{dy}{dx} = \frac{1}{6y - 2}$ so when $y = 3$, the gradient of the tangent is $\frac{1}{16}$.

The equation of the tangent is $y - 3 = \frac{1}{16}(x - 21) \quad \Rightarrow \quad 16y - x - 27 = 0$.

Exercise 3

1 The equation of a curve is $x = 4y^2$. Find $\frac{dy}{dx}$ when $y = 1$.

2 The equation of a curve is $x = y^2 - 3y + 2$. Find $\frac{dy}{dx}$ when $y = 3$.

3 The equation of a curve is $x = 2y^3 + 3y$. Find $\frac{dy}{dx}$ when $y = \frac{1}{2}$.

4 The equation of a curve is $x = y^2 + y$. Find $\frac{dy}{dx}$ when $y = 1$.

Hence find the equation of the tangent to the curve at the point on the curve where $y = 1$.

5 The equation of a curve is $x = e^y + y$. Find $\frac{dy}{dx}$ when $y = 1$.

Hence find the equation of the tangent to the curve at the point on the curve where $y = 2$.

6 The equation of a curve is $x = y - \ln y$. Find $\frac{dy}{dx}$ when $y = 3$.

Hence find the equation of the tangent to the curve at the point on the curve where $y = 3$.

7 The equation of a curve is $x = \ln(y^2)$. Find $\frac{dy}{dx}$ when $y = 2$.

Hence find the equation of the tangent to the curve at the point on the curve where $y = 2$.

8 The equation of a curve is $x = 3y + 4e^y$. Find $\frac{dy}{dx}$ when $y = 1$.

Hence find the equation of the tangent to the curve at the point on the curve where $y = 1$.

5.3 The derivatives of sin x and cos x

When x is measured in radians the gradient function of $\sin x$ is $\cos x$, and the gradient function of $\cos x$ is $-\sin x$.

When $y = \sin x$ then $\dfrac{dy}{dx} = \cos x$.

When $y = \cos x$ then $\dfrac{dy}{dx} = -\sin x$.

These results are found by differentiating from first principles and you can quote them without having to prove them.

It is important to understand that these results are only valid when x is measured in radians and, throughout all the work that follows involving the differentiation of trigonometric functions, the angle is measured in radians unless it is stated otherwise.

Example 6

Question

Find the smallest positive value of x for which $f(x) = x + 2 \cos x$ has a stationary value.

Answer

$f(x) = x + 2 \cos x \quad \Rightarrow \quad f'(x) = 1 - 2 \sin x$

For stationary values, $f'(x) = 0$

so $\quad 1 - 2 \sin x = 0 \quad \Rightarrow \quad \sin x = \dfrac{1}{2}$

The smallest positive angle for which $\sin x = \dfrac{1}{2}$ is $\dfrac{\pi}{6}$ radians.

> **Note**
>
> The answer *must* be given in radians because the rule used to differentiate $\cos x$ is valid only for an angle in radians.

Example 7

Question

Find the smallest positive value of θ for which the curve $y = 2\theta - 3 \sin \theta$ has a gradient of $\dfrac{1}{2}$.

Answer

$y = 2\theta - 3 \sin \theta \quad \Rightarrow \quad \dfrac{dy}{d\theta} = 2 - 3 \cos \theta$

When $\dfrac{dy}{d\theta} = \dfrac{1}{2},\quad 2 - 3 \cos \theta = \dfrac{1}{2} \quad \Rightarrow \quad 3 \cos \theta = \dfrac{3}{2} \quad \Rightarrow \quad \cos \theta = \dfrac{1}{2}$

The smallest positive value of θ for which $\cos \theta = \dfrac{1}{2}$ is $\dfrac{\pi}{3}$ radians.

Exercise 4

1 Write down the derivative of each of these functions with respect to the given variable.

   a $\sin x - \cos x$ **b** $\sin \theta + 4$ **c** $3 \cos \theta$

   d $5 \sin \theta - 6$ **e** $2 \cos \theta + 3 \sin \theta$ **f** $4 \sin x - 5 - 6 \cos x$

2 Write down the derivative of each of these functions with respect to the given variable.

a $x^2 - \sin x$ b $\cos x + \ln x$ c $2 \sin x - e^x$

d $3e^x - \ln x + 4 \cos x$ e $3 + 2 \ln x + 5 \sin x$ f $x^3 - 3\cos x$

3 Find the gradient of each curve at the point whose x-coordinate is given.

a $y = 2 \cos x; \frac{1}{2}\pi$ b $y = \sin x; 0$ c $y = \cos x + \sin x; \pi$

d $y = x - \sin x; \frac{1}{2}\pi$ e $y = 2 \sin x - x^2; -\pi$ f $y = -4 \cos x; \frac{1}{2}\pi$

4 For each of these curves, find the smallest positive value of θ at which the gradient of the curve has the given value.

a $y = 2 \cos \theta; -1$ b $y = \theta + \cos \theta; \frac{1}{2}$

c $y = \sin \theta + \cos \theta; 0$ d $y = 2\theta + \sin \theta; 1$

5 Considering only positive values of x, find the coordinates of the first two stationary points on each of these curves and determine whether they are maximum or minimum points.

a $y = 2 \sin x - x$ b $y = x + 2 \cos x$

6 Find the equation of the tangent to the curve $y = \cos \theta + 3 \sin \theta$ at the point where $\theta = \frac{1}{2}\pi$.

7 Find the equation of the normal to the curve $y = x^2 + \cos x$ at the point where $x = \pi$.

8 Find the coordinates of a point on the curve $y = \sin x + \cos x$ at which the tangent is parallel to the line $y = x$.

5.4 Differentiating products, quotients and composite functions

Differentiating a product

When you are given a is curve $y = u(x)\, v(x)$, where u and v are both functions of x, for example $y = x^2 \sin x$, you need to know that

$\dfrac{dy}{dx}$ is NOT equal to $\dfrac{d}{dx}(x^2) \times \dfrac{d}{dx}(\sin x)$.

Consider $y = u(x)\, v(x)$. A small increase (δx) in x, gives corresponding small increases of δu, δv and δy in the values of u, v and y.

Therefore $y + \delta y = (u + \delta u)(v + \delta v)$

$$= uv + u\delta v + v\delta u + \delta u \delta v$$

Now $y = uv$

therefore $\delta y = u\delta v + v\delta u + \delta u \delta v$

$\Rightarrow$ $\dfrac{\delta y}{\delta x} = u\dfrac{\delta v}{\delta x} + v\dfrac{\delta u}{\delta x} + \delta u\dfrac{\delta v}{\delta x}$

As $\delta x \to 0$, $\dfrac{\delta v}{\delta x} \to \dfrac{dv}{dx}$, $\dfrac{\delta u}{\delta x} \to \dfrac{du}{dx}$ and $\delta u \to 0$

Therefore $\dfrac{dy}{dx}=\lim\limits_{\delta x\to 0}\dfrac{\delta y}{\delta x}$

$$=u\dfrac{dv}{dx}+v\dfrac{du}{dx}+0$$

Therefore $\dfrac{d}{dx}(uv)=v\dfrac{du}{dx}+u\dfrac{dv}{dx}$

Applying this formula to the example $y=(x^2)\sin x$, with $u(x)=x^2$ and $v(x)=\sin x$, gives

$$\dfrac{dy}{dx}=(\sin x)(2x)+(x^2)(\cos x)=2x\sin x+x^2\cos x.$$

Example 8

Question

Differentiate with respect to x

a $\quad x^3\ln x$ 　　　　　　　b $\quad\dfrac{\cos x}{x}$.

Answer

a $\quad u=x^3\quad\Rightarrow\quad\dfrac{du}{dx}=3x^2$

$\quad v=\ln x\quad\Rightarrow\quad\dfrac{dv}{dx}=\dfrac{1}{x}$

$\dfrac{d}{dx}(uv)=v\dfrac{du}{dx}+u\dfrac{dv}{dx}\quad$ gives $\quad\dfrac{dy}{dx}=(3x^2)\ln x+\left(\dfrac{1}{x}\right)x^3=x^2+3x^3\ln x$

b $\quad$ Writing $\dfrac{\cos x}{x}$ as $(\cos x)(x^{-1})$

$\quad$ then $\quad u=\cos x\quad$ gives $\quad\dfrac{du}{dx}=-\sin x$

$\quad$ and $\quad v=x^{-1}\quad$ gives $\quad\dfrac{dv}{dx}=-x^{-2}$

$\quad$ Using $\quad\dfrac{d}{dx}(uv)=v\dfrac{du}{dx}+u\dfrac{dv}{dx}\quad$ gives $\quad\dfrac{dy}{dx}=(-x^{-2})(\cos x)+(-\sin x)(x^{-1})$

$$=-\dfrac{\cos x}{x^2}-\dfrac{\sin x}{x}=-\dfrac{\cos x+x\sin x}{x^2}.$$

Exercise 5

Differentiate each expression with respect to x.

1 $\dfrac{\sin x}{x}$ 　　　　　　**2** $e^x\cos x$ 　　　　　　**3** $(x^3-2)\ln x$

4 $(x+1)\sin x$ 　　　　**5** $\sin x\cos x$ 　　　　**6** $\dfrac{\ln x}{x^2}$

7 $(\cos x)\ln x$ 　　　　**8** $e^x\sin x$ 　　　　　　**9** $x^2\sin x$

10 $x^3\ln 2x$

Differentiating a quotient

To differentiate a function of the form $y=\dfrac{u(x)}{v(x)}$, where u and v are both functions of x, the function can be rewritten as $y=u(x)v^{-1}(x)$ and so the formula to differentiate a product can be used. This method was used in part (b) of Example 8 but it is not always the neatest way to differentiate a quotient. The alternative is to apply the formula derived below.

When a function is of the form $y = \dfrac{u(x)}{v(x)}$, where u and v are both functions of x, a small increase of δx in the value of x gives corresponding small increases of δu and δv in the values u and v. Then, as $\delta x \to 0$, δu and δv also tend to zero.

When $\qquad y = \dfrac{u}{v} \quad$ then $\quad y + \delta y = \dfrac{(u + \delta u)}{(v + \delta v)}$

so $\qquad \delta y = \dfrac{u + \delta u}{v + \delta v} - \dfrac{u}{v} = \dfrac{v\delta u - u\delta v}{v(v + \delta v)}$

Therefore $\quad \dfrac{\delta y}{\delta x} = \dfrac{\left(v\dfrac{\delta u}{\delta x} - u\dfrac{\delta v}{\delta x} \right)}{v(v + \delta v)}$

$\Rightarrow \qquad \dfrac{dy}{dx} = \lim_{\delta x \to 0} \dfrac{\delta y}{\delta x} = \dfrac{v\dfrac{du}{dx} - u\dfrac{dv}{dx}}{v^2}$

Therefore $\qquad \dfrac{dy}{dx} = \dfrac{v\dfrac{du}{dx} - u\dfrac{dv}{dx}}{v^2}$

Example 9

Question

Find $\dfrac{dy}{dx}$ when $y = \dfrac{4x - 3}{\sin x}$.

Answer

Taking $u = 4x - 3$ gives $\dfrac{du}{dx} = 4$ and $v = \sin x$ gives $\dfrac{dv}{dx} = \cos x$

so $\qquad \dfrac{dy}{dx} = \dfrac{\left(v\dfrac{du}{dx} - u\dfrac{dv}{dx} \right)}{v^2}$

$\qquad = \dfrac{4\sin x - (4x - 3)\cos x}{(\sin x)^2}$

Example 10

Question

Find $\dfrac{dy}{dx}$ when $y = \dfrac{\sin x}{\cos x}$.

Answer

Taking $u = \sin x$ gives $\dfrac{du}{dx} = \cos x$ and taking $v = \cos x$ gives $\dfrac{dv}{dx} = -\sin x$

so $\qquad \dfrac{dy}{dx} = \dfrac{\left(v\dfrac{du}{dx} - u\dfrac{dv}{dx} \right)}{v^2}$

$\qquad = \dfrac{\cos^2 x + \sin^2 x}{\cos^2 x} = \dfrac{1}{\cos^2 x} = \sec^2 x$

> **Note**
>
> $\cos^2 x$ means $(\cos x)^2$. Similarly, $\sin^2 x$ means $(\sin x)^2$, and $\tan^2 x$ means $(\tan x)^2$, etc.

The derivative of tan x

In Example 10, you saw how to differentiate $y = \dfrac{\sin x}{\cos x}$. Since $\dfrac{\sin x}{\cos x} = \tan x$, the result gives the derivative of tan x.

When $y = \tan x$, $\dfrac{dy}{dx} = \sec^2 x$.

Exercise 6

Use the quotient rule to differentiate each of these expressions with respect to x.

1 $\dfrac{e^x}{x}$

2 $\dfrac{x^2}{(x+3)}$

3 $\dfrac{(4-x)}{x^2}$

4 $\dfrac{\ln x}{x^3}$

5 $\dfrac{4x}{\sin x + \cos x}$

6 $\dfrac{2x^2}{(x-2)}$

7 $\dfrac{x^{\frac{5}{3}}}{(3x-2)}$

8 $\dfrac{1-\ln x}{x^3}$

9 $\dfrac{\cos x}{\sin x}$

Differentiating a composite function (the chain rule)

The function $\sin(x^2)$ is a **composite function**. That is, when $f(x) = \sin x$ and $g(x) = x^2$ then $fg(x) = \sin(x^2)$.

For any equation of the form $y = gf(x)$ you can make the substitution $u = f(x)$, which means $y = gf(x)$ can be expressed in two simple parts: that is $u = f(x)$ and $y = g(u)$.

A small increase of δx in the value of x gives a corresponding small increase of δu in the value of u.

Then if $\delta x \to 0$, it follows that $\delta u \to 0$.

Hence $\dfrac{dy}{dx} = \lim\limits_{\delta x \to 0}\left(\dfrac{\delta y}{\delta x}\right) = \lim\limits_{\delta x \to 0}\left(\dfrac{\delta y}{\delta u} \times \dfrac{\delta u}{\delta x}\right)$

$\Rightarrow \qquad \dfrac{dy}{dx} = \lim\limits_{\delta x \to 0}\left(\dfrac{\delta y}{\delta u}\right) \times \lim\limits_{\delta x \to 0}\left(\dfrac{\delta u}{\delta x}\right)$

so $\qquad \dfrac{dy}{dx} = \dfrac{dy}{du} \times \dfrac{du}{dx}$

This formula is called the **chain rule**.

Example 11

Question

Find $\dfrac{dy}{dx}$ when $y = \sin(x^2)$.

Answer

Letting $u = x^2$ gives $y = \sin u$.

Using $\dfrac{dy}{dx} = \dfrac{dy}{du} \times \dfrac{du}{dx}$ gives $\dfrac{dy}{dx} =$ either $(\cos u) \times 2x$ or $2x \cos u$

But $u = x^2$, therefore $\dfrac{dy}{dx} = 2x \cos(x^2)$.

Example 12

Question

Given $y = (x^3 + 1)^4$, find $\dfrac{dy}{dx}$.

Answer

Letting $u = x^3 + 1$ gives $y = u^4$

Using $\dfrac{dy}{dx} = \dfrac{dy}{du} \times \dfrac{du}{dx}$ gives

$$\dfrac{dy}{dx} = (4u^3)(3x^2) = 12x^2 u^3$$

Replacing u by $x^3 + 1$ gives $\dfrac{dy}{dx} = 12x^2(x^3 + 1)^3$

Exercise 7

Use the chain rule to differentiate each expression with respect to x.

1 $(3x + 1)^2$ **2** $(3 - x)^4$ **3** $\sin(3x)$

4 e^{2x} **5** $\ln(2x - 1)$ **6** $\cos(5x)$

7 $\sin(x^3)$ **8** $e^{(3x+5)}$ **9** $\sqrt{3x^3 - 4}$

10 $\ln(x^2 - 2x)$ **11** $\cos(3x - 5)$ **12** $\ln(3x + x^2)$

13 $(4 - 2x)^5$ **14** $e^{(x^2 - x)}$ **15** $4\cos(5x - 6)$

16 $\ln(\sin x)$ **17** $e^{(x - x^3)}$ **18** $(2 - x^3)^4$

19 $\sqrt[3]{x^2 - x}$ **20** $(x^5 - 3)^{-\frac{1}{2}}$ **21** $\sin^2 x$

General results

Using the chain rule on some general composite functions gives some standard results that you can quote without having to prove them.

For example, when $y = \sin f(x)$, taking $u = f(x)$ gives $y = \sin u$.

Then $\dfrac{dy}{dx} = \dfrac{dy}{du} \times \dfrac{du}{dx} \Rightarrow \dfrac{dy}{dx} = (\cos u)\dfrac{du}{dx}$.

Notice that   when $y = \sin f(x)$ then $\dfrac{dy}{dx} = f'(x)\cos f(x)$

Similarly   when $y = \cos f(x)$ then $\dfrac{dy}{dx} = -f'(x)\sin f(x)$

In particular $\dfrac{d}{dx}(\sin ax) = a\cos ax$

and $\dfrac{d}{dx}(\cos ax) = -a\sin ax$

When $y = e^{f(x)}$, using $u = f(x)$ gives $y = e^u$, then $\dfrac{dy}{dx} = \dfrac{dy}{du} \times \dfrac{du}{dx} \Rightarrow \dfrac{dy}{dx} = (e^u)\dfrac{du}{dx}$.

So when $y = e^{f(x)}$ then $\dfrac{dy}{dx} = f'(x)e^{f(x)}$.

When $y = \ln f(x)$, using $u = f(x)$ gives $y = \ln u$, then $\dfrac{dy}{dx} = \dfrac{dy}{du} \times \dfrac{du}{dx} \quad \Rightarrow \quad \dfrac{dy}{dx} = \left(\dfrac{1}{u}\right)\dfrac{du}{dx}$.

So when $y = \ln f(x)$ then $\dfrac{dy}{dx} = \dfrac{f'(x)}{f(x)}$.

Example 13

Differentiate $\cos\left(\dfrac{1}{6}\pi - 3x\right)$ with respect to x.

$$\dfrac{d}{dx}\left\{\cos\left(\dfrac{1}{6}\pi - 3x\right)\right\} = -(-3)\sin\left(\dfrac{1}{6}\pi - 3x\right)$$
$$= 3\sin\left(\dfrac{1}{6}\pi - 3x\right)$$

Example 14

Find $\dfrac{dy}{dx}$ when $y = \ln(2x - 3)$.

$$\dfrac{dy}{dx} = \dfrac{2}{2x - 3}$$

Example 15

Find $\dfrac{dy}{d\theta}$ when $y = \cos^3\theta$.

$y = \cos^3\theta = [\cos\theta]^3$

$y = u^3$ where $u = \cos\theta$

$\dfrac{dy}{d\theta} = \dfrac{dy}{du} \times \dfrac{du}{d\theta} = (3u^2)(-\sin\theta) = -3u^2\sin\theta$

Therefore $\quad u = \cos\theta \quad \Rightarrow \quad \dfrac{dy}{d\theta} = -3\cos^2\theta\sin\theta$

Example 15 includes another example of some standard differentiation results:

When $y = \cos^n x$ then $\dfrac{dy}{dx} = -n\cos^{n-1}x \sin x$.

When $y = \sin^n x$ then $\dfrac{dy}{dx} = n\sin^{n-1}x \cos x$

Exercise 8

Differentiate each of these expressions with respect to x.

1 $\sin 4x$

2 $\cos(\pi - 2x)$

3 $\sin\left(\dfrac{1}{2}x + \pi\right)$

4 $\cos^2 x$

5 $e^{\sin x}$

6 $\ln(\cos x)$

7 $\sin x^2$

8 $e^{\cos x}$

9 $\ln(\sin x)$

10 $\cos^4 x$

11 $e^{(x^2 - 2x)}$

12 $\tan 6x$

13 $\ln(2x^2 - 3x)$

14 $\sin(5x - 8)$

5.5 Implicit functions

It is difficult to find y in terms of x in the equation $x^2 - y^2 + y = 1$.

A relationship of this type, where y is not given explicitly as a function of x, is called an **implicit function**, because it is *implied* in the equation that $y = f(x)$.

Differentiating an implicit function

An implicit function can be differentiated term by term. To be able to do this, you need a method which will allow you to differentiate terms like y^2 with respect to x.

When $\quad g(y) = y^2$ and $y = f(x)$

then $\quad g(y) = \{f(x)\}^2 \quad$ is a composite function.

Using the substitution $u = f(x)$ gives $y = u^2$. Differentiating by the chain rule gives

$$\frac{\mathrm{d}}{\mathrm{d}x}y^2 = \frac{\mathrm{d}}{\mathrm{d}x}u^2 = \frac{\mathrm{d}}{\mathrm{d}u}(u^2) \times \frac{\mathrm{d}u}{\mathrm{d}x}$$

$$= 2u \times \frac{\mathrm{d}u}{\mathrm{d}x} = 2y\frac{\mathrm{d}y}{\mathrm{d}x}$$

Therefore $\quad \dfrac{\mathrm{d}}{\mathrm{d}x}(y^2) = 2y\dfrac{\mathrm{d}y}{\mathrm{d}x}$.

In general, $\quad \dfrac{\mathrm{d}}{\mathrm{d}x}g(y) = \left(\dfrac{\mathrm{d}}{\mathrm{d}y}g(y)\right)\left(\dfrac{\mathrm{d}y}{\mathrm{d}x}\right)$.

To differentiate $f(y)$ with respect to x, you must differentiate $f(y)$ with respect to y and then multiply by $\dfrac{\mathrm{d}y}{\mathrm{d}x}$.

For example, $\quad \dfrac{\mathrm{d}}{\mathrm{d}x}y^3 = 3y^2\dfrac{\mathrm{d}y}{\mathrm{d}x} \quad$ and $\quad \dfrac{\mathrm{d}}{\mathrm{d}x}e^y = e^y\dfrac{\mathrm{d}y}{\mathrm{d}x}$.

A term that contains a product of both x and y, for example x^2y^3, can be differentiated using the product rule.

Letting $\quad u = x^2$ and $v = y^3 \quad$ and using $\quad \dfrac{\mathrm{d}}{\mathrm{d}x}(uv) = v\dfrac{\mathrm{d}u}{\mathrm{d}x} + u\dfrac{\mathrm{d}v}{\mathrm{d}x}$

gives $\quad \dfrac{\mathrm{d}}{\mathrm{d}x}(x^2y^3) = y^3\dfrac{\mathrm{d}}{\mathrm{d}x}(x^2) + x^2\dfrac{\mathrm{d}}{\mathrm{d}x}(y^3)$.

$$= 2xy^3 + x^2 3y^2\frac{\mathrm{d}y}{\mathrm{d}x}$$

$$= 2xy^3 + 3x^2y^2\frac{\mathrm{d}y}{\mathrm{d}x}$$

Example 16

Find $\dfrac{dy}{dx}$ in terms of x and y when $x^2 - y^2 + x^2y = 1$.

Differentiating $x^2 - y^2 + x^2y = 1$ term by term gives

$$\frac{d}{dx}(x^2) - \frac{d}{dx}(y^2) + \frac{d}{dx}(x^2y) = \frac{d}{dx}(1)$$

$$\Rightarrow \quad 2x - 2y\frac{dy}{dx} + 2xy + x^2\frac{dy}{dx} = 0$$

Hence $\quad 2x(1+y) = \dfrac{dy}{dx}(2y - x^2) \quad \Rightarrow \quad \dfrac{dy}{dx} = \dfrac{2x(1+y)}{2y - x^2}$.

Example 17

Differentiate $y = xe^y$ with respect to x and hence find $\dfrac{dy}{dx}$ in terms of x and y.

When $\quad y = xe^y \quad$ then $\quad \dfrac{dy}{dx} = \dfrac{d}{dx}(xe^y) = e^y\dfrac{d}{dx}(x) + x\dfrac{d}{dx}(e^y)$

$$\Rightarrow \quad \frac{dy}{dx} = e^y + xe^y\frac{dy}{dx} \quad \Rightarrow \quad \frac{dy}{dx} - xe^y\frac{dy}{dx} = e^y$$

Hence $\quad \dfrac{dy}{dx} = \dfrac{e^y}{1 - xe^y}$.

Example 18

Given that $y = \sin^{-1} x$, show that $\dfrac{dy}{dx} = \dfrac{1}{\sqrt{1-x^2}}$.

$y = \sin^{-1} x \quad \Rightarrow \quad \sin y = x$

Differentiating with respect to x gives

$$\cos y\frac{dy}{dx} = 1 \quad \Rightarrow \quad \frac{dy}{dx} = \frac{1}{\cos y}$$

But $\cos y = \sqrt{1 - \sin^2 y} = \sqrt{1 - x^2}$ so $\dfrac{dy}{dx} = \dfrac{1}{\sqrt{1-x^2}}$.

Exercise 9

In questions 1 to 9, differentiate the equation with respect to x.

1. $x^2 + y^2 = 4$
2. $x^2 + xy + y^2 = 0$
3. $x(x+y) = y^2$
4. $\dfrac{1}{x} + \dfrac{1}{y} = e^y$
5. $\dfrac{1}{x^2} + \dfrac{1}{y^2} = \dfrac{1}{4}$
6. $\dfrac{x^2}{4} - \dfrac{y^2}{9} = 1$

7 $\sin x + \sin y = 1$ **8** $\sin x \cos y = 2$ **9** $xe^y = x + 1$

10 Find $\dfrac{dy}{dx}$ as a function of x when $y^2 = 2x + 1$.

11 Find the gradient of $x^2 + y^2 = 9$ at the points where $x = 1$.

12 Find $\dfrac{dy}{dx}$ given that $y = \tan^{-1} x$.

13 Find the equations of the tangents to $x^2 - 3y^2 = 4y$

    a   at the points where $y = 2$

    b   at the point (x_1, y_1).

14 Find the equations of the tangents to $x^2 + xy + y^2 = 3$ at the points where $x = 1$.

15 Find the equation of the tangent at $\left(1, \dfrac{1}{3}\right)$ to the curve whose equation is $2x^2 + 3y^2 - 3x + 2y = 0$.

5.6 Parametric equations

Look at the equations $x = t^2, y = t - 1$.

A point P(x, y) is on the curve represented by $x = t^2, y = t - 1$ if and only if the coordinates of P are $(t^2, t - 1)$.

The variable t is called a **parameter**.

Substituting some numbers for t gives a pair of corresponding values of x and y.

For example, when $t = 3$, $x = 9$ and $y = 2$, therefore $(9, 2)$ is a point on the curve.

The direct relationship between x and y is found by eliminating t from these two **parametric equations**.

In this case $y = t - 1 \implies t = y + 1$

Substituting $t = y + 1$ in the equation $x = t^2$ gives $x = (y + 1)^2$.

So the Cartesian equation of this curve is $(y + 1)^2 = x$.

Sketching a curve from parametric equations

Look again at the curve whose parametric equations are

     $x = t^2$ and $y = t - 1$.

The table shows the values of x and y that correspond to some values of t.

t	-2	-1	0	1	2
x	4	1	0	1	4
y	-3	-2	-1	0	1

In order to sketch the curve, begin by plotting these points on a graph.

Afterwards, look at what happens to x and to y as t varies:

     $x \geq 0$ for all values of t,

     as $t \to \infty$, $x \to \infty$ and $y \to \infty$,

     as $t \to -\infty$, $x \to \infty$ and $y \to -\infty$,

There are no values of t for which either x or y is undefined so it is reasonable to assume that the curve has no breaks.

There is now enough information to sketch the curve, as shown in the margin.

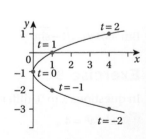

Example 19

Find the Cartesian equation of the curve whose parametric equations are

a $x = t^2$

$y = 2t$

b $x = 2 \cos \theta$

$y = 2 \sin \theta$

c $x = 2t$

$y = \dfrac{2}{t}$.

a $y = 2t \implies t = \dfrac{1}{2}y$

$x = t^2 \implies x = \left(\dfrac{1}{2}y\right)^2 = \dfrac{1}{4}y^2 \implies y^2 = 4x$

b Using $\cos^2 \theta + \sin^2 \theta = 1$ where $2 \cos \theta = x$ and $2 \sin \theta = y$ gives

$x^2 + y^2 = 4\cos^2 \theta + 4\sin^2 \theta = 4$

c $y = \dfrac{2}{t} \implies t = \dfrac{2}{y}$

$x = 2t \implies x = \dfrac{4}{y} \implies xy = 4$

Exercise 10

1 Find the Cartesian equation of each of these curves.

a $x = 2t^2, y = t$
b $x = \cos \theta, y = \sin \theta$
c $x = t, y = \dfrac{4}{t}$

2 Sketch each curve given in question 1.

3 Find the Cartesian equation of the curve given by the parametric equations

$x = \dfrac{t}{1-t}$ and $y = \dfrac{t^2}{1-t}$.

4 Show that the Cartesian equation of the curve given by $x = \dfrac{1}{t}$ and $y = \dfrac{t+1}{t^2}$

is $y = x + x^2$.

5 Find the Cartesian equation of the curve given by $x = t$ and $y = t^3 - t$.

6 Find the Cartesian equation of the curve given by $x = 2 \cos \theta$ and $y = 3 \sin \theta$.

7 a Show that, the parametric equations $x = 4 \cos \theta$ and $y = 4 \sin \theta$ give the equation of a circle.

b State the centre and radius of the circle.

5.7 Finding $\dfrac{dy}{dx}$ using parametric equations

When both x and y are given as functions of t then a small increase of δt in the value of t results in corresponding small increases of δx and δy in the values of x and y.

Therefore $\dfrac{\delta y}{\delta x} = \dfrac{\delta y}{\delta t} \times \dfrac{\delta t}{\delta x}$

As $\delta t \to 0$, δx and δy also approach zero, therefore $\dfrac{dy}{dx} = \dfrac{dy}{dt} \times \dfrac{dt}{dx}$

Using the formula $\dfrac{dt}{dx} = \dfrac{1}{\frac{dx}{dt}}$ gives

$$\frac{dy}{dx} = \frac{\frac{dy}{dt}}{\frac{dx}{dt}}$$

Therefore $\dfrac{dy}{dx}$ is found in terms of t by differentiating both x and y with respect to t and using the formula above.

Example 20

Question

The parametric equations of a curve are $x = t^2$ and $y = t - 1$.

a   Find the gradient of the curve at the point where $t = 1$.

b   Find the coordinates of the point on the curve where the gradient is 2.

c   Show that there are no stationary points on this curve.

Answer

a $x = t^2$ and $y = t - 1$

Using $\dfrac{dy}{dx} = \dfrac{\frac{dy}{dt}}{\frac{dx}{dt}}$ gives

$\dfrac{dy}{dt} = 1$ and $\dfrac{dx}{dt} = 2t$ $\Rightarrow$ $\dfrac{dy}{dx} = \dfrac{1}{2t}$

Therefore when $t = 1$, $\dfrac{dy}{dx} = \dfrac{1}{2}$.

b   When $\dfrac{dy}{dx} = 2$, $\dfrac{1}{2t} = 2$ $\Rightarrow$ $t = \dfrac{1}{4}$

and when $t = \dfrac{1}{4}$, then $x = \dfrac{1}{16}$ and $y = -\dfrac{3}{4}$.

The gradient of the curve is 2 at the point $\left(\dfrac{1}{16}, -\dfrac{3}{4} \right)$.

c   There are no values of t for which $\dfrac{dy}{dx} = 0$, so there are no stationary points on this curve.

Example 21

Question

Find the stationary point on the curve whose parametric equations are $x = t^3$, $y = (t + 1)^2$.

Answer

$\dfrac{dy}{dt} = 2(t + 1)$ and $\dfrac{dx}{dt} = 3t^2$ so $\dfrac{dy}{dx} = \dfrac{\frac{dy}{dt}}{\frac{dx}{dt}} = \dfrac{2(t + 1)}{3t^2}$

(continued)

(continued)

At stationary points $\dfrac{dy}{dx} = 0 \quad \Rightarrow \quad t = -1$

When $t = -1$, then $x = -1$ and $y = 0$.

Therefore the stationary point is $(-1, 0)$.

Example 22

a Find the equation of the normal to the curve $x = t^2$, $y = t + \dfrac{2}{t}$, at the point where $t = 1$.

b Show, without sketching the curve, that this normal does not cross the curve again.

a $x = t^2$ and $y = t + \dfrac{2}{t}$ give $\dfrac{dy}{dt} = 1 - \dfrac{2}{t^2}$ and $\dfrac{dx}{dt} = 2t$

therefore $\dfrac{dy}{dx} = \dfrac{dy}{dt} \div \dfrac{dx}{dt} = \dfrac{1 - \dfrac{2}{t^2}}{2t} = \dfrac{t^2 - 2}{2t^3}$

When $t = 1$, then $x = 1$, $y = 3$ and $\dfrac{dy}{dx} = -\dfrac{1}{2}$.

Therefore the gradient of the normal at $(1, 3)$ is $\dfrac{-1}{-\dfrac{1}{2}} = 2$.

The equation of this normal is $y - 3 = 2(x - 1)$, that is, $y = 2x + 1$.

b All points for which $x = t^2$ and $y = t + \dfrac{2}{t}$ are on the given curve.

For any point that is on both the curve and the normal, these coordinates also satisfy the equation of the normal.

At points common to the curve and the normal,

$$t + \frac{2}{t} = 2t^2 + 1 \quad \Rightarrow \quad 2t^3 - t^2 + t - 2 = 0 \qquad [1]$$

When a cubic equation can be factorised, each factor equated to zero gives a root of the equation.

One point where the curve and normal meet is the point where $t = 1$, so $t = 1$ is a root of [1] and $(t - 1)$ is a factor of the LHS.

that is, $(t - 1)(2t^2 + t + 2) = 0$

Therefore, at any other point where the normal meets the curve, the value of t is a root of the equation $2t^2 + t + 2 = 0$

Checking the value of $b^2 - 4ac$ shows that this equation has no real roots so there are no more points where the normal meets the curve.

Example 23

The parametric equations of a curve are $x = \cos\theta$ and $y = \sin\theta$. Find the equation of the tangent to the curve at the point P $(\cos\alpha, \sin\alpha)$ on the curve.

$\left.\begin{array}{l} x = \cos\theta \\ y = \sin\theta \end{array}\right\}$ are the parametric equations of a circle, centre O and radius 1, as

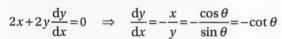

$x^2 + y^2 = \cos^2\theta + \sin^2\theta = 1$

Differentiating $x^2 + y^2 = 1$ implicitly gives

$$2x + 2y\frac{dy}{dx} = 0 \quad \Rightarrow \quad \frac{dy}{dx} = -\frac{x}{y} = -\frac{\cos\theta}{\sin\theta} = -\cot\theta$$

Therefore the gradient of the tangent at P is $-\cot\alpha$.

Use $y - y_1 = m(x - x_1)$ to find the equation of the tangent, with $y_1 = \sin\alpha$, $x_1 = \cos\alpha$ and $m = -\cot\alpha$.

The equation of the tangent at P is

$$y - \sin\alpha = -\cot\alpha(x - \cos\alpha)$$

$$\Rightarrow \quad y = -x\cot\alpha + \text{cosec }\alpha$$

In the example above, α can be *any* of the possible values of θ, so α can be replaced by θ in the equation of the tangent, giving

$$y = -x\cot\theta + \text{cosec }\theta$$

This is the equation of the tangent at *any* point $(\cos\theta, \sin\theta)$ on the curve. It is a *general equation* for a tangent to the curve because, by taking a particular value of θ, you can find the equation of the tangent at that point on the curve. You can also derive this equation directly, without first using α as the parameter.

Example 24

Find, in terms of t, the equation of the tangent to the curve $x = t^2$, $y = t + \dfrac{2}{t}$ at a general point $\left(t^2, t + \dfrac{2}{t}\right)$ on the curve.

From Example 22, the gradient at any point is given by $\dfrac{dy}{dx} = \left(\dfrac{dy}{dt}\right) \div \left(\dfrac{dt}{dx}\right) = \dfrac{t^2 - 2}{2t^3}$.

Hence, the equation of the tangent at any point is $y - \left(t + \dfrac{2}{t}\right) = \left(\dfrac{t^2 - 2}{2t^3}\right)(x - t^2)$.

This simplifies to $2t^3 y - x(t^2 - 2) - (t^4 + 6t^2) = 0$.

Exercise 11

1 Find $\dfrac{dy}{dx}$, in terms of the parameter, for each curve.

   a $x = 2t^2, y = t$ **b** $x = \cos\theta, y = \sin\theta$ **c** $x = t, y = \dfrac{4}{t}$

2 **a**   Given that $x = \dfrac{t}{1-t}$ and $y = \dfrac{t^2}{1-t}$, find $\dfrac{dy}{dx}$ in terms of t.

   b   Find the value of $\dfrac{dy}{dx}$ at the point where $x = 1$.

3 Given that $x = t^2$ and $y = t^3$, find $\dfrac{dy}{dx}$ in terms of t.

4 Find the stationary points of the curve whose parametric equations are $x = t, y = t^3 - t$, and distinguish between them.

5. A curve has parametric equations $x = \theta - \cos\theta$, $y = \sin\theta$.

   Find the smallest positive value of θ at which the gradient of this curve is zero.

6. Find the equation of the tangent to the curve $x = t^2$, $y = 4t$ at the point where $t = -1$.

7. **a** Find the equation of the normal to the curve $x = 2\cos\theta$, $y = 3\sin\theta$ at the point where $\theta = \dfrac{1}{4}\pi$.

   b Find the coordinates of the point where this normal cuts the curve again.

8. **a** Find the equation of the tangent at a general point to each of the curves in question 1 parts **a** and **c**.

   b Find the equation of the normal at a general point to each of the curves in question 1 parts **a** and **c**.

9. **a** Find the equation of the normal at the point $\left(2s, \dfrac{2}{s}\right)$ to the curve whose parametric equations are $x = 2s$, $y = \dfrac{2}{s}$.

   b Find, in terms of s, the coordinates of the point where this normal cuts the curve again.

10. The parametric equations of a curve are $x = t$ and $y = \dfrac{1}{t}$.

    a Find the general equation of the tangent to this curve, that is the equation of the tangent at the point $\left(t, \dfrac{1}{t}\right)$.

    b Find, in terms of t, the coordinates of the points at which the tangent cuts the x and y axes.

11. A curve has parametric equations $x = t^2$, $y = 4t$.

    a Find the equation of the normal to this curve at the point $(t^2, 4t)$.

    b Find the coordinates of the points where this normal cuts the coordinate axes.

Summary

Derivatives of standard functions

$$\frac{d}{dx}(e^x) = e^x$$

$$\frac{d}{dx}(\ln x) = \frac{1}{x}$$

$$\frac{d}{dx}(\sin x) = \cos x$$

$$\frac{d}{dx}(\cos x) = -\sin x$$

$$\frac{d}{dx}(\tan x) = \sec^2 x$$

Formulae for differentiating combinations of functions

$$\frac{dy}{dx} = \frac{1}{\frac{dx}{dy}}$$

$$\frac{d}{dx}(uv) = v\frac{du}{dx} + u\frac{dv}{dx}$$

$$\frac{dy}{dx} = \frac{v\dfrac{du}{dx} - u\dfrac{dv}{dx}}{v^2}$$

$$\frac{dy}{dx} = \frac{dy}{du} \times \frac{du}{dx}$$

Derivatives of some composite functions

$$\frac{d}{dx}(\sin f(x)) = f'(x)\cos f(x)$$

$$\frac{d}{dx}(\cos f(x)) = -f'(x)\sin f(x)$$

$$\frac{d}{dx}(\sin ax) = a\cos ax$$

$$\frac{d}{dx}(\cos ax) = -a\sin ax$$

$$\frac{d}{dx}e^{f(x)} = f'(x)e^{f(x)}$$

$$\frac{d}{dx}(\ln f(x)) = \frac{f'(x)}{f(x)}$$

▶ To differentiate a function of y with respect to x use $\dfrac{d}{dx}g(y) = \left(\dfrac{d}{dy}g(y)\right)\left(\dfrac{dy}{dx}\right)$.

▶ When a curve is defined in terms of a parameter t, the gradient of the curve in terms of t can be found by using $\dfrac{dy}{dx} = \dfrac{\dfrac{dy}{dt}}{\dfrac{dx}{dt}}$.

Review

In questions 1 to 19, find the derivative of each expression.

1   **a** $-\sin 4\theta$ **b** $\theta - \cos\theta$ **c** $\sin^3\theta + \sin 3\theta$

2   **a** $x^3 + e^x$ **b** $e^{(2x+3)}$ **c** $e^x \sin x$

3   **a** $\ln\left(\dfrac{1}{3}x^{-3}\right)$ **b** $\ln\left(\dfrac{2}{x^2}\right)$ **c** $\ln\left(\dfrac{\sqrt{x}}{4}\right)$

4   **a** $3\sin x - e^{-x}$ **b** $\ln x^{\frac{1}{2}} - \dfrac{1}{2}\cos x$

    c $x^4 + 4e^x - \ln 4x$ **d** $\dfrac{1}{2}e^{-x} + x^{-\frac{1}{2}} - \ln\dfrac{1}{2}x$

5 $(x+1)\ln x$ **6** $\sin^2 3x$ **7** $(4x-1)^{\frac{2}{3}}$

8 $(3\sqrt{x} - 2x)^2$ **9** $\dfrac{(x^4-1)}{(x+1)^3}$ **10** $\dfrac{\ln x}{\ln(x-1)}$

11 $\ln(\cot x)$ 　　**12** $x^2 \tan x$ 　　**13** $\dfrac{e^x}{x-1}$

14 $\dfrac{1+\sin x}{1-\sin x}$ 　　**15** $x^2\sqrt{x-1}$ 　　**16** $(1-x^2)(1-x)^2$

17 $\ln\sqrt{\dfrac{(x+3)^3}{x^2+2}}$ 　　**18** $\sin x \cos^3 x$ 　　**19** $e^{\cos^2 x}$

20 Find the value(s) of x for which these expressions have stationary values.

　a $3x - e^x$ 　　**b** $x^2 - 2\ln x$

In questions 21 to 24, find

a the gradient of the curve at the given point

b the equation of the tangent to the curve at that point

c the equation of the normal to the curve at that point.

21 $y = \sin x - \cos x;\ x = \dfrac{1}{2}\pi$

22 $y = x + e^x;\ x = 1$

23 $y = 1 + x + \sin x;\ x = 0$

24 $x = \cos y;\ y = \dfrac{\pi}{2}$

25 Find the coordinates of a point on the curve where the tangent is parallel to the given line.

　a $y = 3x - 2\cos x;\ y = 4x$ 　　**b** $y = 2\ln x - x;\ y = x$

26 Differentiate with respect to x.

　a y^4 　　**b** xy^2 　　**c** $\dfrac{1}{y}$

　d $x\ln y$ 　　**e** $\sin y$ 　　**f** e^y

　g $y\cos x$ 　　**h** $y\cos y$

In questions 27 to 30, find $\dfrac{dy}{dx}$ in terms of x and y.

27 $x^2 - 2y^2 = 4$ 　　**28** $\dfrac{1}{x} + \dfrac{1}{y} = 2$

29 $x^2y^3 = 9$ 　　**30** $x^2y^2 = \dfrac{(y+1)}{(x+1)}$

In questions 31 to 36, find $\dfrac{dy}{dx}$ in terms of the parameter.

31 $x = t^2,\ y = t^3$ 　　**32** $x = (t+1)^2,\ y = t^2 - 1$

33 $x = \sin^2\theta,\ y = \cos^3\theta$ 　　**34** $x = 4t,\ y = \dfrac{4}{t}$

35 $x = e^t,\ y = 1 - t$ 　　**36** $x = \dfrac{t}{1-t},\ y = \dfrac{t^2}{1-t}$

37 Find the equation of the tangent to the curve $x = \cos\theta,\ y = 2\sin\theta$ at the point where $\theta = \dfrac{3}{4}\pi$.

Assessment

1 The equation of a curve is $y = 3 - x^2 + \ln x$.

  a Find $\dfrac{dy}{dx}$.

  b Find the equation of the tangent to the curve at the point where $x = 1$.

  c Find the equation of the normal to the curve at the point where $x = 1$.

2 The equation of a curve is $x = 3y^2 - 2y$.

  a Find $\dfrac{dy}{dx}$ in terms of y.

  b Find the equation of the tangent to the curve at the point where $y = 1$.

  c Find the equation of the normal to the curve at the point where $y = 1$.

3 The equations of a curve are $x = \sin t$ and $y = \cos 2t$.

  a Find $\dfrac{dy}{dx}$ in terms of x.

  b Find the coordinates of the stationary point on the curve.

4 The equations of a curve are $x = e^t - t$ and $y = e^{2t} - 2t$.

  a Show that $\dfrac{dy}{dx} = 2(e^t + 1)$.

  b Hence show that there are no stationary point on the curve.

5 **a** Differentiate $y^2 - 2xy + 3y = 2x$ with respect to x.

  b Find the coordinates of the points on the curve where $x = 1$.

  c Find the equations of the tangent to the curve $y^2 - 2xy + 3y = 2x$ at the points where $x = 1$.

6 A curve is given by the parametric equations $x = t$, $y = \dfrac{1}{t}$.

  a Find the equation of the normal to the curve at the point $\left(a, \dfrac{1}{a} \right)$.

  b This normal cuts the curve again at the point with parameter b. Find a relationship between a and b.

7 The parametric equations of a curve are

    $x = 2 \cos \theta, y = 3 \sin \theta$.

  Find the equation of the tangent to this curve at the point $(2 \cos \theta, 3 \sin \theta)$.

8 The parametric equations of a curve are $x = \cos \theta$ and $y = 1 + \sin \theta$. Find the equation of the normal to the curve at the point $(\cos \theta, 1 + \sin \theta)$.

9 The equation of a curve is $y = \ln \dfrac{1}{x} + 4x$.

  Find the coordinate of the stationary point on the curve.

10 The curve shown in the diagram is called an asteroid.

  The parametric equations of this curve are

  $x = 27 \cos^3 \theta$ and $y = 27 \sin^3 \theta$ for $0 \leq \theta \leq 2\pi$.

  a Find the coordinates of the point on the curve where $\theta = \dfrac{\pi}{2}$.

  b Show that the Cartesian equation of the curve is $x^{\frac{2}{3}} + y^{\frac{2}{3}} = 9$.

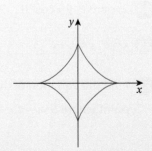

11 Given that $x = \dfrac{1}{\sin\theta}$, use the quotient rule to show that $\dfrac{dy}{d\theta} = -\operatorname{cosec}\theta\cot\theta$.

AQA MPC3 January 2012 (part question)

12 a Find $\dfrac{dy}{dx}$ when $y = e^{3x} + \ln x$

    b i Given that $u = \dfrac{\sin x}{1 + \cos x}$, show that $\dfrac{du}{dx} = \dfrac{1}{1 + \cos x}$

      ii Hence show that if $y = \ln\left(\dfrac{\sin x}{1 + \cos x}\right)$, then $\dfrac{dy}{dx} = \operatorname{cosec} x$

AQA MPC3 January 2013

13 A curve is defined by the parametric equations $x = \cos 2t$, $y = \sin t$.

The point P on the curve is where $t = \dfrac{\pi}{6}$,

    a Find the gradient at P.

    b Find the equation of the normal to the curve at P in the form $y = mx + c$.

    c The normal at P intersects the curve again at the point Q $(\cos 2q, \sin q)$.

    Use the equation of the normal to form a quadratic equation in $\sin q$ and hence find the x-coordinate of Q.

AQA MPC4 June 2015

14 A curve is defined by the equation $9x^2 - 6xy + 4y^2 = 3$.

Find the coordinates of the two stationary points of this curve.

AQA MPC4 June 2012

6 Integration

Introduction

In the history of calculus, integration was studied before differentiation as a process of summation – for example, to find areas under curves. This chapter looks at the other interpretation of integration as the reverse of differentiation.

It also demonstrates how you can use products and fractions to integrate some expressions that cannot be integrated by recognition or by substitution. The chapter ends with an application for finding volumes.

Recap

You will need to remember...

▶ The derivatives of polynomial, exponential, logarithmic and trigonometric functions.
▶ How to integrate polynomial functions.
▶ How to evaluate a definite integral.
▶ How to express a rational function as partial fractions.
▶ How to use the trigonometric double angle formulae.
▶ The product rule and the chain rule for differentiating.

Objectives

By the end of this chapter, you should know how to...

▶ Find the integrals of functions by recognition of their derivatives.
▶ Use substitution to find the integrals of functions that are not immediately recognised as derivatives of known functions.
▶ Integrate products of functions using integration by parts.
▶ Express a rational function as partial fractions that can be integrated.
▶ Use trigonometric formulae to convert functions involving powers of $\sin x$, $\cos x$ or $\tan x$ to forms that can be integrated.
▶ Find volumes formed when the region between a curve and the x- or y-axis is rotated about that axis.

6.1 Standard integrals

Whenever you recognise a function $f'(x)$ as the derivative of another function $f(x)$, then you can integrate $f'(x)$ 'by sight':

$$\frac{d}{dx}f(x) = f'(x) \quad \Rightarrow \quad \int f'(x)dx = f(x) + K$$

Integrating exponential functions

You have seen that $\frac{d}{dx}e^x = e^x$, so therefore

$$\int e^x \, dx = e^x + K$$

You have seen that $\frac{d}{dx}(ce^x) = ce^x$, so therefore

$$\int ce^x \, dx = ce^x + K$$

You have seen that $\frac{d}{dx}(e^{f(x)}) = f'(x)e^{f(x)}$, so therefore,

$$\int f'(x)e^{f(x)}dx = e^{f(x)} + K$$

For example, $\int 2e^{2x-1} \, dx = e^{2x-1} + K$.

Example 1

Write down the integral of $2e^x$ with respect to x.

$$\int 2e^x\,dx = 2\int e^x\,dx = 2e^x + K$$

Exercise 1

In questions 1 to 10, integrate each expression with respect to x.

1 e^{4x}

2 $4e^{-x}$

3 $e^{(3x-2)}$

4 $2e^{(1-5x)}$

5 $6e^{-2x}$

6 $5e^{(x-3)}$

7 $e^{\left(2+\frac{x}{2}\right)}$

8 $e^{(2+x)}$

9 $e^{2x} + \dfrac{1}{e^{2x}}$

In questions 10 to 13, evaluate

10 $\displaystyle\int_0^2 e^{2x}\,dx$

11 $\displaystyle\int_{-1}^1 2e^{(x+1)}\,dx$

12 $\displaystyle\int_2^3 e^{(2-x)}\,dx$

13 $\displaystyle\int_0^2 e^{-x}\,dx.$

Integrating $\dfrac{1}{x}$

Trying to integrate $\dfrac{1}{x} = x^{-1}$ using the rule $\displaystyle\int x^n\,dx = \dfrac{1}{n+1}x^{(n+1)} + K$ fails because

when $n = -1$, then $\dfrac{1}{n+1} = \dfrac{1}{0}$ which is meaningless.

However $\dfrac{1}{x}$ can be *recognised* as the derivative of $\ln x$, but $\ln x$ is defined only when $x > 0$.

Therefore, provided that $x > 0 \quad \dfrac{d}{dx}(\ln x) = \dfrac{1}{x} \quad \Leftrightarrow \quad \displaystyle\int \dfrac{1}{x}\,dx = \ln x + K.$

When $x < 0$ the statement $\displaystyle\int \dfrac{1}{x}\,dx = \ln x$ is not valid because the logarithm of a negative number does not exist.

However, the function $\dfrac{1}{x}$ exists for negative values of x, as the graph of $y = \dfrac{1}{x}$ shows.

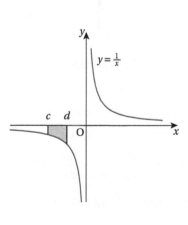

Also, the definite integral $\displaystyle\int_c^d \dfrac{1}{x}\,dx$, which is represented by the shaded area on the graph region, clearly exists.

So it must be possible to integrate $\dfrac{1}{x}$ when x is negative.

When $x < 0$ then $-x > 0$ so $\displaystyle\int \dfrac{1}{x}\,dx = \int \dfrac{-1}{(-x)}\,dx = \ln(-x) + K$

Therefore when $x < 0$, $\displaystyle\int \dfrac{1}{x}\,dx = \ln(-x) + K$

and when $x > 0$, $\displaystyle\int \dfrac{1}{x}\,dx = \ln x + K.$

These two results can be combined using $|x|$ so that, for both positive and negative values of x,

$$\int \dfrac{1}{x}\,dx = \ln|x| + K$$

Also $\dfrac{\mathrm{d}}{\mathrm{d}x}(\ln x^a)=\dfrac{\mathrm{d}}{\mathrm{d}x}(a\ln x)=\dfrac{a}{x}$.

Therefore, $\displaystyle\int\dfrac{a}{x}\,\mathrm{d}x=a\ln|x|+K$

For example, $\displaystyle\int\dfrac{4}{x}\,\mathrm{d}x=4\ln|x|+K.$

Example 2

Question

Find $\displaystyle\int\dfrac{2x-1}{2x}\,\mathrm{d}x.$

Answer

First express $\dfrac{2x-1}{2x}$ as a proper fraction.

$$\dfrac{2x-1}{2x}=1-\dfrac{1}{2x}=1-\left(\dfrac{1}{2}\right)\left(\dfrac{1}{x}\right)$$

Therefore $\displaystyle\int\dfrac{2x-1}{2x}\,\mathrm{d}x=\int\left(1-\dfrac{1}{2}\left(\dfrac{1}{x}\right)\right)\mathrm{d}x=x-\dfrac{1}{2}\ln|x|+K.$

Exercise 2

Integrate with respect to x.

1 $\dfrac{2}{x}$

2 $\dfrac{1}{4x}$

3 $\dfrac{3}{2x}$

4 $\dfrac{x+1}{x}$

5 $\dfrac{x^2+x-1}{x}$

6 $e^x+\dfrac{2}{x}$

Evaluate these integrals.

7 $\displaystyle\int_1^2\dfrac{1}{3x}\,\mathrm{d}x$

8 $\displaystyle\int_1^3\left(1-\dfrac{1}{x}\right)\mathrm{d}x$

9 $\displaystyle\int_1^2\left(\dfrac{1-x}{x}\right)\mathrm{d}x$

10 $\displaystyle\int_2^3\left(e^x-\dfrac{1}{x}\right)\mathrm{d}x$

11 $\displaystyle\int_4^5\left(\dfrac{2-x}{3x}\right)\mathrm{d}x$

12 $\displaystyle\int_{-3}^{-2}\left(2x^2-\dfrac{1}{x}\right)\mathrm{d}x$

Integrating $(ax+b)^n$

Look at the function $f(x)=(2x+3)^4$.

To differentiate $f(x)$ using the chain rule, make the substitution

$$u=2x+3\Rightarrow f(x)=u^4$$

and using $\dfrac{\mathrm{d}y}{\mathrm{d}x}=\dfrac{\mathrm{d}y}{\mathrm{d}u}\times\dfrac{\mathrm{d}u}{\mathrm{d}x}$ gives $\dfrac{\mathrm{d}}{\mathrm{d}x}(2x+3)^4=4u^3\times2=(4)(2)(2x+3)^3$

Therefore $\displaystyle\int(4)(2)(2x+3)^3\,\mathrm{d}x=(2x+3)^4+K$

Hence $\displaystyle\int(2x+3)^3\,\mathrm{d}x=\dfrac{1}{(2)(4)}(2x+3)^4+K$

Working with $f(x)=(ax+b)^{n+1}$ in a similar way gives the general result

$$\int(ax+b)^n\,\mathrm{d}x=\dfrac{1}{(a)(n+1)}(ax+b)^{n+1}+K.$$

Exercise 3

Integrate these functions with respect to x.

1 $(2x-3)^3$ **2** $(3x+1)^4$ **3** $(5x-2)^4$ **4** $(2-x)^{-2}$

5 $(x+3)^{-2}$ **6** $(1+x)^{\frac{1}{2}}$ **7** $(1+3x)^5$ **8** $(2-5x)^4$

Integrating trigonometric functions

As you have seen the derivatives $\sin x$, $\cos x$ and $\tan x$, you will recognise these integrals.

$$\frac{d}{dx}(\sin x)=\cos x \quad \Leftrightarrow \quad \int \cos x\,dx=\sin x+K$$

$$\frac{d}{dx}(\cos x)=-\sin x \quad \Leftrightarrow \quad \int \sin x\,dx=-\cos x+K$$

$$\frac{d}{dx}(\tan x)=\sec^2 x \quad \Leftrightarrow \quad \int \sec^2 x\,dx=\tan x+K$$

Also, when $y=\sin(f(x))$ then $\dfrac{dy}{dx}=f'(x)\cos f(x)$

so $\qquad \int f'(x)\cos f(x)\,dx=\sin f(x)+K.$

Similarly, when $y=\cos(f(x))$ then $\dfrac{dy}{dx}=-f'(x)\sin f(x)$

so $\qquad \int -f'(x)\sin f(x)\,dx=\cos f(x)+K.$

In particular, $\quad \int \cos(ax+b)\,dx=\dfrac{1}{a}\sin(ax+b)+K$

and $\qquad \int \sin(ax+b)\,dx=-\dfrac{1}{a}\cos(ax+b)+K.$

Example 3

Find the integral of $\sec^2(3x-\pi)$ with respect to x.

Answer

When $y=\tan(f(x))$ then $\dfrac{dy}{dx}=f'(x)\sec^2 f(x)$

So $\quad \int f'(x)\sec^2 f(x)\,dx=\tan f(x)+K$

$\qquad \int \sec^2(3x-\pi)\,dx=\dfrac{1}{3}\tan(3x-\pi)+K$

> **Note**
>
> Check your answer by differentiation:
> $$\frac{d}{dx}\left(\frac{1}{3}\tan(3x-\pi)\right)$$
> $$=\left(\frac{1}{3}\right)(3)\sec^2(3x-\pi)$$
> $$=\sec^2(3x-\pi)$$

Exercise 4

In questions 1 to 10, integrate the expression with respect to x.

1 $\sin 2x$ **2** $\cos 7x$ **3** $\sec^2 4x$

4 $\sin\left(\dfrac{1}{4}\pi+x\right)$ **5** $3\cos\left(4x-\dfrac{1}{2}\pi\right)$ **6** $\sec^2\left(\dfrac{1}{3}\pi+2x\right)$

7 $2\sin(3x-\alpha)$ **8** $5\cos\left(\alpha-\dfrac{1}{2}x\right)$ **9** $\cos 3x-\cos x$

10 $\sec^2 2x$

Evaluate these integrals.

11 $\int_0^{\frac{\pi}{6}} \sin 3x \, dx$

12 $\int_{\frac{\pi}{6}}^{\frac{\pi}{4}} \cos\left(2x - \frac{1}{2}\pi\right) dx$

13 $\int_0^{\frac{\pi}{2}} 2\sin\left(2x - \frac{1}{2}\pi\right) dx$

14 $\int_0^{\frac{\pi}{8}} \sec^2 2x \, dx$

Questions 15 to 35 contain a variety of functions, including some integrals covered in the AS book. Integrate each function with respect to x.

You can check your answers are correct by mentally differentiating them. You should get the original function.

15 $\sin\left(\frac{1}{2}\pi - 2x\right)$

16 $e^{(4x-1)}$

17 $\sec^2 7x$

18 $\dfrac{1}{2x-3}$

19 $\dfrac{1}{\sqrt{2x-3}}$

20 $\dfrac{1}{(3x-2)^2}$

21 e^{5x}

22 $\dfrac{x-1}{x}$

23 $(3x-5)^2$

24 $e^{(4x-5)}$

25 $\sqrt{4x-5}$

26 $\sin\left(5x - \frac{1}{3}\pi\right)$

27 $\dfrac{3}{2(1-x)}$

28 $\dfrac{4}{(x-6)^4}$

29 $\cos\left(3x - \frac{1}{3}\pi\right)$

30 $(x-2)(2x+4)$

31 $x(x-3)^2$

32 $3 - 2x(1-3x)$

33 $\dfrac{2}{3x^3}$

34 $\dfrac{3}{2x}$

35 $\dfrac{2}{3}(x-3)^2$

6.2 Integrating products by substitution

Section 6.1 demonstrated how you need to be able to recognise some functions as being the derivatives of others, in order to be able to to integrate them. Recognition is equally important to help you avoid errors in integration.

For example, look at the derivative of the product $x^2 \sin x$.

Using the product formula gives $\dfrac{d}{dx}(x^2 \sin x) = 2x\sin x + x^2 \cos x$

The derivative is not a simple product, therefore in general,

$\int uv \, dx$ is NOT $\left(\int u \, dx\right)\left(\int v \, dx\right)$.

However, using the chain rule to differentiate the composite function $(1 + x^2)^3$ gives

$$\frac{d}{dx}(1+x^2)^3 = 6x(1+x^2)^2$$

This time the derivative *is* a product, so clearly the integral of a product *may* be a composite function.

First look at the function e^u where u is a function of x.

Differentiating as a composite function gives

$$\frac{d}{dx}(e^u) = \left(\frac{du}{dx}\right)(e^u)$$

Therefore any product of the form $\left(\dfrac{du}{dx}\right)e^u$ can be integrated by recognition,

since $\displaystyle\int\left(\dfrac{du}{dx}\right)e^u\,dx=e^u+K.$

For example $\displaystyle\int 2xe^{x^2}\,dx=e^{x^2}+K$ (recognising that $u=x^2$)

$\displaystyle\int \cos x e^{\sin x}\,dx=e^{\sin x}+K$ (recognising that $u=\sin x$)

$\displaystyle\int x^2e^{x^3}\,dx=\dfrac{1}{3}\int 3x^2e^{x^3}\,dx=\dfrac{1}{3}e^{x^3}+K$ (recognising that $u=x^3$)

In these simple cases the substitution of u for $f(x)$ can be done mentally. All the results can be checked by differentiating them mentally.

Similar, but slightly less simple functions, can also be integrated by changing the variable when the substitution is written down.

Integration by substitution

Look at the general function $g(u)$ where u is a function of x.

$$\dfrac{d}{dx}g(u)=\dfrac{du}{dx}g'(u)=g'(u)\dfrac{du}{dx}$$

Therefore $\displaystyle\int g'(u)\dfrac{du}{dx}dx=g(u)+K$ \hfill [1]

and $\displaystyle\int g'(u)\,du=g(u)+K$ \hfill [2]

Comparing [1] and [2] gives $\displaystyle\int g'(u)\dfrac{du}{dx}dx=\int g'(u)\,du$

so $\displaystyle\cdots\dfrac{du}{dx}dx=\cdots dx$ \hfill [3]

Therefore integrating $f(u)\dfrac{du}{dx}$ with respect to x is *equivalent* to integrating $f(u)$ with respect to u.

So the relationship in [3] is neither an equation nor an identity, but is a pair of equivalent operations.

For example, to find $\displaystyle\int 2x(x^2+1)^5\,dx,$

making the substitution $u=x^2+1$ gives

$$\int (x^2+1)^5 2x\,dx=\int u^5(2x)\,dx$$

But $\dfrac{du}{dx}=2x$ and as $\cdots\dfrac{du}{dx}dx=\cdots du$

then $\cdots 2x\,dx=\cdots du$

so $\displaystyle\int(x^2+1)^5 2x\,dx=\int u^5\,du=\dfrac{1}{6}u^6+K=\dfrac{1}{6}(x^2+1)^6+K$

To use this method, the integral needs to be expressed as an integral only in terms of u and du.

In practice, you can go direct from $\dfrac{du}{dx}=2x$ to $\cdots 2x\,dx=\cdots du$ by 'separating the variables'.

Products which can be integrated by this method are those in which one factor is basically the derivative of the function in the other factor, so substitute u for this function.

Example 4

Question

Integrate $x^2 \sqrt{x^3+5}$ with respect to x.

Answer

In this product x^2 is basically the derivative of $x^3 + 5$, so use the substitution $u = x^3 + 5$.

When $u = x^3 + 5$, $\dfrac{du}{dx} = 3x^2$

$\Rightarrow \qquad \ldots du = \ldots 3x^2\, dx$

Therefore $\displaystyle\int x^2 \sqrt{x^3+5}\, dx = \frac{1}{3}\int (x^3+5)^{\frac{1}{2}}(3x^2\, dx) = \frac{1}{3}\int u^{\frac{1}{2}}\, du$

$$= \left(\frac{1}{3}\right)\left(\frac{2}{3}\right)u^{\frac{3}{2}} + K = \frac{2}{9}(x^3+5)^{\frac{3}{2}} + K$$

Example 5

Question

Find $\displaystyle\int \cos x \sin^3 x\, dx$.

Answer

Writing the given integral in the form $\cos x\,(\sin x)^3$ suggests substituting $u = \sin x$.

When $u = \sin x$, $\dfrac{du}{dx} = \cos x \Rightarrow \ldots du = \ldots \cos x\, dx$

$\Rightarrow \quad \displaystyle\int \cos x \sin^3 x\, dx = \int (\sin x)^3 \cos x\, dx = \int u^3\, du$

$$= \frac{1}{4}u^4 + K = \frac{1}{4}\sin^4 x + K$$

> **Note**
>
> The integrals in Examples 4 and 5 can also be done by inspection.

Example 6

Question

Find $\displaystyle\int \frac{\ln x}{x}\, dx$.

Answer

This looks like a fraction but writing $\dfrac{(\ln x)}{x}$ as $\left(\dfrac{1}{x}\right)(\ln x)$ and since $\dfrac{1}{x} = \dfrac{d}{dx}(\ln x)$ you can make the substitution $u = \ln x$.

When $u = \ln x$, $\dfrac{du}{dx} = \dfrac{1}{x} \Rightarrow \ldots du = \ldots \dfrac{1}{x}\, dx$

Hence $\displaystyle\int \ln x \frac{1}{x}\, dx = \int u\, du = \frac{1}{2}u^2 + K$

So $\displaystyle\int \frac{\ln x}{x}\, dx = \frac{1}{2}(\ln x)^2 + K$

> **Note**
>
> $(\ln x)^2$ is *not* the same as $\ln x^2$.

Exercise 5

In questions 1 to 8, integrate the function with respect to x.

1 $4x^3 e^{x^4}$

2 $\sin x e^{\cos x}$

3 $(\sec^2 x)e^{\tan x}$

4 $(2x+1)e^{(x^2+x)}$

5 $\sec^2 x e^{(1-\tan x)}$

6 $(1+\cos x)e^{(x+\sin x)}$

7 $2xe^{(1+x^2)}$

8 $(3x^2-2)e^{(x^3-2x)}$

In questions 9 to 18, integrate by making the suggested substitution.

9 $\displaystyle\int x(x^2-3)^4\,dx;\, u=x^2-3$

10 $\displaystyle\int x\sqrt{1-x^2}\,dx;\, u=1-x^2$

11 $\displaystyle\int \cos 2x(\sin 2x+3)^2\,dx;\, u=\sin 2x+3$

12 $\displaystyle\int x^2(1-x^3)\,dx;\, u=1-x^3$

13 $\displaystyle\int e^x\sqrt{1+e^x}\,dx;\, u=1+e^x$

14 $\displaystyle\int \cos x\sin^4 x\,dx;\, u=\sin x$

15 $\displaystyle\int \sec^2 x\tan^3 x\,dx;\, u=\tan x$

16 $\displaystyle\int x^n(1+x^{n+1})^2\,dx;\, u=1+x^{n+1};\, n\neq-1$

17 $\displaystyle\int 3\sin x(1-\cos x)^4\,dx;\, u=1-\cos x$

18 $\displaystyle\int \sqrt{x}\sqrt{1+x^{\frac{3}{2}}}\,dx;\, u=1+x^{\frac{3}{2}}$

In questions 19 to 24, use a suitable substitution to find the integral or integrate by sight.

19 $\displaystyle\int x^3(x^4+4)^2\,dx$

20 $\displaystyle\int e^x(1-e^x)^3\,dx$

21 $\displaystyle\int \sin\theta\sqrt{1-\cos\theta}\,d\theta$

22 $\displaystyle\int (x+1)\sqrt{x^2+2x+3}\,dx$

23 $\displaystyle\int xe^{x^2+1}\,dx$

24 $\displaystyle\int \sec^2 x(1+\tan x)\,dx$

6.3 Integration by parts

Some functions (which are products of simpler functions) cannot be expressed in the form $f(u)\dfrac{du}{dx}$, so you cannot integrate these functions by using a substitution.

Using the product rule to differentiate a product $u(x)v(x)$ gives

$$\frac{d}{dx}(uv)=v\frac{du}{dx}+u\frac{dv}{dx} \quad\Rightarrow\quad v\frac{du}{dx}=\frac{d}{dx}(uv)-u\frac{dv}{dx}$$

Using $v\dfrac{du}{dx}$ to represent a product which is to be integrated with respect to x gives

$$\int v\frac{du}{dx}\,dx=\int \frac{d}{dx}(uv)\,dx-\int u\frac{dv}{dx}\,dx,\text{ and hence}$$

$$\int v\frac{du}{dx}\,dx=uv-\int u\frac{dv}{dx}\,dx$$

To give the right-hand side of the formula, one factor, $\dfrac{du}{dx}$, has to be integrated to give u. The other factor, v, has to be differentiated to give $\dfrac{dv}{dx}$. If either factor can be chosen, choose the one that gives the simpler expression when differentiated.

This method for integrating a product is called **integrating by parts**.

Example 7

Integrate xe^x with respect to x.

Taking $\quad v = x$ and $\dfrac{du}{dx} = e^x$ gives $\quad \dfrac{dv}{dx} = 1$ and $u = e^x$

Then $\quad \displaystyle\int v\frac{du}{dx}\,dx = uv - \int u\frac{dv}{dx}\,dx$

gives $\quad \displaystyle\int xe^x\,dx = (e^x)(x) - \int(e^x)(1)\,dx$

$$= xe^x - e^x + K$$

Example 8

Find $\displaystyle\int x^4 \ln x \, dx$.

Because $\ln x$ can be differentiated but *not integrated,* use $v = \ln x$.

Taking $\quad v = \ln x$ and $\dfrac{du}{dx} = x^4$

gives $\quad \dfrac{dv}{dx} = \dfrac{1}{x}$ and $u = \dfrac{1}{5}x^5$

Integrating by parts then gives

$$\int x^4 \ln x \, dx = \left(\frac{1}{5}x^5\right)(\ln x) - \int\left(\frac{1}{5}x^5\right)\left(\frac{1}{x}\right)dx = \frac{1}{5}x^5 \ln x - \frac{1}{5}\int x^4 \, dx$$

$$\Rightarrow \qquad \int x^4 \ln x \, dx = \frac{1}{5}x^5 \ln x - \frac{1}{25}x^5 + K$$

Example 9

Find $\displaystyle\int \ln x \, dx$.

$\displaystyle\int \ln x \, dx$ has not yet been found, but integration by parts can be used when $\ln x$ is regarded as the product of 1 and $\ln x$ as follows.

Taking $\qquad v = \ln x$ and $\dfrac{du}{dx} = 1$ gives $\quad \dfrac{dv}{dx} = \dfrac{1}{x}$ and $u = x$

Then $\qquad \displaystyle\int v\frac{du}{dx}\,dx = uv - \int u\frac{dv}{dx}\,dx$

becomes $\quad \displaystyle\int \ln x \, dx = x\ln x - \int x\left(\frac{1}{x}\right)dx = x\ln x - x + K$

Therefore $\quad \displaystyle\int \ln x \, dx = x(\ln x - 1) + K$

Exercise 6

In questions 1 to 16, integrate each expression with respect to x.

1 $x \cos x$

2 $x e^x$

3 $x \ln 3x$

4 $x e^{-x}$

5 $3x \sin x$

6 $x \sin 2x$

7 $x e^{2x}$

8 $x^2 e^{4x}$

9 $x \sin x$

10 $\ln 2x$

11 $e^x(x+1)$

12 $x(1+x)^7$

13 $x \sin\left(x + \dfrac{1}{6}\pi\right)$

14 $x \cos nx; \; n \neq 0$

15 $x \ln x$

16 $3x \cos 2x$

17 By writing $\cos^3 \theta$ as $(\cos^2 \theta)(\cos \theta)$, use integration by parts to find $\displaystyle\int \cos^3 \theta \, d\theta$.

Each product in questions 18 to 26 can be integrated either:

a by immediate recognition, or

b by a suitable substitution, or

c by parts.

Choose the best method in each case and integrate each function.

18 $(x-1)e^{(x^2-2x+4)}$

19 $(x+1)^2 e^x$

20 $\sin x(4 + \cos x)^3$

21 $\cos x \, e^{\sin x}$

22 $x^4 \sqrt{1+x^5}$

23 $e^x(e^x+2)^4$

24 $(x-1)e^{(2x-1)}$

25 $x(1-x^2)^9$

26 $\cos x \sin^5 x$

6.4 Integrating fractions

There are several different ways to integrate a fraction.

Integrating fractions using recognition

Using the chain rule to differentiate $y = \ln f(x)$ with $u = f(x)$ gives

$$\frac{d}{dx}\ln u = \left(\frac{1}{u}\right)\left(\frac{du}{dx}\right) = \frac{\frac{du}{dx}}{u}$$

so $\quad \dfrac{d}{dx}\ln f(x) = \dfrac{f'(x)}{f(x)}$

Therefore $\quad \displaystyle\int \frac{f'(x)}{f(x)}\,dx = \ln|f(x)| + K$

As such, all fractions of the form $\dfrac{f'(x)}{f(x)}$ can be integrated *immediately* by recognition.

For example $\displaystyle\int \frac{\cos x}{1+\sin x}\,dx = \ln|1+\sin x| + K \quad$ because $\quad \dfrac{d}{dx}(1+\sin x) = \cos x$

and $\quad \displaystyle\int \frac{e^x}{e^x+4}\,dx = \ln|e^x+4| + K \quad$ because $\quad \dfrac{d}{dx}(e^x+4) = e^x$.

Recognition works only for an integral whose numerator is basically the derivative of the *complete denominator*.

Example 10

Question

Find $\displaystyle\int \frac{x^2}{1+x^3}\,dx$.

Answer

$$\int \frac{x^2}{1+x^3}\,dx = \frac{1}{3}\int \frac{3x^2}{1+x^3}\,dx$$

The integral is now in the form $\displaystyle\int \frac{f'(x)}{f(x)}\,dx$ so   use $\displaystyle\int \frac{f'(x)}{f(x)}\,dx = \ln|f(x)|$

Hence $\displaystyle\int \frac{x^2}{1+x^3}\,dx = \frac{1}{3}\ln|1+x^3| + K$

Example 11

Question

By writing $\tan x$ as $\dfrac{\sin x}{\cos x}$,   find $\displaystyle\int \tan x\,dx$.

Answer

$$\int \tan x\,dx = \int \frac{\sin x}{\cos x}\,dx = -\int \frac{f'(x)}{f(x)}\,dx \quad \text{where} \quad f(x) = \cos x$$

So $\displaystyle\int \frac{\sin x}{\cos x}\,dx = -\ln|\cos x| + K$

Therefore $\displaystyle\int \tan x\,dx = K - \ln|\cos x| \quad \text{or} \quad K + \ln|\sec x|$

> **Note**
>
> Similarly,
> $$\int \cot x\,dx = \ln|\sin x| + K$$

Integrating fractions using substitution

An integral whose numerator is the derivative not of the complete denominator, but of an expression *within* the denominator, belongs to the next type of integral. For example

$$\int \frac{2x}{\sqrt{x^2+1}}\,dx.$$

As $2x$ is the derivative of x^2+1, use the substitution $u = x^2 + 1$.

Then $\dfrac{du}{dx} = 2x$, which gives ... $du = ...\,2x\,dx$.

This substitution converts the given integral into the form $\displaystyle\int \frac{1}{\sqrt{u}}\,du$.

Example 12

Question

Find $\displaystyle\int \frac{e^x}{(1-e^x)^2}\,dx$.

Answer

e^x is basically the derivative of $1 - e^x$, but not of $(1 - e^x)^2$, so make the substitution $u = 1 - e^x$.

When $u = 1 - e^x$ then ...$du = ...-e^x\,dx$

So $\displaystyle\int \frac{e^x}{(1-e^x)^2}\,dx = \int \frac{-1}{u^2}\,du = \frac{1}{u} + K$

$\Rightarrow \displaystyle\int \frac{e^x}{(1-e^x)^2}\,dx = \frac{1}{1-e^x} + K$

Example 13

Find $\int \dfrac{\sec^2 x}{\tan^3 x}\,dx$.

$\sec^2 x$ is the derivative of $\tan x$ but not of $\tan^3 x$.

Taking $u = \tan x$ gives $\ldots du = \ldots \sec^2 x\,dx$

Then $\qquad \int \dfrac{\sec^2 x}{\tan^3 x}\,dx = \int \dfrac{1}{u^3}\,du = -\dfrac{1}{2}u^{-2} + K$

Therefore $\qquad \int \dfrac{\sec^2 x}{\tan^3 x}\,dx = \dfrac{-1}{2\tan^2 x} + K$

Exercise 7

In questions 1 to 18 integrate the expression with respect to x.

1 $\dfrac{\cos x}{4 + \sin x}$

2 $\dfrac{e^x}{3e^x - 1}$

3 $\dfrac{x}{(1-x^2)^3}$

4 $\dfrac{\sin x}{\cos^3 x}$

5 $\dfrac{x^3}{1+x^4}$

6 $\dfrac{2x+3}{x^2+3x-4}$

7 $\dfrac{x^2}{\sqrt{2+x^3}}$

8 $\dfrac{\cos x}{(\sin x - 2)^2}$

9 $\dfrac{1}{x\ln x}$ $\left(\text{this is equivalent to } \dfrac{\frac{1}{x}}{\ln x} \right)$

10 $\dfrac{\cos x}{\sin^6 x}$

11 $\dfrac{2x}{1-x^2}$

12 $\dfrac{e^x}{\sqrt{1-e^x}}$

13 $\dfrac{x-1}{3x^2-6x+1}$

14 $\dfrac{\cos x}{\sin^n x}\, n \neq 0$

15 $\dfrac{\sin x}{\cos^n x}\, n \neq 0$

16 $\dfrac{\sin x}{4+\cos x}$

17 $\dfrac{x-1}{x(x-2)}$

18 $\dfrac{e^x-1}{(e^x-x)^2}$

Evaluate these integrals.

19 $\displaystyle\int_1^2 \dfrac{2x+1}{x^2+x}\,dx$

20 $\displaystyle\int_0^1 \dfrac{x}{x^2+1}\,dx$

21 $\displaystyle\int_2^3 \dfrac{2x}{(x^2-1)^3}\,dx$

22 $\displaystyle\int_0^1 \dfrac{e^x}{(1+e^x)^2}\,dx$

23 $\displaystyle\int_{\frac{\pi}{6}}^{\frac{\pi}{3}} \dfrac{\sin(2x-n)}{\cos(2x-\pi)}\,dx$

24 $\displaystyle\int_2^4 \dfrac{1}{x(\ln x)}\,dx$

Using partial fractions

When a rational function cannot be integrated by inspection or by substitution, it may be easy to integrate when expressed in partial fractions. Remember, though, that only proper fractions can be converted directly into partial fractions. An improper fraction must first be rearranged so that it is made up of non-fractional terms and a proper fraction.

Example 14

Find $\int \dfrac{2x-3}{(x-1)(x-2)}\,dx$.

First decompose into partial fractions

$$\frac{2x-3}{(x-1)(x-2)}=\frac{A}{x-1}+\frac{B}{x-2} \quad\Rightarrow\quad 2x-3=A(x-2)+B(x-1) \quad\Rightarrow\quad A=1 \text{ and } B=1$$

So $\quad\dfrac{2x-3}{(x-1)(x-2)}=\dfrac{1}{x-1}+\dfrac{1}{x-2}$

Therefore $\quad\displaystyle\int\frac{2x-3}{(x-1)(x-2)}\,dx=\int\frac{1}{x-1}\,dx+\int\frac{1}{x-2}\,dx$

$$= \ln|x-1|+\ln|x-2|+K$$
$$= \ln|(x-1)(x-2)|+K$$

Example 15

Find $\int \dfrac{x^2+1}{x^2-1}\,dx$.

This fraction is improper, so first express it as a sum of non-fractional terms and a proper fraction. Then decompose the proper fraction into partial fractions.

$$\frac{x^2+1}{x^2-1}=\frac{(x^2-1)+2}{x^2-1}=1+\frac{2}{x^2-1}=1+\frac{2}{(x-1)(x+1)}$$

$$=1+\frac{1}{x-1}-\frac{1}{x+1}$$

Then $\quad\displaystyle\int\frac{x^2+1}{x^2-1}\,dx=\int 1\,dx+\int\frac{1}{x-1}\,dx-\int\frac{1}{x+1}\,dx$

$$=x+\ln|x-1|-\ln|x+1|+K$$

When you are asked to integrate a fraction, look at the fraction carefully, as fractions needing different integration techniques often *look* very similar. Example 16 shows this clearly.

Example 16

Integrate with respect to x.

a $\quad\dfrac{x+1}{x^2+2x-8}$
b $\quad\dfrac{x+1}{(x^2+2x-8)^2}$
c $\quad\dfrac{x+2}{x^2+2x-8}$.

a This fraction is basically of the form $\dfrac{f'(x)}{f(x)}$

$$\int\frac{x+1}{x^2+2x-8}\,dx=\frac{1}{2}\int\frac{2x+2}{x^2+2x-8}\,dx=\frac{1}{2}\ln|x^2+2x-8|+K$$

b This time the numerator is basically the derivative of the function *within* the denominator.

Let $u=x^2+2x-8 \quad\Rightarrow\quad \ldots du=\ldots(2x+2)\,dx=\ldots 2(x+1)\,dx$

(continued)

(continued)

$$\Rightarrow \int \frac{x+1}{(x^2+2x-8)^2}\,\mathrm{d}x = \frac{1}{2}\int \frac{1}{u^2}\,\mathrm{d}u = -\frac{1}{2u}+K = -\frac{1}{2(x^2+2x-8)}+K$$

c   In this fraction the numerator is not related to the derivative of the denominator so, since the denominator factorises, use partial fractions.

$$\int \frac{x+2}{x^2+2x-8}\,\mathrm{d}x = \int \frac{\frac{1}{3}}{x+4}\,\mathrm{d}x + \int \frac{\frac{2}{3}}{x-2}\,\mathrm{d}x$$

$$= \frac{1}{3}\ln|x+4| + \frac{2}{3}\ln|x-2| + K$$

Exercise 8

In questions 1 to 12, integrate the expression with respect to x.

1 $\dfrac{2}{x(x+1)}$

2 $\dfrac{4}{(x-2)(x+2)}$

3 $\dfrac{x}{(x-1)(x+1)}$

4 $\dfrac{x-1}{x(x+2)}$

5 $\dfrac{x-1}{(x-2)(x-3)}$

6 $\dfrac{1}{x(x-1)(x+1)}$

7 $\dfrac{x}{x+1}$

8 $\dfrac{x+4}{x}$

9 $\dfrac{x}{x+4}$

10 $\dfrac{3x-4}{x(1-x)}$

11 $\dfrac{x^2-2}{x^2-1}$

12 $\dfrac{x^2}{(x+1)(x+2)}$

In questions 13 to 18, choose the best method to integrate the expression with respect to x.

13 $\dfrac{x}{x^2-1}$

14 $\dfrac{2x}{(x^2-1)^2}$

15 $\dfrac{2}{x^2-1}$

16 $\dfrac{2x-5}{x^2-5x+6}$

17 $\dfrac{2x}{x^2-5x+6}$

18 $\dfrac{2x-3}{x^2-5x+6}$

In questions 19 to 24, evaluate the definite integral.

19 $\displaystyle\int_0^4 \frac{x+2}{x+1}\,\mathrm{d}x$

20 $\displaystyle\int_{-1}^1 \frac{5}{x^2+x-6}\,\mathrm{d}x$

21 $\displaystyle\int_1^2 \frac{x+2}{x(x+4)}\,\mathrm{d}x$

22 $\displaystyle\int_0^1 \frac{2}{3+2x}\,\mathrm{d}x$

23 $\displaystyle\int_{\frac{1}{2}}^3 \frac{2}{(3+2x)^2}\,\mathrm{d}x$

24 $\displaystyle\int_1^2 \frac{2x}{3+2x}\,\mathrm{d}x$

6.5 Special techniques for integrating some trigonometric functions

Integrating a function containing an odd power of sin x or cos x

When the power of $\sin x$ or $\cos x$ is an *odd power* other than 1, the formula $\cos^2 x + \sin^2 x = 1$ can convert the given function to one that can be integrated.

For example, $\sin^3 x$ is converted to $(\sin^2 x)(\sin x) \Rightarrow (1-\cos^2 x)(\sin x)$

$$\Rightarrow \sin x - \cos^2 x \sin x$$

Example 17

Integrate with respect to x.

a $\quad \cos^5 x$

b $\quad \sin^3 x \cos^2 x$.

a $\quad \cos^5 x = (\cos^2 x)^2 \cos x = (1 - \sin^2 x)^2 \cos x = (1 - 2\sin^2 x + \sin^4 x)\cos x$

Therefore $\quad \displaystyle\int \cos^5 x \, dx = \int \cos x \, dx - 2\int \sin^2 x \cos x \, dx + \int \sin^4 x \cos x \, dx$

Hence $\quad \displaystyle\int \cos^5 x \, dx = \sin x - 2\left(\frac{1}{3}\right)\sin^3 x + \left(\frac{1}{5}\right)\sin^5 x + K$

$$= \sin x - \frac{2}{3}\sin^3 x + \frac{1}{5}\sin^5 x + K$$

b $\quad \sin^3 x \cos^2 x = \sin x (1 - \cos^2 x)\cos^2 x$

so $\quad \displaystyle\int \sin^3 x \cos^2 \, dx = \int \cos^2 x \sin x \, dx - \int \cos^4 x \sin x \, dx$

$$= -\frac{1}{3}\cos^3 x + \frac{1}{5}\cos^5 x + K$$

> **Note**
>
> For any value of n,
>
> $$\int \cos x \sin^n x \, dx$$
>
> $$= \frac{1}{n+1}\sin^{n+1} x + K.$$

Integrating a function containing only even powers of sin x or cos x

When integrating even powers of $\sin x$ or $\cos x$, the double angle identities are useful. For example

$$\cos^4 x = (\cos^2 x)^2 = \left[\frac{1}{2}(1 + \cos 2x)\right]^2 = \frac{1}{4}[1 + 2\cos 2x + \cos^2 2x]$$

Then use a double angle identity again, which gives

$$\cos^4 x = \frac{1}{4}(1 + 2\cos 2x) + \frac{1}{4}\left[\frac{1}{2}(1 + \cos 4x)\right] = \frac{3}{8} + \frac{1}{2}\cos 2x + \frac{1}{8}\cos 4x$$

Now each of these terms can be integrated.

Example 18

Integrate with respect to x.

a $\quad \sin^2 x$

b $\quad 16\sin^2 x \cos^2 x$.

a $\quad \displaystyle\int \sin^2 x \, dx = \int \frac{1}{2}(1 - \cos 2x) \, dx = \frac{1}{2}x - \frac{1}{4}\sin 2x + K$

b $\quad \displaystyle 16\int \sin^2 x \cos^2 x \, dx = 16\int (\sin x \cos x)^2 \, dx$

$$= \frac{16}{4}\int \sin^2 2x \, dx$$

$$= 4\int \frac{1}{2}(1 - 4\cos 4x) \, dx$$

$$= 2\left(x - \frac{1}{4}\sin 4x\right) + k$$

Integrating any power of tan x

The formula $\tan^2 x = \sec^2 x - 1$ is useful when integrating a power of $\tan x$.

For example, $\tan^3 x$ becomes $\tan x(\sec^2 x - 1) = \sec^2 x \tan x - \tan x$

then $\int \tan^3 x \, dx$

$$= \int \sec^2 x \tan x - \int \tan x \, dx$$

$$= \frac{1}{2}\tan^2 x + \ln |\cos x| + c$$

Example 19

Question

Integrate these expressions with respect to x.

a $\tan^4 x$ 　　　　　　　　　　b $\tan^5 x$

Answer

a $\int \tan^4 x \, dx = \int \tan^2 x(\sec^2 x - 1)\, dx = \int \sec^2 x \tan^2 x \, dx - \int \tan^2 x \, dx$

$$= \frac{1}{3}\tan^3 x - \int (\sec^2 x - 1)\, dx$$

$$= \frac{1}{3}\tan^3 x - \tan x + x + K$$

b $\int \tan^5 x \, dx = \int \tan^3 x(\sec^2 x - 1)\, dx$

$$= \int \sec^2 x \tan^3 x - \int \tan x(\sec^2 x - 1)\, dx$$

$$= \int \sec^2 x \tan^3 x \, dx - \int \sec^2 x \tan x \, dx + \int \tan x \, dx$$

$$= \frac{1}{4}\tan^4 x - \frac{1}{2}\tan^2 x + \ln |\sec x| + K$$

Note

To integrate *any* power of tan$x$, the formula $\tan^2 x = \sec^2 x - 1$ is used to convert tan$^2 x$ *only, one step at a time*. Converting tan$^4 x$ to $(1 - \sec^2 x)^2$ does not help.

Exercise 9

Integrate each expression with respect to x.

1. $\cos^2 x$ 　　　　　　2. $\cos^3 x$
3. $\sin^5 x$ 　　　　　　4. $\tan^2 x$
5. $\sin^4 x$ 　　　　　　6. $\tan^3 x$
7. $\cos^4 x$ 　　　　　　8. $\sin^3 x$

6.6 Volume of revolution

When a region is rotated about a straight line, the three-dimensional object formed is called a **solid of revolution**, and its volume is a **volume of revolution**.

The line about which rotation takes place is always a line of symmetry for the solid of revolution. Also, any cross-section of the solid which is perpendicular to the axis of rotation is circular.

The diagram shows the solid of revolution formed when the shaded region is rotated about the x-axis.

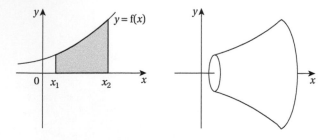

To calculate the volume of this solid, it can be divided it into 'slices' by making cuts perpendicular to the axis of rotation.

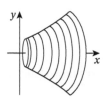

If the cuts are reasonably close together, each slice is approximately a cylinder and the approximate volume of the solid can be found by summing the volumes of these cylinders.

Look at an element formed by one cut through the point $P(x, y)$ and the other cut a distance δx from the first.

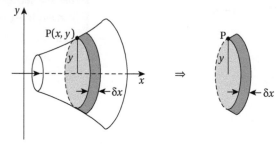

The volume, δV, of this element is approximately that of a cylinder of radius y and 'height' δx,

so $\delta V \approx \pi y^2 \delta x$

Then the total volume of the solid is V, where $V \approx \sum_{x_1}^{x_2} \pi y^2 \, \delta x$

The smaller δx is, the closer is this approximation to V,

Therefore $V = \lim_{\delta x \to 0} \sum_{x_1}^{x_2} \pi y^2 \, \delta x = \pi \int_{x_1}^{x_2} y^2 \, \mathrm{d}x$

When the equation of the rotated curve is known, this integral can be evaluated and the volume of the solid of revolution found.

For example, to find the volume generated when the area between part of curve $y = \mathrm{e}^x$ and the x-axis is rotated about the x-axis, use $\pi \int (\mathrm{e}^x)^2 \, \mathrm{d}x = \pi \int \mathrm{e}^{2x} \, \mathrm{d}x$.

When a region rotates about the y-axis we can use a similar method based on slices perpendicular to the y-axis, giving

$$V = \pi \int_{y_1}^{y_2} x^2 \, dy$$

Example 20

Find the volume generated when the region bounded by the x- and y-axes, the line $x = 1$ and the curve $y = e^x$ is rotated through one revolution about the x-axis.

$$V = \pi \int_0^1 y^2 \, dx = \pi \int_0^1 (e^x)^2 \, dx$$

$$= \pi \int_0^1 e^{2x} \, dx = \pi \left[\frac{1}{2} e^{2x} \right]_0^1 = \frac{1}{2} \pi (e^2 - e^0)$$

Therefore the volume is $\frac{1}{2} \pi (e^2 - 1)$ cubic units.

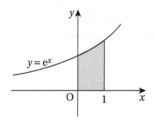

> **Note**
>
> $\int y \, dx$ means integrate y with respect to x, so y must be expressed in terms of x.

Example 21

The region enclosed by the curve $y = x^2 + 1$, the line $y = 2$ and the y-axis is rotated completely about the y-axis. Find the volume of the solid generated.

Rotating the shaded region about the y-axis gives the solid shown.

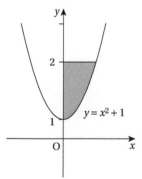
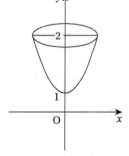

$$V = \pi \int_1^2 x^2 \, dy$$

The integral is with respect to y, so x^2 must be expressed in terms of y.

Using the equation $y = x^2 + 1$ gives $x^2 = y - 1$

$$V = \pi \int_1^2 (y - 1) \, dy = \pi \left[\frac{1}{2} y^2 - y \right]_1^2$$

$$= \pi \left[(2 - 2) - \left(\frac{1}{2} - 1 \right) \right]$$

The volume of the solid is $\frac{1}{2} \pi$ cubic units.

Exercise 10

In questions 1 to 5, find the volume generated when the region enclosed by the equations is rotated completely about the x-axis.

1 The x-axis and $y = x(4-x)$

2 $y = e^x$, the y-axis, the x-axis and $x = 3$

3 $y = \dfrac{1}{x}$, $x = 1$, $x = 2$ and the x-axis

4 $y = x^2$, the y-axis, $x = -2$ and $x = 2$

5 $y = \cos x$, the y-axis, the x-axis, $x = 0$ and $x = \dfrac{\pi}{2}$

In questions 6 to 9, the region enclosed by the curve and line(s) given is rotated about the y-axis to form a solid. Find the volume generated.

6 $y = x^2$, $y = 4$

7 $y = 4 - x^2$, $y = 0$

8 $y = x^3$, $y = 1$, $y = 2$, for $x \geq 0$

9 $y = \ln x$, $x = 0$, $y = 0$, $y = 1$

Summary

Integrals that can be found by inspection

$$\int e^x \, dx = e^x + K, \text{ and } \int f'(x)e^{f(x)} \, dx = e^{f(x)} + K$$

$$\int \frac{1}{x} \, dx = \ln|x| + K, \text{ and } \int \frac{f'(x)}{f(x)} \, dx = \ln|f(x)| + K$$

$$\int (ax+b)^n \, dx = \frac{1}{(a)(n+1)}(ax+b)^{n+1} + K$$

$$\int \cos x \, dx = \sin x + K, \text{ and } \int f'(x)\cos f(x) \, dx = -\sin f(x) + K$$

$$\int \sin x \, dx = -\cos x + K, \text{ and } \int -f'(x)\sin f(x) \, dx = \cos f(x) + K$$

$$\int \sec^2 x \, dx = \tan x + K, \text{ and } \int f'(x)\sec^2 f(x) x \, dx = \tan f(x) + K$$

Integrals that can be found by substitution

▶ Functions of the form $f'(x)g(f(x))$ or $\dfrac{f'(x)}{g(f(x))}$ can be integrated using the substitution $u = f(x)$ and

$$\dots \frac{du}{dx} \, dx = \dots \, du$$

Integration by parts

▶ A product, $f(x)g(x)$, may be integrated using $\displaystyle\int v \frac{du}{dx} \, dx = uv - \int u \frac{dv}{dx} \, dx$ where

$$v = f(x) \quad \text{and} \quad \frac{du}{dx} = g(x).$$

Integrating partial fractions

▶ A rational function that cannot be integrated by recognition or by substitution may be integrated by first expressing it in partial fractions.

Integrals containing powers of sin x, cos x or tan x

▶ When the power of sin x or cos x is an *odd power* other than 1, use the formula $\cos^2 x + \sin^2 x = 1$.

▶ When only even powers of sin x or cos x are involved, the double angle formulae can transform the function into one that can be integrated.

▶ When any power of tan x is involved, the formula $\tan^2 x = \sec^2 x - 1$ can be used to transform the function into one that can be integrated.

Volume of revolution

▶ The volume generated when the area enclosed by the x-axis, the ordinates at a and b, and part of the curve $y = f(x)$ rotates completely about the x-axis, is given by $\pi \int_{x=a}^{x=b} y^2 \, dx$.

▶ The volume generated when the region enclosed by the y-axis and the lines $y = a$ and $y = b$ and the curve $y = f(x)$ is rotated completely about the y-axis, is given by $\pi \int_{y=a}^{y=b} x^2 \, dy$.

Review

In questions 1 to 11, integrate the expression with respect to x.

1 $3e^{-x}$

2 $2xe^{x^2}$

3 $\sec^2 x(3 \tan x - 4)$

4 $\dfrac{4}{3x}$

5 $\sec^2 x \tan^3 x$

6 $(3x+4)^5$

7 $(\sin x)e^{\cos x}$

8 $\cos\left(4x + \dfrac{\pi}{7}\right)$

9 $(x-1)e^{(x^2-2x+3)}$

10 $x^2(1-x^3)^9$

11 $\sqrt{2x+1}$

12 Find $\int x\sqrt{x^2+1}\,dx$ using the substitution $u = x^2 + 1$.

13 Find $\int \cos x(2-\sin x)^2\,dx$ using the substitution $u = 2 - \sin x$.

In questions 14 to 16 use integration by parts to integrate the function.

14 x^2e^{2x}

15 $(x+1)\ln(x+1)$

16 $x^2 \cos x$

In questions 17 to 32 choose the best method and find the integral in each case.

17 $\int e^{2x+3}dx$

18 $\int x\sqrt{2x^2-5}\,dx$

19 $\int xe^x\,dx$

20 $\int \ln x\,dx$

21 $\int \sin^2 3x\,dx$

22 $\int xe^{-x^2}\,dx$

23 $\displaystyle\int \cos x \sin^2 x \, dx$

24 $\displaystyle\int u(u+7)^9 \, du$

25 $\displaystyle\int \frac{x^2}{(x^3+9)^5} \, dx$

26 $\displaystyle\int \frac{\sin 2y}{1-\cos 2y} \, dy$

27 $\displaystyle\int \frac{1}{2x+7} \, dx$

28 $\displaystyle\int \sin 3x \sqrt{1+\cos 3x} \, dx$

29 $\displaystyle\int \frac{x+2}{x^2+4x-5} \, dx$

30 $\displaystyle\int \frac{x+1}{x^2+4x-5} \, dx$

31 $\displaystyle\int x \cos 3x \, dx$

32 $\displaystyle\int \ln 5x \, dx$

33 The diagram shows part of the curve $y = 1 - \cos^2 x$ between $x = 0$ and $x = \pi$.

Find the volume generated when the region shown is rotated about the x-axis.

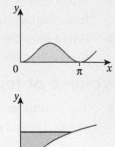

34 The region shown in the diagram is enclosed by the curve $y = \ln x$, the x-axis, the y-axis and the line $y = 1$.

Show that the volume of the solid formed when the area is rotated completely about the y-axis is $\frac{1}{2}\pi(e^2 - 1)$.

Assessment

1 Integrate

  a $(1-x)e^{(1-x)^2}$

  b $\dfrac{2x}{x^2+1}$

  c $\cos x \sin^4 x$

2 Integrate

  a $\sin x(4+\cos x)^3$

  b $x(2x+3)^7$

  c $xe^{(2-3x)}$

3 Find $\displaystyle\int \frac{3x}{\sqrt{x^2-3}} \, dx$ using the substitution $u = x^2 - 3$.

4 **a** Express $\dfrac{2}{(x^2-2x-3)}$ in partial fractions.

  b Find $\displaystyle\int \frac{2}{(x^2-2x-3)} \, dx$.

5 **a** Find $\displaystyle\int 3y\sqrt{9-y^2} \, dy$.

  b Find $\displaystyle\int_0^{\frac{\pi}{2}} x \sin 4x \, dx$.

6 The diagram shows the region enclosed by the curve $y = x^2 - 2x + 1$, the x-axis and the lines $x = 0$ and $x = 3$.

Find the volume generated when these regions are rotated about the x-axis.

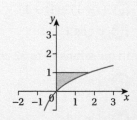

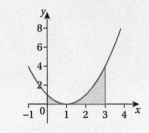

7 The diagram shows the region enclosed by the curve $y = \ln(x+1)$, the y-axis and the line $y = 1$.

Find the volume generated when the region is rotated about the y-axis.

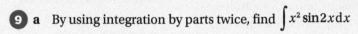

8 a Given that $\dfrac{4x^3 - 2x^2 + 16x - 3}{2x^2 - x + 2}$ can be expressed as $Ax + \dfrac{B(4x-1)}{2x^2 - x + 2}$, find the values of the constants A and B.

b The gradient of a curve is given by $\dfrac{dy}{dx} = \dfrac{4x^3 - 2x^2 + 16x - 3}{2x^2 - x + 2}$

The point $(-1, 2)$ lies on the curve. Find the equation of the curve.

AQA MPC4 June 2014

9 a By using integration by parts twice, find $\displaystyle\int x^2 \sin 2x \, dx$

b A curve has equation $y = x\sqrt{\sin 2x}$ for $0 \le x \le \dfrac{\pi}{2}$.

The region bounded by the curve and the x-axis is rotated through 2π radians about the x-axis to generate a solid. Find the exact value of the volume of the solid generated.

AQA MPC3 June 2014

7 Differential Equations

Introduction

Any equation that contains a derivative, for example $x\dfrac{dy}{dx}=2$, is called a differential equation. This chapter looks at how to find a direct relationship between x and y. There are many different types of differential equation, with each type requiring a specific technique to find its solution. This chapter looks at first order linear differential equations where you can solve the equation by separating the variables.

Objectives

By the end of this chapter, you should know how to...
- Solve first order linear differential equations by separating the variables.
- Form differential equations from given information.

Recap

You will need to remember...
- When two quantities, x and y, are such that y is proportional to x, then $y \propto x \Rightarrow y = kx$ where k is a constant.
- When two quantities x and y, are such that y is inversely proportional to x, then $y \propto \dfrac{1}{x} \Rightarrow y = \dfrac{k}{x}$ where k is a constant.

Applications

Differential equations occur whenever the variation in one quantity, p, depends upon the changing value of another quantity, q,

the rate of increase of p with respect to q can be expressed as $\dfrac{dp}{dq}$.

There are many everyday situations where such relationships exist, for example:
- Liquid expands when heated so, if V is the volume of a quantity of liquid and T is the temperature, then the rate at which the volume increases with temperature can be written $\dfrac{dV}{dT}$.
- When the profit, P, made by a company selling phones depends upon the number, n, of phones sold, then $\dfrac{dP}{dn}$ represents the rate at which profit increases with respect to sales.

7.1 First order differential equations with separable variables

An equation with at least one term containing $\dfrac{dy}{dx}, \dfrac{d^2y}{dx^2}, \ldots$ is called a **differential equation**. When the equation contains only $\dfrac{dy}{dx}$ it is a **first order linear differential equation**.

For example, $x + 2\dfrac{dy}{dx} = 3y$ is a first order linear differential equation.

A differential equation represents a relationship between two variables. The same relationship can often be expressed in a form that does not contain a derivative, for example, $\dfrac{dy}{dx} = 2x$ and $y = x^2 + K$ expresses the same relationship between x and y.

Converting a differential equation into a direct one is called solving the differential equation and this process involves integration.

Look at the differential equation $3y\dfrac{dy}{dx} = 5x^2$. \hfill [1]

Integrating both sides of the equation with respect to x gives $\displaystyle\int 3y\dfrac{dy}{dx}\,dx = \int 5x^2\,dx$.

Using $\ldots\dfrac{dy}{dx}dx = \ldots dy$ gives

$$\int 3y\,dy = \int 5x^2\,dx \qquad [2]$$

Temporarily removing the integral signs from this equation gives

$$3y\,dy = 5x^2\,dx \qquad [3]$$

This relationship can be found directly from equation [1] by separating the variables – that is, by separating dy from dx and collecting on one side all the terms involving y together with dy, while collecting all the x terms along with dx on the other side.

The relationship shown in [3] does not have any meaning, and it *should not be written down as a step in the solution*. It only provides a way of making a quick mental conversion from the differential equation [1] to the form [2] in which each side can be integrated separately.

Returning to equation [2] and integrating each side gives

$$\frac{3}{2}y^2 = \frac{5}{3}x^3 + A$$

A constant of integration does not need to be added on both sides. A constant on one side only is enough.

When solving differential equations, the constant of integration is usually denoted by A, B, etc. and is called the arbitrary constant.

Example 1

Solve the differential equation $\dfrac{1}{x}\dfrac{dy}{dx} = \dfrac{2y}{x^2 + 1}$.

$$\frac{1}{x}\frac{dy}{dx} = \frac{2y}{x^2+1} \quad \Rightarrow \quad \frac{1}{y}\frac{dy}{dx} = \frac{2x}{x^2+1}$$

Separating the variables gives

$$\int \frac{1}{y}\,dy = \int \frac{2x}{x^2+1}\,dx$$

$$\Rightarrow \quad \ln|y| = \ln|x^2+1| + A$$

Exercise 1

Solve each differential equation.

1 $y\dfrac{dy}{dx}=\sin x$

2 $x^2\dfrac{dy}{dx}=y^2$

3 $\dfrac{1}{x}\dfrac{dy}{dx}=\dfrac{1}{y^2-2}$

4 $\tan y\dfrac{dy}{dx}=\dfrac{1}{x}$

5 $\dfrac{dy}{dx}=y^2$

6 $\dfrac{1}{x}\dfrac{dy}{dx}=\dfrac{1}{1-x^2}$

7 $(x-3)\dfrac{dy}{dx}=y$

8 $\tan y\dfrac{dx}{dy}=4$

9 $u\dfrac{du}{dv}=v+2$

10 $\dfrac{y^2}{x^3}\dfrac{dy}{dx}=\ln x$

11 $e^x\dfrac{dy}{dx}=\dfrac{x}{y}$

12 $\sec x\dfrac{dy}{dx}=e^y$

13 $r\dfrac{dr}{d\theta}=\sin^2\theta$

14 $\dfrac{dv}{du}=\dfrac{v+1}{u+2}$

15 $xy\dfrac{dy}{dx}=\ln x$

16 $y(x+1)=(x^2+2x)\dfrac{dy}{dx}$

17 $v^2\dfrac{dv}{dt}=(2+t)^3$

18 $x\dfrac{dy}{dx}=\dfrac{1}{y}+y$

19 $r\dfrac{d\theta}{dr}=\cos^2\theta$

20 $y\sin^3 x\dfrac{dy}{dx}=\cos x$

21 $\dfrac{uv}{u-1}=\dfrac{du}{dv}$

22 $e^x\dfrac{dy}{dx}=e^{y-1}$

23 $\tan x\dfrac{dy}{dx}=2y^2\sec^2 x$

24 $\dfrac{dy}{dx}=\dfrac{x(y^2-1)}{(x^2+1)}$

Calculation of the constant of integration

Look again at $3y\dfrac{dy}{dx}=5x^2 \iff \dfrac{3}{2}y^2=\dfrac{5}{3}x^3+A$

The value of A cannot be found from the differential equation alone, but further information is needed – for example a point on the curve.

To find the equation of the curve that satisfies the differential equation $2\dfrac{dy}{dx}=\dfrac{\cos x}{y}$ and which passes through the point $(0, 2)$, first solve the differential equation.

Separating the variables gives $\displaystyle\int 2y\,dy=\int\cos x\,dx \Rightarrow y^2=\sin x+A$

When $x=0$ and $y=2$, $\quad 4=A+0 \Rightarrow A=4$

Hence the equation of the specified curve is $y^2=4+\sin x$.

Example 2

Find the equation of the curve satisfied by the differential equation $y=x\dfrac{dy}{dx}$ and which passes through the point $(1, 2)$.

Separating the variables in the equation $y=x\dfrac{dy}{dx}$ gives $\displaystyle\int\dfrac{1}{y}\,dy=\int\dfrac{1}{x}\,dx$

$$\Rightarrow \quad \ln|y|=\ln|x|+A$$

When $x=1$ and $y=2$ then $\quad \ln 2=A+\ln 1=A+0, \Rightarrow A=\ln 2$

So $\qquad\qquad\qquad\qquad \ln y=\ln x+\ln 2 \Rightarrow \ln y=\ln 2x$

Therefore $\qquad\qquad\qquad y=2x$

Exercise 2

In questions 1 to 6, find the equation of the curve satisfied by the differential equation and passing through the given point.

1 $y^2 \dfrac{dy}{dx} = x^2 + 1$; $(2, 1)$

2 $e^t \dfrac{ds}{dt} = \sqrt{s}$; $(0, 4)$

3 $\dfrac{y}{x} \dfrac{dy}{dx} = \dfrac{y^2 - 1}{x^2 - 1}$; $(2, 3)$

4 $e^{-x} \dfrac{dy}{dx} = 1$; $(0, -1)$

5 $\dfrac{2y}{3} \dfrac{dy}{dx} = e^{-3x}$; $(1, 2)$

6 $\dfrac{dy}{dx} = \dfrac{y+1}{x^2 - 1}$; $(-3, 1)$

7 Solve the differential equation $(1 + x^2) \dfrac{dy}{dx} - y(y+1)x = 0$, given that $y = 1$ when $x = 0$.

7.2 Natural occurrence of differential equations

Differential equations often arise when a physical situation is interpreted mathematically (that is, when a mathematical model is made to represent a physical situation). For example:

▶ A body falls from rest in a medium which causes the velocity to decrease at a rate proportional to the velocity.

As the velocity is *decreasing* with time, its rate of increase is *negative*.

Using v for velocity and t for time, the rate of change of velocity with respect to time is negative so $\dfrac{dv}{dt} \propto -v$.

The motion of the body satisfies the differential equation $-\dfrac{dv}{dt} = kv$.

▶ During the first stages of the growth of yeast cells in a culture, the number of cells present increases in proportion to the number already formed.

Therefore the rate of change of the number of cells, n, with respect to time is such that $\dfrac{dn}{dt} \propto n$, and n can be found from the differential equation $\dfrac{dn}{dt} = kn$.

> **Note**
>
> To form (and then solve) differential equations from naturally occurring information, you do not need to understand the background to the situation or experiment.

Exercise 3

In questions 1 to 6 form, but *do not solve,* the differential equations representing the given information.

1 A body moves with a velocity v which is inversely proportional to its displacement s from a fixed point.

2 The rate at which the height h of a plant increases is proportional to the natural logarithm of the difference between its present height and its final height H.

3 When water enters a tank, the rate at which the depth of water, h metres, in the tank is increasing is proportional to the volume, V cubic metres, which is already in the tank.

4 The rate at which a radioactive material loses mass is proportional to the mass, m kilograms, at time t seconds.

5 The depth of a sump in an engine is H cm. At time t hours the depth of oil in the sump is h cm. Oil is leaking from the sump so that h is changing at a rate proportional to $H - h$.

6 In freezing weather, the ice on a lake is d cm thick t hours after freezing starts. The depth of ice is increasing at a rate that is inversely proportional to itself.

7 The manufacturers of soap powder are concerned that the number, n, of people buying their product at any time t months has remained constant for some months. An advertisment results in the number of customers increasing at a rate proportional to the square root of n.

Express as differential equations the progress of sales

  a before advertising

  b after advertising.

8 In a community, the number, n, of people suffering from an infectious disease is N_1 at a particular time. The disease then becomes epidemic and spreads so that the number of sick people increases at a rate proportional to n, until the total number of sufferers is N_2. The rate of increase then becomes inversely proportional to n until N_3 people have the disease. After this, the total number of sick people decreases at a constant rate. Write down the differential equation governing the incidence of the disease

  a for $N_1 \le n \le N_2$

  b for $N_2 \le n \le N_3$

  c for $n \ge N_3$.

Solving naturally occurring differential equations

We have seen that when one naturally occurring quantity varies with another, the relationship between them often involves a constant of proportion. Consequently, a differential equation that represents the relationship contains a constant of proportion whose value is not always known. So the initial solution of the differential equation contains both this constant and the arbitrary constant. Extra given information may allow either or both constants to be evaluated.

Example 3

A particle moves in a straight line with an acceleration that is inversely proportional to its velocity. (Acceleration is the rate of increase of velocity.)

  a Form a differential equation to represent this data.

  b Given that the acceleration is $2\ \mathrm{ms^{-2}}$ when the velocity is $5\ \mathrm{ms^{-1}}$, solve the differential equation.

a Using $\dfrac{\mathrm{d}v}{\mathrm{d}t}$ for acceleration gives $\dfrac{\mathrm{d}v}{\mathrm{d}t} \propto \dfrac{1}{v} \quad \Rightarrow \quad \dfrac{\mathrm{d}v}{\mathrm{d}t} = \dfrac{k}{v}$

b $v = 5$ when $\dfrac{\mathrm{d}v}{\mathrm{d}t} = 2$, so $\quad 2 = \dfrac{k}{5} \quad \Rightarrow \quad k = 10$

Therefore $\dfrac{\mathrm{d}v}{\mathrm{d}t} = \dfrac{10}{v}$

Separating the variables gives $\displaystyle\int v\ \mathrm{d}v = \int 10\ \mathrm{d}t \quad \Rightarrow \quad \frac{1}{2}v^2 = 10t + A$

Natural growth and decay

A naturally occurring relationship arises when the rate of change of a quantity Q is proportional to the value of Q. Often (but not always) the rate of change is with respect to time, so the relationship can be expressed as

$$\frac{dQ}{dt} \propto Q.$$

This relationship of proportionality can be represented by the differential equation

$$\frac{dQ}{dt} = kQ \quad \text{where } k \text{ is a constant of proportion.}$$

Solving this differential equation by separating the variables gives

$$\int \frac{1}{Q}\, dQ = \int k\, dt \quad \Rightarrow \quad \ln Q = kt + A$$

giving $Q = e^{kt+A} = e^{A}e^{kt}$ or $Q = Be^{kt}$ (letting $B = e^{A}$ which is a constant).

This equation shows that Q varies exponentially with time.

The diagram shows a sketch of the corresponding graph.

Quantities that behave in this way undergo exponential growth. For example, the number of yeast cells undergoes exponential growth when the rate of increase of the number of cells is proportional to the number of cells present.

When it is the rate of *decrease* of Q that is proportional to Q, then

$$\frac{dQ}{dt} = -kQ \quad \Rightarrow \quad Q = Be^{-kt}$$

This graph is typical of a quantity undergoing exponential decay.

If, when $t = 0$, the value of Q is Q_0, the equations representing exponential growth and exponential decay respectively become

$$Q = Q_0 e^{kt} \quad \text{and} \quad Q = Q_0 e^{-kt}.$$

Half-life

When a substance is decaying exponentially, the time taken for one-half of the original quantity to decay is called the **half-life** of the substance. So if the original amount is Q_0, the half-life is given by

$$\frac{1}{2}Q_0 = Q_0 e^{-kt} \quad \Rightarrow \quad e^{kt} = 2 \quad \Rightarrow \quad t = \frac{1}{k}\ln 2$$

therefore the value of the half-life is $\frac{1}{k}\ln 2$.

Example 4

When a uniform rod is heated it expands so that the rate of increase of its length, l, with respect to the temperature, θ° C, is proportional to the length.

a Form the differential equation that models this information.

When the temperature is 0° C the length of the rod is L.

b Solve the differential equation using the information given.

c Given that the length of the rod has increased by 1% when the temperature is 20° C, find the value of θ at which the length of the rod has increased by 5%.

a From the given information, $\dfrac{dl}{d\theta} = kl$.

b $\dfrac{dl}{d\theta} = kl \quad \Rightarrow \quad \displaystyle\int \dfrac{1}{l}\,dl = \int k\,d\theta$

Therefore $\ln l = k\theta + A \quad \Rightarrow \quad l = Be^{k\theta}$

When $\theta = 0$, $l = L$, so $B = L$

Therefore $l = Le^{k\theta}$

c When $\theta = 20$, the length has increased by 1% so $l = L + 0.01L = 1.01L$

Then $\qquad 1.01L = Le^{20k}$

$\qquad\qquad e^{20k} = 1.01 \quad \Rightarrow \quad 20k = \ln 1.01$

$\Rightarrow \qquad\qquad k = 0.0004975\ldots$

Therefore $l = Le^{k\theta} \quad \Rightarrow \quad 0.0004975\,\theta = \ln(l/L)$

When $\qquad l = L + 0.05\,L = 1.05\,L, \quad 0.0004975\,\theta = \ln 1.05$

$\Rightarrow \qquad\qquad \theta = 98°$ (to the nearest degree)

Example 5

The rate at which the atoms in a mass of radioactive material are disintegrating is proportional to N, the number of atoms present at any time. Initially the number of atoms is M.

a Form and solve the differential equation that represents this data.

b Given that half of the original mass disintegrates in 152 days, evaluate the constant of proportion in the differential equation.

a The rate at which the atoms are disintegrating is $-\dfrac{dN}{dt}$.

Therefore $\qquad -\dfrac{dN}{dt} = kN$

Separating the variables gives $\displaystyle\int \dfrac{1}{N}\,dN = -\int k\,dt$

Therefore $\ln N = -kt + A \quad \Rightarrow \quad N = Be^{-kt}$

When $\qquad t = 0, N = M \quad$ so $\quad B = M$

Therefore $N = Me^{-kt}$

b When $\qquad N = \dfrac{1}{2}M, t = 152$

So $\qquad \dfrac{1}{2}M = Me^{-152k} \quad \Rightarrow \quad \ln\left(\dfrac{1}{2}\right) = -152k$

Therefore $152k = \ln 2 \quad \Rightarrow \quad k = \dfrac{\ln 2}{152} = 0.00456$ (to 3 significant figures)

Exercise 4

1 Grain is pouring on to a barn floor where it forms a pile whose height h is increasing at a rate that is inversely proportional to h^3.

   a Form a differential equation from the information given.

   b The initial height of the pile is 1m and the height doubles after 3 minutes. Use this information to solve the differential equation.

   c Find the time after which the height of the pile has grown to 3 m.

2 The gradient of any point of a curve is proportional to the square root of the x-coordinate.

  a Form a differential equation from this information.

  b Given that the curve passes through the point $(1, 2)$ and at that point the gradient is 0.6, solve the differential equation.

  c Show that the curve passes through the point $(4, 4.8)$ and find the gradient at this point.

3 The number of micro-organisms in a liquid is growing at a rate proportional to the number n of organisms present at any time t. Initially there are N organisms.

  a Form a differential equation that models the growth in the size of the colony.

  b Given that the colony increases by 50% in 10 hours, find the time that elapses from the start of the reaction before the size of the colony doubles.

4 The half-life of a radioactive element that is decaying exponentially is 500 years.

  a Form a differential equation that models the mass remaining, m g, of a sample which initially contained 1000 g, at time t years after decay began.

  b Solve the differential equation.

  c How many years it will be before the original mass of the element is reduced by 75%?

5 In a chemical reaction, a substance is transformed into a compound. The mass of the substance after any time t is m, and the substance is being transformed at a rate that is proportional to the mass of the substance at that time. Given that the original mass is 50 g and that 20 g is transformed after 200 seconds

  a Form and solve the differential equation relating m and t.

  b Find the mass of the substance transformed in 300 seconds.

Summary

▶ An equation with at least one term containing $\dfrac{dy}{dx}$ or $\dfrac{d^2y}{dx^2}$, ... is called a differential equation.

▶ Solving a differential equation means converting a differential equation into a relationship that does not contain a derivative.

▶ You can solve a differential equation with separable variables by: separating dy from dx; and collecting on one side all the terms involving y together with dy, while collecting all the x terms, along with dx, on the other side.

Review

1 The radius, r metres, of a circular patch of oil pollution is increasing at a rate that is proportional to r^2.

Form, but do not solve, a differential equation for this situation.

2 Solve the differential equation $\dfrac{dy}{dx} = 3x^2y^2$ given that $y = 1$ when $x = 0$.

3 Solve the differential equation $\dfrac{dy}{dx} = x(y^2 + 1)$ given that $y = 0$ when $x = 2$.

4 **a** Find the equation of the curve which passes through the point $\left(\dfrac{1}{2}, 1\right)$ and is defined by the differential equation $ye^{y^2}\dfrac{dy}{dx} = e^{2x}$.

b Show that the curve also passes through the point $(2, 2)$.

5 A virus has infected the population of rabbits on an isolated island and the evidence suggests that the growth in the number of rabbits infected is proportional to the number already infected. Initially 20 rabbits were recorded as being infected.

a Form a differential equation that models the growth in the number infected.

b Thirty days after the initial evidence was collected, 60 rabbits were infected. After how many further days does the model predict that 200 rabbits will be infected?

c In the event, only 100 rabbits were infected by that time. Give one reason why the model turned out to be unreliable.

6 A student models the spending power of a given sum of money over several years. The student assumes that the rate of decrease in spending power is proportional to the spending power at any given time.

a Use the information that \$100 in January 2000 buys only \$90 worth of goods in January 2010. Find and solve the differential equation connecting the value of goods, \$$y$, that \$100 in January 2000 will buy in January x years later.

b Give one reason why this model might be unsuitable for predicting the spending power that \$100 in January 2000 will have in January 2050.

Assessment

1 Solve the differential equation $\dfrac{dy}{dx} = \dfrac{4x}{y}$ given that $y = 3$ when $x = 1$.

2 Find the equation of the curve defined by $\dfrac{dy}{dx} = \dfrac{\sin x}{\cos y}$ which passes through the point $\left(\dfrac{\pi}{2}, \dfrac{3\pi}{2}\right)$.

3 The rate of increase of y with respect to x is inversely proportional to xy.

a Form a differential equation relating x and y.

b Solve the differential equation given that $y = 2$ when $x = 4$ and that $y = 10$ when $x = 1$.

4 The rate of decrease of the temperature of a liquid is proportional to the amount by which this temperature exceeds the temperature of its surroundings. (This is Newton's Law of Cooling.)

Taking θ as the excess temperature at any time t, and θ_0 as the initial excess

a show that $\theta = \theta_0 e^{-kt}$.

A pan of water at 65°C is standing in a kitchen whose temperature is a steady 15°C.

b Show that, after cooling for t minutes, the water temperature, ϕ, can be modelled by the equation $\phi = 15 + 50e^{-kt}$ where k is a constant.

c Given that after 10 minutes the temperature of the water has fallen to 50°C, find the value of k.

d Find the temperature of the water after 15 minutes.

5 Use the knowledge gained from the last question to undertake the following piece of detective work.

You are a forensic doctor called to a murder scene. When the victim was discovered, the body temperature was measured and found to be 20°C. You arrive one hour later and find the body temperature at that time to be 18°C. Assuming that the ambient temperature remained constant at 17°C in that intervening hour, give the police an estimate of the time of death. (Take 37°C as normal body temperature. You may assume also that the human body is largely made up of water.)

6 Solve the differential equation $\dfrac{dy}{dx} = y^2 x \sin 3x$ given that $y = 1$ when $x = \dfrac{\pi}{6}$.

Give your answer in the form $y = \dfrac{9}{f(x)}$.

AQA MPC4 January 2012

7 a A pond is initially empty and is then filled gradually with water. After t minutes, the depth of the water, x metres, satisfies the differential equation $\dfrac{dx}{dt} = \dfrac{\sqrt{4+5x}}{5(1+t)^2}$

Solve this differential equation to find x in terms of t.

b Another pond is gradually filling with water. After t minutes, the surface of the water forms a circle of radius r metres. The rate of change of the radius is inversely proportional to the area of the surface of the water.

i Write down a differential equation, in the variables r and t and a constant of proportionality, which represents how the radius of the surface of the water is changing with time.

ii When the radius of the pond is 1 metre, the radius is increasing at a rate of 4.5 metres per second. Find the radius of the pond when the radius is increasing at a rate of 0.5 metres per second.

AQA MPC4 June 2015

8 Numerical Methods

Introduction

There are many situations when it's not possible to find an exact solution to a problem. For example, the roots of many equations cannot be found exactly. There are also many integrals that cannot be found exactly. This chapter gives some methods for finding approximate values for roots of equations and approximate values of definite integrals.

Objectives

By the end of this chapter, you should know how to...

► Approximately locate the roots of an equation.

► Use an iteration formula to find a root of an equation to a given number of decimal places.

► Use the mid-ordinate rule and Simpson's rule to find an approximate value for the area under a curve.

Recap

You will need to remember...

► The shapes of the graphs of quadratic and cubic functions.

► How to find the stationary values on a curve.

► The shapes of the graphs of trigonometric functions, exponential functions and logarithmic functions.

► The shape of the graph of $f(x) = \dfrac{1}{x}$ shown in the diagram.

► The value of the definite integral
$$\int_{x_1}^{x_2} f'(x)\,dx = \left[f(x)\right]_{x_1}^{x_2} = f(x_2) - f(x_1).$$

► The trapezium rule for finding the approximate value of the area under a curve.

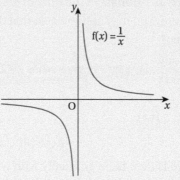

8.1 Approximately locating the roots of an equation

When the roots of an equation cannot be found exactly, you can locate the roots roughly. You can do this by sketching graphs.

For example, look at the equation $e^x = 4x$.

The roots of this equation are the values of x where the curve $y = e^x$ and the line $y = 4x$ intersect.

The sketch shows that there is one root between $x = 0$ and $x = 1$ and another root somewhere near $x = 2$.

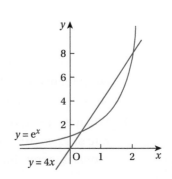

To see how to locate the roots more accurately look at the graph of a general curve $y = f(x)$.

The roots of the equation $f(x) = 0$ are the values of x where this curve crosses the x-axis.

As you can see from the graph each time that the curve crosses the x-axis, the sign of y changes.

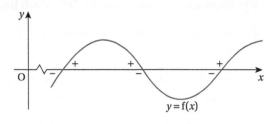

So when one root only of the equation $f(x) = 0$ lies between x_1 and x_2 and when the curve $y = f(x)$ is unbroken between the points where $x = x_1$ and $x = x_2$ then $f(x_1)$ and $f(x_2)$ are opposite in sign.

The condition that the curve $y = f(x)$ must be unbroken in the interval between x_1 and x_2 is essential.

This curve in the diagram crosses the x-axis between x_1 and x_2 but $f(x_1)$ and $f(x_2)$ have the same sign, because the curve is broken between these values.

Returning to the equation $e^x = 4x$, the larger root can be located a little more precisely.

First write the equation in the form $f(x) = 0$, that is $e^x - 4x = 0$. This curve is unbroken for all real x. Then find where there is a change in the sign of $f(x)$.

There is a root near $x = 2$, so see if it lies between 1.8 and 2.2.

Using $f(x) = e^x - 4x$, gives $f(1.8) = e^{1.8} - 4(1.8) = -1.1...$

and $\qquad\qquad\qquad f(2.2) = e^{2.2} - 4(2.2) = 0.2...$

Therefore the larger root of the equation lies between 1.8 and 2.2 (and is likely to be nearer to 2.2 as $f(2.2)$ is nearer to zero than $f(1.8)$ is).

Example 1

Question

Show that the equation $x^3 - 2x^2 + x + 1 = 0$ has a root between $x = -1$ and $x = 0$.

Answer

Using $f(x) = x^3 - 2x^2 + x + 1$ gives $f(-1) = -3$ and $f(0) = 1$.

$f(-1)$ and $f(0)$ are opposite in sign, so a root of $x^3 - 2x^2 + x + 1 = 0$ lies between $x = -1$ and $x = 0$.

Exercise 1

1. Show that the equation $\cos x = x^2 - 1$ has a root between $x = 1$ and $x = 2$.
2. Show that the equation $(x^2 - 4) = \dfrac{1}{x}$ has a root between $x = 2$ and $x = 3$.
3. Show that the equation $x(2^x) = 1$ has a root between $x = 0$ and $x = 1$.
4. Show that the equation $x \ln x = 1$ has a root between $x = 1$ and $x = 2$.
5. Show that the equation $\sin x = x^2$ has a root between $x = 0$ and $x = 1$.

6 Show that the equation $\ln x + 2^x = 0$ has a root between $x = 0.1$ and $x = 0.5$.

7 Show that the equation $4 + 5x^2 - x^3 = 0$ has a root between 5 and 5.2.

8 Show that the equation $x^4 - 4x^3 - x^2 + 4x - 10 = 0$ has a root between 4 and 4.2.

9 Sketch the curves $y = x^3$ and $y = 3x^2 - 1$ and use your sketch to find the number of real roots of the equation $x^3 - 3x^2 + 1 = 0$.

10 a Sketch graphs to show that the equation $3^{-x} = x^2 + 2$ has just one root.

   b Show that this root is exactly -1.

11 Show that one root of the equation $2^x = \dfrac{1}{2}(x + 3)$ lies between $x = -3$ and $x = -2$.

12 Sketch the graphs of $y = x^3$ and $y = 5 - x^2$. Hence find two consecutive integer values of x between which there is a root of the equation $x^3 + x^2 - 5 = 0$.

13 Sketch graphs to show that there is only one positive root of the equation $3 \sin x - x = 0$, where the angle is measured in radians.
   Show that, if this root is α, then $2 < \alpha < 3$.

14 On the same axes, sketch the graphs of $y = 4x$ and $y = 2^x$ for values of x from -1 to 4. Hence show that one root of the equation $2^x - 4x = 0$ is 4, and that there is another root between 0 and 1.

8.2 Using the iteration $x_{n+1} = g(x_n)$

This method of iteration, can often be used to find successive approximations to a root of an equation $f(x) = 0$. It can be written in the form $x = g(x)$. The roots of the equation $x = g(x)$ are the values of x at the points of intersection of the line $y = x$ and the curve $y = g(x)$.

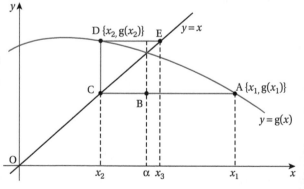

Taking x_1 as a first approximation to a root α then, as you can see in the diagram:

▶ A is the point on the *curve* where $x = x_1$ and $y = g(x_1)$

▶ B is the point where $x = \alpha$ and $y = g(x_1)$

▶ C is the point on the *line* where $y = g(x_1)$ and $x = x_2$.

When, in the region of α, the slope of $y = g(x)$ is less steep than that of the line $y = x$ (that is provided that $|g'(x)| < 1$) then CB < BA.

Therefore, x_2 is closer to α than x_1 is, so x_2 is a better approximation to α.

The point C is on the line $y = x$ and therefore $x_2 = g(x_1)$.

Next, taking the point D on the curve where $x = x_2$, $y = g(x_2)$ and repeating the argument above, shows that x_3 is a better approximation to α than x_2 is, where $x_3 = g(x_2)$.

This process can be repeated as often as necessary to achieve the required degree of accuracy and is called **iteration**.

The rate at which these approximations converge to α depends on the value of $|g'(x)|$ near α. The smaller $|g'(x)|$ is, the more rapid the convergence. However this method fails if $|g'(x)| > 1$ near α, as $x_1, x_2, \ldots$ diverge from α.

These diagrams illustrate some of the factors which determine the success, or failure, of this iterative method to find a solution to the equation $x = g(x)$.

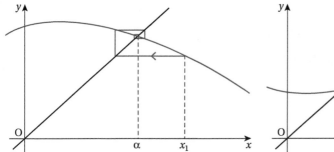
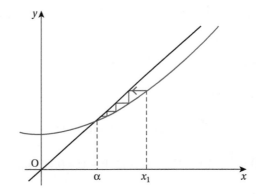

Rapid rate of convergence ($|g'(x)|$ small).

> **Note**
>
> The diagrams on the left are called cobweb diagrams. The diagrams on the right are called staircase diagrams.

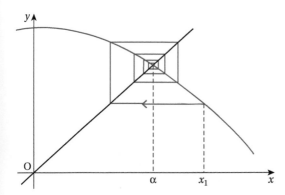

Slow rate of convergence ($|g'(x)| < 1$ but close to 1).

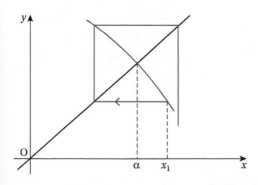
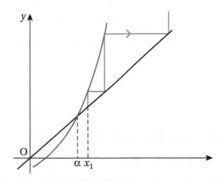

Divergence, that is, the method fails to converge ($|g'(x)| > 1$).

Example 2

The curve $y = x^3$ intersects the curve $y = 1 - 2x^2 - 5x$ at a single point where $x = \alpha$.

a Show that α lies between 0 and 0.2.

b Show that the equation $x^3 + 2x^2 + 5x - 1 = 0$ can be written in the form
$$x = -\frac{1}{5}(x^3 + 2x^2 - 1).$$

c Show that iteration $x_{n+1} = -\frac{1}{5}(x_n^3 + 2x_n^2 - 1)$ converges.

d Use the iteration $x_{n+1} = -\frac{1}{5}(x_n^3 + 2x_n^2 - 1)$ with $x_1 = 0$ to find the value of α correct to 3 decimal places.

a At the point of intersection $x^3 = 1 - 2x^2 - 5x$.

Therefore α is a root of the equation $x^3 = 1 - 2x^2 - 5x \implies x^3 + 2x^2 + 5x - 1 = 0$

Using $f(x) = x^3 + 2x^2 + 5x - 1 \implies f(0) = -1$ and $f(0.2) = 0.088$.

As $f(0) < 0$ and $f(0.2) > 0$, α lies between 0 and 0.2.

b $x^3 + 2x^2 + 5x - 1 = 0 \implies 5x = 1 - 2x^2 - x^3$

Therefore the equation can be written in the form
$$x = -\frac{1}{5}(x^3 + 2x^2 - 1).$$

c Using $g(x) = -\frac{1}{5}(x^3 + 2x^2 - 1)$ gives $g'(x) = -\frac{1}{5}(3x^2 + 4x)$ and

$g'(0.2) = -\frac{1}{5}(0.12 + 0.8) = -0.184$ so $|g'(x)| < 1$ near the root.

Therefore the iteration converges.

d Using $x_{n+1} = -\frac{1}{5}(x_n^3 + 2x_n^2 - 1)$ with $x_1 = 0$

then $x_2 = g(x_1) = -\frac{1}{5}[0^3 + 2(0)^2 - 1] = 0.2$

and $x_3 = g(x_2) = -\frac{1}{5}[(0.2)^3 + 2(0.2)^2 - 1] = 0.1824$

$x_4 = g(x_3) = \frac{1}{5}[(0.1824)^3 + 2(0.1824)^2 - 1] = 0.1854\ldots$

$x_5 = g(x_4) = -\frac{1}{5}[(0.1854\ldots)^3 + 2(0.1854\ldots)^2 - 1] = 0.1849\ldots$

It looks as if $\alpha = 0.185$ is correct to 3 decimal places, that is $0.1846 < \alpha < 0.1854$. The degree of accuracy can be checked by finding the signs of $f(0.1846)$ and $f(0.1854)$.

$f(0.1846)$ is negative and $f(0.1854)$ is positive, so $x = 0.185$ correct to 3 decimal places.

Note

This does *not* show that $x = 0.1850$ to 4 d.p.

Exercise 2

In questions 1 to 4

a show that each equation has a root between $x = 0$ and $x = 1$

b express each equation in the form $x = g(x)$

c determine, without doing the iteration, whether your form for $x = g(x)$ gives an iteration formula that converges.

1 $x^3 - x^2 + 10x - 2 = 0$ 2 $3x^3 - 2x^2 - 9x + 2 = 0$

3 $2x^3 + x^2 + 6x - 1 = 0$ 4 $x^2 + 8x - 8 = 0$

5 The equation $x(2^x) = 1$ has a root between $x = 0$ and $x = 1$.

   a Show that the equation can be written in the form $x = 2^{-x}$.

   b Use the iteration $x_{n+1} = 2^{-x_n}$ with $x_1 = 0.5$ to find x_2 and x_3, giving your answers to 2 decimal places.

6 The equation $x \ln (x + 2) = 1$ has a root between $x = 0$ and $x = 1$.

   a Show that the equation can be written in the form $x = \dfrac{1}{\ln(x+2)}$.

   b Use the iteration $x_{n+1} = \dfrac{1}{\ln(x_n+2)}$ with $x_1 = 0.8$ to find x_2 and x_3, giving your answers to 2 decimal places.

7 a Show that the equation $x^3 - x^2 - 5x + 1 = 0$ can be written in the form $x = \dfrac{1}{5}(x^3 - x^2 + 1)$.

   b The equation $x^3 - x^2 - 5x + 1 = 0$ has a root between 0.1 and 0.2. Use the iteration $x_{n+1} = \dfrac{1}{5}(x_n^3 - x_n^2 + 1)$ with $x_1 = 0.1$ to find x_2 and x_3 giving your answers to 3 decimal places.

   c The equation $x^3 - x^2 - 5x + 1 = 0$ also has a root between 2.7 and 2.8. Use the iteration $x_{n+1} = \dfrac{1}{5}(x_n^3 - x_n^2 + 1)$ with $x_1 = 2.7$ to find four successive values of x_n and hence show that this iteration does not converge to this root.

8 a Show that the equation $e^{x+5} = \sqrt{x^2 - 1}$ has a root α between -4 and -3.5.

   b Show that the equation $e^{x+5} = \sqrt{x^2 - 1}$ can be written as $x = -5 + \dfrac{1}{2}\ln(x^2 - 1)$.

   c Use the iteration $x_{n+1} = -5 + \dfrac{1}{2}\ln(x_n^2 - 1)$ to show that $\alpha = -3.72$ correct to 2 decimal places.

8.3 Rules to find the approximate value of an area under a curve

The mid-ordinate rule

The **mid-ordinate rule** is similar to the trapezium rule but it uses rectangular strips instead of trapeziums. The area is divided into equal width strips where the top boundary of each strip is a horizontal line through the point on the curve at the centre of the strip.

The sum of the areas of these rectangles gives an approximate value for the area under the curve, so

$$A \approx h\left[y_{\frac{1}{2}} + y_{\frac{3}{2}} + \cdots + y_{\frac{(2n-1)}{2}} \right]$$

The mid-ordinate value is usually more accurate than the trapezium rule. This is because the top of the trapezium is often entirely on one side of the curve, so that part of the area under the curve is not allowed for. The top line of the rectangular strip, however, usually cuts across the curve this means that the part of the area included in the rectangle, but which is not under the curve, tends to balance the area under the curve that is not included in the rectangle.

The diagrams showing the tops of two enlarged strips show this.

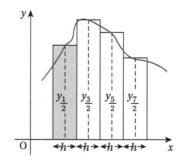

The trapezium rule

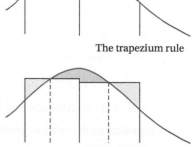

The mid-ordinate rule

Simpson's rule

Simpson's rule is a formula that gives a better approximation to the area under a curve than that obtained from the trapezium and the mid-ordinate rules.

Using the same notation as before, Simpson's rule states that

$$A \approx \frac{1}{3}h\left[y_0 + 4y_1 + 2y_2 + 4y_3 + 2y_4 + \ldots + 2y_{n-2} + 4y_{n-1} + y_n\right], \text{ so}$$

$$A \approx \frac{1}{3}h\left[(y_0 + y_n) + 4(y_1 + y_3 + \cdots) + 2(y_2 + y_4 + \cdots)\right]$$

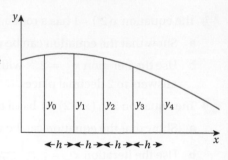

This formula is based on dividing the required area into equal width strips and, for each *pair* of strips, finding a parabola which passes through the top of the three ordinates bounding the two strips.

Because this formula is based on pairs of strips it follows that it can be used only when the number of strips is even, that is, when the number of ordinates is odd.

Simpson's rule gives an even more accurate approximation than the other rules because, as it is based on a parabola, the tops of the strips are even nearer to the shape of the curve.

The degree of accuracy of an area under a curve, given by either of these rules, depends upon the number of strips into which the required area is divided. This is because the narrower the strip, the nearer its shape at the top becomes to the shape of the curve.

Example 3

Question

Use four strips to find an approximate value for the definite integral $\int_1^5 x^3 \, dx$ using

a   the mid-ordinate rule

b   Simpson's rule.

Answer

The integral represents the area of the region bounded by the x-axis, the lines $x = 1$ and $x = 5$, and the curve $y = x^3$.

Five ordinates are used when there are four strips whose widths must all be the same.

From $x = 1$ to $x = 5$ there are four units, so the width of each strip must be 1 unit. Hence the five ordinates are where $x = 1$, $x = 2$, $x = 3$, $x = 4$ and $x = 5$.

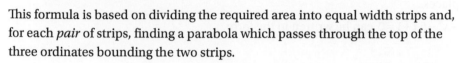

a   The values of the mid-ordinates are given by

$$y_1 = 1.5^3, \, y_2 = 2.5^3, \, y_3 = 3.5^3, \, y_4 = 4.5^3$$

The width of each strip is 1 unit so using the mid-ordinate rule gives

$$A = (1)[3.375 + 15.63 + 42.88 + 91.13] = 153.0$$

The area is approximately 153 square units.

b   There is an odd number of ordinates, so Simpson's rule can be used.

$$A \approx \frac{1}{3}(1)[(1 + 125) + 4(8 + 64) + 2(27)] = 156$$

The area is approximately 156 square units.

The exact value of the area is given by $\int_1^5 x^3 \, dx = \left[\frac{1}{4}x^4\right]_1^5 = 156.25 - 0.25 = 156$

so Simpson's rule gave an accurate answer for this area.

Exercise 3

1 Estimate the value of $\int_0^4 x^2 \, dx$ using 5 ordinates and

   a the mid-ordinate rule

   b Simpson's rule.

2 Estimate the value of $\int_1^3 \frac{1}{x^2} \, dx$ using 5 ordinates and

   a the mid-ordinate rule

   b Simpson's rule.

3 Estimate the value of $\int_0^{\frac{2\pi}{3}} \sqrt{\sin x} \, dx$ using 3 ordinates and

   a the mid-ordinate rule

   b Simpson's rule.

4 Estimate the value of $\int_1^3 \ln x \, dx$ using 3 ordinates and

   a the mid-ordinate rule

   b Simpson's rule.

5 Estimate the value of $\int_0^\pi 2 + (\cos x)\ln(x+1) \, dx$ using 3 ordinates and

   a the mid-ordinate rule

   b Simpson's rule.

Summary

Locating the root of an equation between two values

▶ When a root of the equation $f(x) = 0$ lies between x_1 and x_2 and when the curve $y = f(x)$ is unbroken between the points where $x = x_1$ and $x = x_2$ then $f(x_1)$ and $f(x_2)$ are opposite in sign.

Using the iteration $x_{n+1} = g(x_n)$

▶ The iteration formula $x_{n+1} = g(x_n)$ can be used to find a root of the equation $f(x) = 0$ provided that the equation can be rearranged as $x = g(x)$ for some function $g(x)$, and that $|g'(x)| < 1$ near the root.

The mid-ordinate rule

▶ $A = \int_a^b f(x)\,dx \approx h\left[y_{\frac{1}{2}} + y_{\frac{3}{2}} + \cdots + y_{\frac{(2n-1)}{2}} \right]$

where the values of y are heights of the rectangles, that is the y-coordinates of the points on the curve at the centre of each strip.

Simpson's Rule

▶ To use this rule there must be an odd number of ordinates.

$A = \int_a^b f(x)\,dx \approx \frac{1}{3}h[(y_0 + y_n) + 4(y_1 + y_3 + \ldots) + 2(y_2 + y_4 + \ldots)]$

Review

1 The function f is given by $f(x) = 12 \ln x - x^{\frac{3}{2}}$ for positive values of x. The curve $y = f(x)$ crosses the x-axis at the point A. Show that the value of x at the point A lies between 1.1 and 1.2.

2 **a** Show that the equation $x^3 - 2x^2 - 1 = 0$ has a root between 2 and 3.

   b Show that $x^3 - 2x^2 - 1 = 0$ can be written as $x = 2 + \dfrac{1}{x^2}$.

   c Taking $x_1 = 2$ as the first approximation to this root, use the iteration $x_{n+1} = 2 + \dfrac{1}{x_n^2}$ twice to obtain a better approximation.

3 By sketching graphs in the interval $0 < x < 4$, find two consecutive integers between which there is a root of the equation $\ln(x - 2) = \dfrac{1}{x}$. Use the change of sign of $f(x) = \ln(x - 2) - \dfrac{1}{x}$ to verify that your answer is correct.

4 Use the mid-ordinate rule with 5 ordinates to find an approximate value of $\displaystyle\int_1^5 (e^{2x} - 1)\, dx$.

5 Use Simpson's rule with 3 ordinates to find an approximate value of $\displaystyle\int_1^5 (\ln x)(\cos x)\, dx$.

6 Use the mid-ordinate rule with 3 ordinates to find an approximate value of $\displaystyle\int_0^{\frac{2\pi}{3}} \dfrac{1}{\sin x + \cos x}\, dx$.

Assessment

1 **a** Sketch graphs of $y = e^x$ and $y = x^2 + 2$ to show that the curves intersect at a single point where $x = \alpha$.

   b Show that α lies between $x = 1.2$ and $x = 1.4$.

   c Show that the equation $e^x = x^2 + 2$ can be written in the form $x = \ln(x^2 + 2)$.

   d Use the iteration $x_{n+1} = \ln(x_n^2 + 2)$ with $x_1 = 1.2$ to find x_2 and x_3.

2 **a** Show that the equation $x^4 + x^2 - x = 0$ has one root exactly equal to 0 and another root between 0.6 and 0.7.

   b By expressing the equation in part **a** in the form $x = \dfrac{1}{x^2 + 1}$ show that the iteration formula $x_{n+1} = \dfrac{1}{(x_n^2 + 1)}$ converges to the larger root and find this root correct to 3 significant figures.

3 **a** Use the mid-ordinate rule with 4 ordinates to find an approximate value for $\displaystyle\int_1^4 \left(x + \dfrac{4}{\sqrt{x}}\right)^2 dx$.

   b Find the value of $\displaystyle\int_1^4 \left(x + \dfrac{4}{\sqrt{x}}\right)^2 dx$.

4 a The equation of a curve is $y = x\ln(x^2 + 1)$.

Use Simpson's rule with 3 ordinates to find an approximate value of

$$\int_0^2 x\ln(x^2+1)\,dx.$$

b Use the substitution $u = x^2 + 1$ to find $\int x\ln(x^2+1)\,dx$.

c Hence find the exact value of $\int_0^2 x\ln(x^2+1)\,dx$.

5 Use Simpson's rule, with five ordinates (four strips), to calculate an estimate

for $\int_0^\pi x^{\frac{1}{2}}\sin x\,dx$.

Give your answer to four significant figures.

AQA MPC3 June 2014

6 Use the mid-ordinate rule with four strips to find an estimate for

$$\int_{1.5}^{5.5} e^{2-x}\ln(3x-2)\,dx,$$ giving your answer to three decimal places.

AQA MPC3 June 2015 (part question)

7 A curve is defined by the equation $y = (x^2 - 4)\ln(x+2)$ for $x > 3$.

The curve intersects the line $y = 15$ at a single point, where $x = \alpha$.

a Show that α lies between 3.5 and

b Show that the equation $(x^2 - 4)\ln(x+2) = 15$ can be arranged into the

form $x = \pm\sqrt{4 + \dfrac{15}{\ln(x+2)}}$.

c Use the iteration $x_{n+1} = \sqrt{4 + \dfrac{15}{\ln(x_n+2)}}$ with $x_1 = 3.5$ to find the values of

x_2 and x_3, giving your answers to three decimal places.

AQA MPC3 January 2011

9 Vectors

Introduction

Many quantities need a size and a direction to define them. For example, to give directions to get from a point A to a point B, giving the distance between the points does not help. The direction of B from A is also needed. An aircraft flying in a wind needs to know how to set its course to take into account not only the speed and but also the direction of the wind. This chapter shows how to work with quantities that need both size and direction to define them.

Recap

You will need to remember...

- The modulus of a quantity means its magnitude irrespective of its sign, for example when $x = 2$ or -2, $|x| = 2$.
- Pythagoras' theorem.
- A parameter is a variable that can take any value.
- $\cos^{-1} x$ means the angle whose cosine is x.

Objectives

By the end of this chapter, you should know how to...

- Find the position vector of the midpoint of a line.
- Locate a vector in space using x, y and z coordinates.
- Add and subtract vectors in three dimensions.
- Find the magnitude of a vector.
- Understand and find a vector equation of a line.
- Determine whether two lines are parallel, intersect or are skew.
- Find the scalar product of two vectors.
- Find the coordinates of the foot of the perpendicular from a point to a line.
- Find the perpendicular distance of a point from a line.

9.1 Properties of vectors

A vector is a quantity which has both magnitude and a specific direction in space.

A **scalar** is a quantity that is fully defined by magnitude alone.

Vector representation

A vector can be represented by a section of a straight line, whose length represents the magnitude of the vector and whose direction, indicated by an arrow, represents the direction of the vector.

The modulus of a vector

The **modulus** of a vector **a** is its magnitude and is written $|\mathbf{a}|$ or a. In other words, $|\mathbf{a}|$ is the length of the line representing **a**.

Equal vectors

Two vectors with the same magnitude and the same direction are equal.

$$\mathbf{a} = \mathbf{b} \quad \Leftrightarrow \quad \begin{cases} |\mathbf{a}| = |\mathbf{b}| \text{ and} \\ \text{the direction of } \mathbf{a} \text{ and } \mathbf{b} \text{ are the same} \end{cases}$$

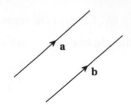

Therefore a vector can be represented by *any* line of the right length and direction, regardless of the line's position, so each of the lines in the diagram represents the vector **c**.

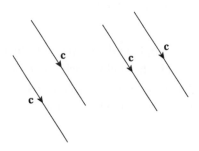

Equal magnitude and opposite direction

When two vectors, **a** and **b**, have the same magnitude but opposite directions then $\mathbf{b} = -\mathbf{a}$.

$-\mathbf{a}$ is a vector of magnitude $|\mathbf{a}|$ in the opposite direction to **a**.

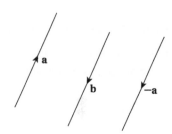

Multiplication of a vector by a scalar

When λ is a positive real number, then $\lambda\mathbf{a}$ is a vector in the same direction as **a** with magnitude $\lambda|\mathbf{a}|$. It follows that $-\lambda\mathbf{a}$ is a vector in the opposite direction to **a**, with magnitude $\lambda|\mathbf{a}|$.

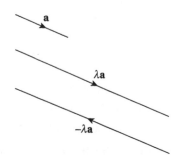

Addition of vectors

When the sides $\overrightarrow{AB}$ and $\overrightarrow{BC}$ of a triangle ABC represent the vectors **p** and **q** then the third side $\overrightarrow{AC}$ represents the vector sum, or **resultant**, of **p** and **q**, which is denoted by $\mathbf{p} + \mathbf{q}$.

Note that **p** and **q** follow each other round the triangle (in this case in the clockwise sense), whereas the resultant, $\mathbf{p} + \mathbf{q}$, goes the opposite way round (anticlockwise in the diagram).

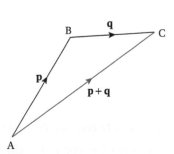

This is called the **triangle law** for addition of vectors. It can be extended to cover the addition of more than two vectors.

Let $\overrightarrow{AB}, \overrightarrow{BC}, \overrightarrow{CD}$ and $\overrightarrow{DE}$ represent the vectors **a**, **b**, **c** and **d** respectively.

The triangle law gives $\overrightarrow{AB} + \overrightarrow{BC} = \mathbf{a} + \mathbf{b} = \overrightarrow{AC}$

then $\overrightarrow{AC} + \overrightarrow{CD} = (\mathbf{a} + \mathbf{b}) + \mathbf{c} = \overrightarrow{AD}$

and $\overrightarrow{AD} + \overrightarrow{DE} = (\mathbf{a} + \mathbf{b} + \mathbf{c}) + \mathbf{d} = \overrightarrow{AE}$

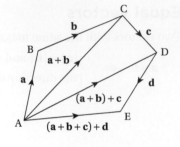

Note that the vectors **a**, **b**, **c** and **d** may not be coplanar so the polygon may not be two-dimensional.

AE completes the polygon of which AB, BC, CD and DE are four sides taken in order (that is, they follow each other round the polygon in the *same sense*). Again, the side representing the resultant closes the polygon in the *opposite* sense. The vectors **a**, **b**, **c** and **d** may not be coplanar so the polygon may not be two-dimensional.

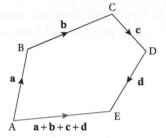

Order of vector addition

The order in which the addition is performed does not matter as is shown by looking at the parallelogram ABCD.

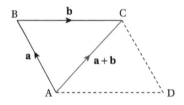
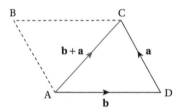

Because the opposite sides of a parallelogram are equal and parallel, $\overrightarrow{AB}$ and $\overrightarrow{DC}$ both represent **a** and $\overrightarrow{BC}$ and $\overrightarrow{AD}$ both represent **b**.

In $\triangle ABC$, $\overrightarrow{AC} = \mathbf{a} + \mathbf{b}$ and in $\triangle ADC$, $\overrightarrow{AC} = \mathbf{b} + \mathbf{a}$.

Therefore $\mathbf{a} + \mathbf{b} = \mathbf{b} + \mathbf{a}$.

The angle between two vectors

There are two angles between two lines, α and $180° - \alpha$.

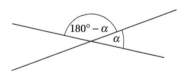

The angle between two vectors, however, is defined uniquely.

It is the angle between their directions, when the lines representing them *both converge* or *both diverge* (see diagrams below).

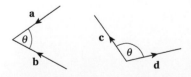

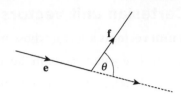

In some cases one of the lines may have to be produced in order to mark the correct angle (see diagram below).

9.2 Position vectors

Usually a vector has no specific location in space and is called a **free vector**. Some vectors, however, are constrained to a specific position, such as the vector $\overrightarrow{OA}$ where O is a fixed origin.

$\overrightarrow{OA}$ is called the position of A relative to O.

This displacement is unique and *cannot* be represented by any other line of equal length and direction.

Vectors such as $\overrightarrow{OA}$, representing quantities that have a specific location, are called **position vectors**.

The position vector of the midpoint of a line

Look at the line AB where the position vectors of A and B relative to O are **a** and **b** respectively, and C is the midpoint of AB.

In the diagram, $\overrightarrow{AB} = \mathbf{b} - \mathbf{a}$

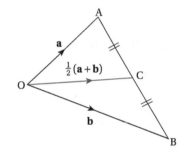

Therefore $\overrightarrow{AC} = \dfrac{1}{2}(\mathbf{b} - \mathbf{a})$

Hence $\overrightarrow{OC} = \overrightarrow{OA} + \overrightarrow{AC} = \mathbf{a} + \dfrac{1}{2}(\mathbf{b} - \mathbf{a}) = \dfrac{1}{2}\mathbf{a} + \dfrac{1}{2}\mathbf{b}$

Therefore the position vector of C is $\dfrac{1}{2}(\mathbf{a} + \mathbf{b})$.

Exercise 1

The position vectors, relative to O, of A, B, C and D are **a**, **b**, **c** and **d** respectively. P, Q and R are the midpoints of AB, BC and CD respectively.

1 **a** Find the position vector of the midpoint of AC.

   b Find the position vector of the midpoint of BD.

2 **a** Find the position vector of the midpoint of PQ.

   b Find the position vector of the midpoint of QR.

3 Show that PQ is parallel to AC.

9.3 The location of a point in space

Any point P in a plane can be located by giving its distances from a fixed point O, in each of two perpendicular directions. These distances are the Cartesian coordinates of the point.

In three-dimensional space, when O is a fixed point, any other point P can be located by giving its distances from O in each of *three* mutually perpendicular directions, so three coordinates are needed to locate a point in space. The familiar x- and y-axes are used, together with a third z-axis. Then any point has coordinates (x, y, z) relative to the origin O.

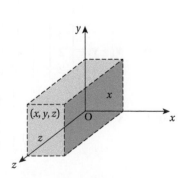

Cartesian unit vectors

A **unit vector** is a vector whose magnitude is one unit.

When **i** is a unit vector in the direction of the x-axis

   j is a unit vector in the direction of the y-axis

   k is a unit vector in the direction of the z-axis

then the position vector, relative to O, of any point P can be given in terms of **i**, **j** and **k**.

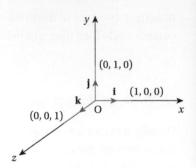

The point P distant

   3 units from O in the x direction

   4 units from O in the y direction

   5 units from O in the z direction

has coordinates (3, 4, 5) and $\overrightarrow{OP} = 3\mathbf{i} + 4\mathbf{j} + 5\mathbf{k}$.

This can also be written as $\overrightarrow{OP} = \begin{bmatrix} 3 \\ 4 \\ 5 \end{bmatrix}$ and this is the form used in this book.

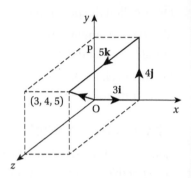

In general, when P is a point, (x, y, z) and $\overrightarrow{OP} = \mathbf{r}$, then $\mathbf{r} = \begin{bmatrix} x \\ y \\ z \end{bmatrix}$ is the position vector of P.

Free vectors can be given in the same form. For example, the vector $\begin{bmatrix} 3 \\ 4 \\ 5 \end{bmatrix}$ can

represent the position vector of the point P(3, 4, 5) but it can equally well represent *any* vector of length and direction equal to those of $\overrightarrow{OP}$.

> **Note**
>
> Note that, unless a vector is *specified* as a position vector, it is taken to be free.

9.4 Operations on cartesian vectors

Addition and subtraction

To add or subtract vectors given in the form $\begin{bmatrix} x \\ y \\ z \end{bmatrix}$ the coordinates are added separately. For example

when $\mathbf{v}_1 = \begin{bmatrix} 3 \\ 2 \\ 2 \end{bmatrix}$ and $\mathbf{v}_2 = \begin{bmatrix} 1 \\ 2 \\ -3 \end{bmatrix}$

then $\mathbf{v}_1 + \mathbf{v}_2 = \begin{bmatrix} 3 \\ 2 \\ 2 \end{bmatrix} + \begin{bmatrix} 1 \\ 2 \\ -3 \end{bmatrix} = \begin{bmatrix} 3+1 \\ 2+2 \\ 2-3 \end{bmatrix} = \begin{bmatrix} 4 \\ 4 \\ -1 \end{bmatrix}$

And $\mathbf{v}_1 - \mathbf{v}_2 = \begin{bmatrix} 3-1 \\ 2-2 \\ 2-(-3) \end{bmatrix} = \begin{bmatrix} 2 \\ 0 \\ 5 \end{bmatrix}$

Modulus

The magnitude or the modulus of $\mathbf{v}$, where $\mathbf{v} = \begin{bmatrix} 12 \\ -3 \\ 4 \end{bmatrix}$ is the length of OP
where P is the point $(12, -3, 4)$.

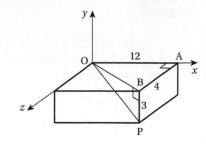

Using Pythagoras twice we have

$$OB^2 = OA^2 + AB^2 = 12^2 + 4^2$$

$$OP^2 = OB^2 + BP^2 = (12^2 + 4^2) + (-3)^2$$

$$\therefore \quad OP = \sqrt{12^2 + 4^2 + 3^2} = 13$$

In general, when $\mathbf{v} = \begin{bmatrix} a \\ b \\ c \end{bmatrix}$ then $|\mathbf{v}| = \sqrt{a^2 + b^2 + c^2}$.

Parallel vectors

Two vectors $\mathbf{v}_1$ and $\mathbf{v}_2$ are parallel when $\mathbf{v}_1 = \lambda \mathbf{v}_2$ where λ is a constant.

For example, $\begin{bmatrix} 2 \\ -3 \\ -1 \end{bmatrix}$ is parallel to $\begin{bmatrix} 4 \\ -6 \\ -2 \end{bmatrix}$ (by taking $\lambda = 2$), and

$\begin{bmatrix} 1 \\ 1 \\ 1 \end{bmatrix}$ is parallel to $\begin{bmatrix} -3 \\ -3 \\ -3 \end{bmatrix}$ (by taking $\lambda = -3$).

Equal vectors

When two vectors $\mathbf{v}_1 = \begin{bmatrix} a_1 \\ b_1 \\ c_1 \end{bmatrix}$ and $\mathbf{v}_2 = \begin{bmatrix} a_2 \\ b_2 \\ c_2 \end{bmatrix}$ are equal then $a_1 = a_2$ and $b_1 = b_2$ and $c_1 = c_2$.

Example 1

Given the vector $\mathbf{v} = \begin{bmatrix} 5 \\ -2 \\ 4 \end{bmatrix}$ state whether

each of these vectors is parallel to $\mathbf{v}$, equal to $\mathbf{v}$ or neither.

a $\begin{bmatrix} 10 \\ -4 \\ 8 \end{bmatrix}$
b $-\dfrac{1}{2}\begin{bmatrix} -10 \\ 4 \\ -8 \end{bmatrix}$

c $\begin{bmatrix} -5 \\ 2 \\ -4 \end{bmatrix}$
d $\begin{bmatrix} 4 \\ -2 \\ 5 \end{bmatrix}$

a $\begin{bmatrix} 10 \\ -4 \\ 8 \end{bmatrix} = 2\begin{bmatrix} 5 \\ -2 \\ 4 \end{bmatrix}$ (taking $\lambda = 2$), therefore $\begin{bmatrix} 10 \\ -4 \\ 8 \end{bmatrix}$ is parallel to $\mathbf{v}$.

(continued)

(continued)

Answer

b $-\dfrac{1}{2}\begin{bmatrix} -10 \\ 4 \\ -8 \end{bmatrix} = \begin{bmatrix} 5 \\ -2 \\ 4 \end{bmatrix}$ therefore $-\dfrac{1}{2}\begin{bmatrix} -10 \\ 4 \\ -8 \end{bmatrix}$ is equal to **v**.

c $\begin{bmatrix} -5 \\ 2 \\ -4 \end{bmatrix} = -\begin{bmatrix} 5 \\ -2 \\ 4 \end{bmatrix}$ (taking $\lambda = -1$), therefore $\begin{bmatrix} 5 \\ -2 \\ 4 \end{bmatrix}$ is parallel to **v**.

d $\begin{bmatrix} 4 \\ -2 \\ 5 \end{bmatrix}$ is not a multiple of $\begin{bmatrix} 5 \\ -2 \\ 4 \end{bmatrix}$ therefore $\begin{bmatrix} 4 \\ -2 \\ 5 \end{bmatrix}$ is not equal or parallel to **v**.

Example 2

Question

A triangle ABC has its vertices at the points A(2, –1, 4), B(3, –2, 5) and C(–1, 6, 2). Find, in the form $\begin{bmatrix} a \\ b \\ c \end{bmatrix}$ the vectors $\overrightarrow{AB}, \overrightarrow{BC}$ and $\overrightarrow{CA}$ and hence find the lengths of the sides of the triangle.

Answer

The coordinate axes are not drawn in the diagram, as they can cause confusion when two or more points are illustrated. The origin should always be included as it provides a reference point.

$\overrightarrow{AB} = \overrightarrow{OB} - \overrightarrow{OA} = \begin{bmatrix} 3 \\ -2 \\ 5 \end{bmatrix} - \begin{bmatrix} 2 \\ -1 \\ 4 \end{bmatrix} = \begin{bmatrix} 1 \\ -1 \\ 1 \end{bmatrix}$

$\overrightarrow{BC} = \overrightarrow{OC} - \overrightarrow{OB} = \begin{bmatrix} -1 \\ 6 \\ 2 \end{bmatrix} - \begin{bmatrix} 3 \\ -2 \\ 5 \end{bmatrix} = \begin{bmatrix} -4 \\ 8 \\ -3 \end{bmatrix}$

$\overrightarrow{CA} = \overrightarrow{OA} - \overrightarrow{OC} = \begin{bmatrix} 2 \\ -1 \\ 4 \end{bmatrix} - \begin{bmatrix} -1 \\ 6 \\ 2 \end{bmatrix} = \begin{bmatrix} 3 \\ -7 \\ 2 \end{bmatrix}$

Therefore $AB = |\overrightarrow{AB}| = \sqrt{(1)^2 + (-1)^2 + (1)^2} = \sqrt{3}$

$BC = |\overrightarrow{BC}| = \sqrt{(-4)^2 + (8)^2 + (-3)^2} = \sqrt{89}$

$CA = |\overrightarrow{CA}| = \sqrt{(3)^2 + (-7)^2 + (2)^2} = \sqrt{62}$

Exercise 2

1 Write down, in the form $\begin{bmatrix} a \\ b \\ c \end{bmatrix}$ the vector represented by $\overrightarrow{OP}$ if P is a point with coordinates.

    a (3, 6, 4)   **b** (1, –2, –7)   **c** (1, 0, –3).

2 $\overrightarrow{OP}$ represents a vector **r**. Write down the coordinates of P when

a $r = \begin{bmatrix} 5 \\ -7 \\ 2 \end{bmatrix}$ **b** $r = \begin{bmatrix} 1 \\ 4 \\ 0 \end{bmatrix}$ **c** $r = \begin{bmatrix} 0 \\ 1 \\ -1 \end{bmatrix}$.

3 Find the length of the line OP when P is the point

a $(2, -1, 4)$ **b** $(3, 0, 4)$ **c** $(-2, -2, 1)$.

4 Find the modulus of the vector **v** when

a $v = \begin{bmatrix} 2 \\ -4 \\ 4 \end{bmatrix}$ **b** $v = \begin{bmatrix} 6 \\ 2 \\ -3 \end{bmatrix}$ **c** $v = \begin{bmatrix} 11 \\ -7 \\ -6 \end{bmatrix}$.

5 When $p = \begin{bmatrix} 1 \\ 1 \\ 1 \end{bmatrix}$, $q = \begin{bmatrix} 2 \\ -1 \\ 3 \end{bmatrix}$ and $r = \begin{bmatrix} -1 \\ 3 \\ -1 \end{bmatrix}$ find

a $p + q$ **b** $p - r$ **c** $p + q + r$ **d** $p - 2q + 3r$.

6 Given $\overrightarrow{OA} = a = \begin{bmatrix} 4 \\ -12 \\ 0 \end{bmatrix}$, $\overrightarrow{OB} = b = \begin{bmatrix} 1 \\ 6 \\ 0 \end{bmatrix}$ and $\overrightarrow{OD} = \lambda \overrightarrow{OA}$,

find the value of λ for which $\overrightarrow{OD} + \overrightarrow{OB}$ is parallel to the x-axis.

7 State which of the following vectors are parallel to $\begin{bmatrix} 3 \\ -1 \\ -2 \end{bmatrix}$.

a $\begin{bmatrix} 6 \\ -3 \\ -4 \end{bmatrix}$ **b** $\begin{bmatrix} -9 \\ 3 \\ 6 \end{bmatrix}$ **c** $\begin{bmatrix} -3 \\ -1 \\ -2 \end{bmatrix}$

d $-2 \begin{bmatrix} 3 \\ 1 \\ 2 \end{bmatrix}$ **e** $\begin{bmatrix} \frac{3}{2} \\ -\frac{1}{2} \\ -1 \end{bmatrix}$ **f** $\begin{bmatrix} -1 \\ \frac{1}{3} \\ \frac{2}{3} \end{bmatrix}$

8 Given that $a = \begin{bmatrix} 4 \\ 1 \\ -6 \end{bmatrix}$, state whether each of these vectors

is parallel to **a**, equal to **a** or neither parallel or equal to **a**.

a $\begin{bmatrix} 8 \\ 2 \\ -10 \end{bmatrix}$ **b** $\begin{bmatrix} -4 \\ -1 \\ 6 \end{bmatrix}$ **c** $2\begin{bmatrix} 2 \\ \frac{1}{2} \\ -3 \end{bmatrix}$

9 The triangle ABC has its vertices at the points A$(-1, 3, 0)$, B$(-3, 0, 7)$ and C$(-1, 2, 3)$.

Find, in the form $\begin{bmatrix} a \\ b \\ c \end{bmatrix}$, the vectors representing

a $\overrightarrow{AB}$ **b** $\overrightarrow{AC}$ **c** $\overrightarrow{CB}$.

10 Find the lengths of the sides of the triangle described in question 9.

11. Find $|\mathbf{a} - \mathbf{b}|$ where $\mathbf{a} = \begin{bmatrix} 1 \\ -1 \\ 2 \end{bmatrix}$ and $\mathbf{b} = \begin{bmatrix} 2 \\ -1 \\ 0 \end{bmatrix}$.

12. A, B, C and D are the points $(0, 0, 2)$, $(-1, 3, 2)$, $(1, 0, 4)$ and $(-1, 2, -2)$ respectively. Find the vectors representing $\overrightarrow{AB}, \overrightarrow{BD}, \overrightarrow{CD}$ and $\overrightarrow{AD}$.

9.5 Properties of a line joining two points

The points A and B have coordinates (x_1, y_1, z_1) and (x_2, y_2, z_2) respectively.

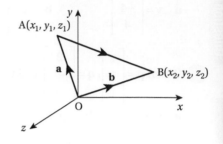

$$\overrightarrow{OA} = \begin{bmatrix} x_1 \\ y_1 \\ z_1 \end{bmatrix} \quad \text{and} \quad \overrightarrow{OB} = \begin{bmatrix} x_2 \\ y_2 \\ z_2 \end{bmatrix} \quad \text{so} \quad \overrightarrow{AB} = \overrightarrow{AO} + \overrightarrow{OB} = \overrightarrow{OB} - \overrightarrow{OA}$$

Therefore $\overrightarrow{AB} = \begin{bmatrix} x_2 - x_1 \\ y_2 - y_1 \\ z_2 - z_1 \end{bmatrix}$

The length of AB

Since $AB = \left\| \begin{bmatrix} x_2 - x_1 \\ y_2 - y_1 \\ z_2 - z_1 \end{bmatrix} \right\|$ then

the length of the line joining (x_1, y_1, z_1) and (x_2, y_2, z_2) is

$$\sqrt{(x_2 - x_1)^2 + (y_2 - y_1)^2 + (z_2 - z_1)^2}$$

The position vector of the midpoint of AB

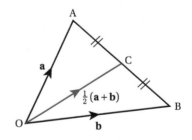

When C is the midpoint of AB then the position vector of C is $\frac{1}{2}(\overrightarrow{OA} + \overrightarrow{OB})$ (you saw this in section 9.2).

So when A is the point (x_1, y_1, z_1) and B is the point (x_2, y_2, z_2)

the coordinates of C are $\left(\frac{1}{2}(x_1 + x_2), \frac{1}{2}(y_1 + y_2), \frac{1}{2}(z_1 + z_2) \right)$

Therefore the coordinates of the midpoint are the arithmetic means of the respective coordinates of the end points.

Example 3

The coordinates of the midpoint R of a line PQ are $(3, -2, 6)$. P is the point $(4, 1, -3)$ and Q is the point (a, b, c). Find the values of a, b and c.

R is the point $(3, -2, 6)$.

But R is the midpoint of PQ, so R is the point $\left(\frac{1}{2}(4+a), \frac{1}{2}(1+b), \frac{1}{2}(-3+c)\right)$

Therefore $\frac{1}{2}(4+a)=3, \quad \frac{1}{2}(1+b)=-2, \quad \frac{1}{2}(-3+c)=6$

So $\qquad a=2, b=-5, c=15$.

Example 4

Find the length of the median through O of the triangle OAB, where A is the point $(2, 7, -1)$ and B is the point $(4, 1, 2)$.

The median of a triangle is the line joining a vertex to the midpoint of the opposite side.

The coordinates of M, the midpoint of AB, are

$$\left(\frac{1}{2}(2+4), \frac{1}{2}(7+1), \frac{1}{2}(-1+2)\right) = \left(3, 4, \frac{1}{2}\right)$$

So the length of OM is $\sqrt{3^2 + 4^2 + \frac{1}{2}^2} = \frac{1}{2}\sqrt{101}$.

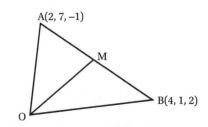

Example 5

The points A, B and C have coordinates $(3, -1, 5)$, $(7, 1, 3)$ and $(-5, 9, -1)$ respectively. L is the midpoint of AB and M is the midpoint of BC. Find the length of LM.

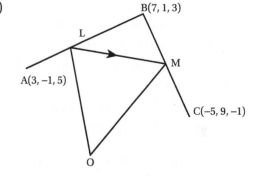

L is the point $\left(\frac{1}{2}(3+7), \frac{1}{2}(-1+1), \frac{1}{2}(5+3)\right) = (5, 0, 4)$

M is the point $\left(\frac{1}{2}(7-5), \frac{1}{2}(1+9), \frac{1}{2}(3-1)\right) = (1, 5, 1)$

Therefore LM $= \sqrt{(1-5)^2 + (5-0)^2 + (1-4)^2} = \sqrt{50} = 5\sqrt{2}$.

Example 6

A, B and C are the points with position vectors $\begin{bmatrix} 2 \\ -1 \\ 5 \end{bmatrix}$, $\begin{bmatrix} 1 \\ -2 \\ 1 \end{bmatrix}$ and $\begin{bmatrix} 3 \\ 1 \\ -2 \end{bmatrix}$ respectively.

D and E are the respective midpoints of BC and AC. Show that DE is parallel to AB.

Using $\mathbf{a} = \begin{bmatrix} 2 \\ -1 \\ 5 \end{bmatrix}$, $\mathbf{b} = \begin{bmatrix} 1 \\ -2 \\ 1 \end{bmatrix}$ and $\mathbf{c} = \begin{bmatrix} 3 \\ 1 \\ -2 \end{bmatrix}$ then

In $\triangle$ OBC, $\overrightarrow{OD} = \frac{1}{2}(\mathbf{b}+\mathbf{c}) = \frac{1}{2}\begin{bmatrix} 4 \\ -1 \\ -1 \end{bmatrix}$

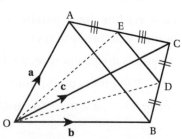

(continued)

and in $\triangle OAC$, $\overrightarrow{OE} = \frac{1}{2}(\mathbf{a}+\mathbf{c}) = \frac{1}{2}\begin{bmatrix} 5 \\ 0 \\ 3 \end{bmatrix}$

Therefore $\overrightarrow{DE} = \overrightarrow{OE} - \overrightarrow{OD} = \frac{1}{2}\begin{bmatrix} 1 \\ 1 \\ 4 \end{bmatrix}$

Also $\overrightarrow{AB} = \mathbf{b} - \mathbf{a} = \begin{bmatrix} -1 \\ -1 \\ -4 \end{bmatrix}$

So $\overrightarrow{AB} = -2\overrightarrow{DE}$ therefore AB and DE are parallel.

Exercise 3

In this exercise A, B, C and D are the points with position vectors

$\begin{bmatrix} 1 \\ 1 \\ -1 \end{bmatrix}$, $\begin{bmatrix} 1 \\ -1 \\ 2 \end{bmatrix}$, $\begin{bmatrix} 0 \\ 1 \\ 1 \end{bmatrix}$ and $\begin{bmatrix} 2 \\ 1 \\ 0 \end{bmatrix}$ respectively.

1 Find $|\overrightarrow{AB}|$ and $|\overrightarrow{BD}|$.

2 Determine whether any of these pairs of lines are parallel.

   a AB and CD

   b AC and BD

   c AD and BC

3 When L and M are the position vectors of the midpoints of AD and BD respectively, show that $\overrightarrow{LM}$ is parallel to $\overrightarrow{AB}$.

4 When H and K are the midpoints of AC and CD respectively, show that $\overrightarrow{HK} = \frac{1}{2}\overrightarrow{AD}$.

5 When L, M, N and P are the midpoints of AD, BD, BC and AC respectively, show that $\overrightarrow{LM}$ is parallel to $\overrightarrow{NP}$.

9.6 The equation of a straight line

A particular line is uniquely located in space if

▶ it has a known direction and passes through a known fixed point, or

▶ it passes through two known fixed points.

A line with known direction passing through a fixed point

Look at a line that is parallel to a vector **m** and which passes through a fixed point A with position vector **a**.

When **r** is the position vector, $\overrightarrow{OP}$, of a point P then

     P is a point on this line $\Leftrightarrow$ $\overrightarrow{AP} = \lambda\mathbf{m}$

where λ is a parameter (a variable scalar).

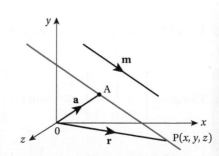

Now $\quad \overrightarrow{OP} = \overrightarrow{OA} + \overrightarrow{AP}$

So $\quad\quad \mathbf{r} = \mathbf{a} + \lambda\mathbf{m}$

Therefore P is on the line $\quad\Leftrightarrow\quad \mathbf{r} = \mathbf{a} + \lambda\mathbf{m}.$

In general, if a line passes through A(x_1, y_1, z_1) and is parallel to $\begin{bmatrix} a \\ b \\ c \end{bmatrix}$ its

equation is $\mathbf{r} = \begin{bmatrix} x_1 \\ y_1 \\ z_1 \end{bmatrix} + \lambda \begin{bmatrix} a \\ b \\ c \end{bmatrix}.$

The point (x_1, y_1, z_1) is only one of an infinite number of points on the line.
Therefore the equation representing a given line is *not* unique.

Example 7

A line passes through the point with position vector $\begin{bmatrix} 2 \\ -1 \\ 4 \end{bmatrix}$

and is parallel to the vector $\begin{bmatrix} 1 \\ 1 \\ -2 \end{bmatrix}.$

Find a vector equation of the line.

The vector equation of a line is $\mathbf{r} = \mathbf{a} + \lambda\mathbf{m}$ where $\mathbf{a}$ is the position vector
of a point on the line and $\mathbf{m}$ is parallel to the line.

For this line, $\mathbf{a} = \begin{bmatrix} 2 \\ -1 \\ 4 \end{bmatrix}$ and $\mathbf{m} = \begin{bmatrix} 1 \\ 1 \\ -2 \end{bmatrix}.$

A vector equation of the line is $\mathbf{r} = \begin{bmatrix} 2 \\ -1 \\ 4 \end{bmatrix} + \lambda \begin{bmatrix} 1 \\ 1 \\ -2 \end{bmatrix}.$

Example 8

Find a vector equation for the line through the points A(3, 4, −7) and B(1, −1, 6).

To find a vector equation of a line, you need to know a point on the line (you
could use either A or B) and a vector parallel to the line.
As A and B are on the line, $\overrightarrow{AB}$ is parallel to the line and $\overrightarrow{AB} = \overrightarrow{OB} - \overrightarrow{OA}.$

$$\overrightarrow{OA} = \begin{bmatrix} 3 \\ 4 \\ -7 \end{bmatrix} \text{ and } \overrightarrow{OB} = \begin{bmatrix} 1 \\ -1 \\ 6 \end{bmatrix}$$

Hence $\quad \overrightarrow{AB} = \overrightarrow{OB} - \overrightarrow{OA} = \begin{bmatrix} 1 \\ -1 \\ 6 \end{bmatrix} - \begin{bmatrix} 3 \\ 4 \\ -7 \end{bmatrix} = \begin{bmatrix} -2 \\ -5 \\ 13 \end{bmatrix}$

(continued)

A vector equation of the line is $\mathbf{r} = \begin{bmatrix} 3 \\ 4 \\ -7 \end{bmatrix} + \lambda \begin{bmatrix} -2 \\ -5 \\ 13 \end{bmatrix}$.

This equation is not unique; we could have used $\overrightarrow{OB}$ instead of $\overrightarrow{OA}$ as the position vector. Furthermore, you could use any multiple of $\overrightarrow{OB} - \overrightarrow{OA}$ as the direction vector, since all are parallel to the line.

$\mathbf{r} = \begin{bmatrix} 1 \\ -1 \\ 6 \end{bmatrix} + \lambda \begin{bmatrix} -2 \\ -5 \\ 13 \end{bmatrix}$ is an equally valid vector equation for this line.

Example 9

Find the coordinates of the point where the line $\mathbf{r} = \begin{bmatrix} 2 \\ -3 \\ 2 \end{bmatrix} + s \begin{bmatrix} 1 \\ -1 \\ 4 \end{bmatrix}$ cuts the xy-plane.

The z-coordinate of any point P on the xy-plane is zero.

Rearranging the equation of the line as $\mathbf{r} = \begin{bmatrix} 2+s \\ -3-s \\ 2+4s \end{bmatrix}$ shows that

it cuts the xy-plane where $2 + 4s = 0$, that is where $s = -\dfrac{1}{2}$.

When $s = -\dfrac{1}{2}$, then $\mathbf{r} = \begin{bmatrix} \frac{3}{2} \\ -\frac{7}{2} \\ 0 \end{bmatrix}$.

Therefore the line cuts the xy-plane at the point $\left(\dfrac{3}{2}, -\dfrac{7}{2}, 0 \right)$.

Exercise 4

1 Write down a vector which is parallel to each of these lines.

a $\mathbf{r} = \begin{bmatrix} 1 \\ -2 \\ 4 \end{bmatrix} + t \begin{bmatrix} 2 \\ -1 \\ -5 \end{bmatrix}$
b $\mathbf{r} = \begin{bmatrix} 2 \\ 0 \\ -1 \end{bmatrix} + s \begin{bmatrix} 0 \\ 3 \\ -5 \end{bmatrix}$

c $\mathbf{r} = \begin{bmatrix} 1-2s \\ 4s-3 \\ 1-s \end{bmatrix}$

2 Write down an equation in vector form for the line through a point A with position vector **a** and in the direction of vector **b** where

a $\mathbf{a} = \begin{bmatrix} 1 \\ -3 \\ 2 \end{bmatrix}, \mathbf{b} = \begin{bmatrix} 5 \\ 4 \\ -1 \end{bmatrix}$
b $\mathbf{a} = \begin{bmatrix} 2 \\ 1 \\ 0 \end{bmatrix}, \mathbf{b} = \begin{bmatrix} 0 \\ 3 \\ -1 \end{bmatrix}$

c A is the origin, $\mathbf{b} = \begin{bmatrix} 1 \\ -1 \\ -1 \end{bmatrix}$

3 The points A(4, 5, 10), B(2, 3, 4) and C(1, 2, −1) are three vertices of a parallelogram ABCD. Find vector equations for the sides AB, BC and AD.

4 **a** Write down a vector equation for the line through A and B when

i $\overrightarrow{AB} = \begin{bmatrix} 3 \\ 1 \\ -4 \end{bmatrix}$ and $\overrightarrow{OB} = \begin{bmatrix} 1 \\ 7 \\ 8 \end{bmatrix}$

ii A and B have coordinates (1, 1, 7) and (3, 4, 1).

b Find, for each line in part **a**, the coordinates of the points where the line crosses the *xy*-plane, the *yz*-plane and the *zx*-plane.

9.7 Pairs of lines

The location of two lines in space can be such that

a the lines are parallel

b the lines are not parallel and intersect

c the lines are not parallel and do not intersect (such lines are called **skew**).

Parallel lines

If two lines are parallel, the vector giving the direction of one line will be a multiple of the vector giving the direction of the other line.

Non-parallel lines

When two lines whose vector equations are $\mathbf{r}_1 = \mathbf{a}_1 + \lambda\mathbf{b}_1$ and $\mathbf{r}_2 = \mathbf{a}_2 + \mu\mathbf{b}_2$ intersect, there must be unique values of λ and μ such that $\mathbf{a}_1 + \lambda\mathbf{b}_1 = \mathbf{a}_2 + \mu\mathbf{b}_2$.

If no such values can be found, the lines do not intersect.

Example 10

Find out whether these pairs of lines are parallel, intersecting or skew.

a $\mathbf{r}_1 = \begin{bmatrix} 1 \\ 1 \\ 2 \end{bmatrix} + \lambda \begin{bmatrix} 3 \\ -2 \\ 4 \end{bmatrix}$ and $\mathbf{r}_2 = \begin{bmatrix} 2 \\ -1 \\ 3 \end{bmatrix} + \mu \begin{bmatrix} -6 \\ 4 \\ -8 \end{bmatrix}$

b $\mathbf{r}_1 = \begin{bmatrix} 1 \\ -1 \\ 3 \end{bmatrix} + s \begin{bmatrix} 1 \\ -1 \\ 1 \end{bmatrix}$ and $\mathbf{r}_2 = \begin{bmatrix} 2 \\ 4 \\ 6 \end{bmatrix} + t \begin{bmatrix} 2 \\ 1 \\ 3 \end{bmatrix}$

c $\mathbf{r}_1 = \begin{bmatrix} 1 \\ 0 \\ 1 \end{bmatrix} + \lambda \begin{bmatrix} 1 \\ 3 \\ 4 \end{bmatrix}$ and $\mathbf{r}_2 = \begin{bmatrix} 2 \\ 3 \\ 0 \end{bmatrix} + \mu \begin{bmatrix} 4 \\ -1 \\ 1 \end{bmatrix}$

a Check first whether the lines are parallel by comparing their directions.

The first line is parallel to $\begin{bmatrix} 3 \\ -2 \\ 4 \end{bmatrix}$.

The second line is parallel to $\begin{bmatrix} -6 \\ 4 \\ -8 \end{bmatrix} = -2 \begin{bmatrix} 3 \\ -2 \\ 4 \end{bmatrix}$.

Therefore these two lines are parallel.

b The directions of the lines are $\begin{bmatrix} 1 \\ -1 \\ 1 \end{bmatrix}$ and $\begin{bmatrix} 2 \\ 1 \\ 3 \end{bmatrix}$.

These are not multiples of one another, so the two lines are not parallel.

If the lines intersect it will be at a point where $\mathbf{r}_1 = \mathbf{r}_2$ that is, where

$$\begin{bmatrix} 1+s \\ -1-s \\ 3+s \end{bmatrix} = \begin{bmatrix} 2+2t \\ 4+t \\ 6+3t \end{bmatrix}.$$

Equating the coordinates of x and y gives

$1 + s = 2(1 + t)$ and $-(1+s) = 4 + t$

so $t = -2$, $s = -3$.

With these values for s and t, the z-coordinates are

first line $\quad 3 + s = 0$

second line $\ 6 + 3t = 0$ $\Big\}$ equal values.

So $\mathbf{r}_1 = \mathbf{r}_2$ when $s = -3$ and $t = -2$.

Therefore the lines *do* intersect at the point with position vector

$$\begin{bmatrix} 1-3 \\ -1+3 \\ 3-3 \end{bmatrix} = \begin{bmatrix} -2 \\ 2 \\ 0 \end{bmatrix} \ (\text{using } s = -3 \text{ in } \mathbf{r}_1).$$

c The directions of these two lines are not multiples of one another, so the lines are not parallel.

If the lines intersect it will be where $\mathbf{r}_1 = \mathbf{r}_2$ that is, where

$$\begin{bmatrix} 1+\lambda \\ 3\lambda \\ 1+4\lambda \end{bmatrix} = \begin{bmatrix} 2+4\mu \\ 3-\mu \\ \mu \end{bmatrix}.$$

Equating the coordinates of x and y gives

$1 + \lambda = 2 + 4\mu$

$3\lambda = 3 - \mu$ $\Big\}$ $\Rightarrow$ $\mu = 0, \lambda = 1$.

With these values of λ and μ, the z-coordinates are

first line $\quad 1 + 4\lambda = 5$

second line $\qquad \mu = 0$ $\Big\}$ not equal values.

So there are no values of λ and μ for which $\mathbf{r}_1 = \mathbf{r}_2$. Hence these lines do not intersect and are therefore skew.

Exercise 5

1 Find whether the following pairs of lines are parallel, intersecting or skew. In the case of intersection state the position vector of the common point.

a $\mathbf{r}_1 = \begin{bmatrix} 1 \\ -1 \\ 1 \end{bmatrix} + \lambda \begin{bmatrix} 3 \\ -4 \\ 1 \end{bmatrix}$ and $\mathbf{r}_2 = \mu \begin{bmatrix} -9 \\ 12 \\ -3 \end{bmatrix}$

b $\mathbf{r}_1 = \begin{bmatrix} 4 - t \\ 8 - 2t \\ 3 - t \end{bmatrix}$ and $\mathbf{r}_2 = \begin{bmatrix} 7 + 6s \\ 6 + 4s \\ 5 + 5s \end{bmatrix}$

c $\mathbf{r}_1 = \begin{bmatrix} 1 \\ 0 \\ 3 \end{bmatrix} + \lambda \begin{bmatrix} 2 \\ 1 \\ 1 \end{bmatrix}$ and $\mathbf{r}_2 = \begin{bmatrix} 2 \\ -1 \\ 1 \end{bmatrix} + \mu \begin{bmatrix} 1 \\ -2 \\ 0 \end{bmatrix}$

2 Two lines which intersect have equations

$$\mathbf{r} = \begin{bmatrix} 2 \\ 9 \\ 13 \end{bmatrix} + \lambda \begin{bmatrix} 1 \\ 2 \\ 3 \end{bmatrix} \text{ and } \mathbf{r} = \begin{bmatrix} a \\ 7 \\ -2 \end{bmatrix} + \mu \begin{bmatrix} -1 \\ 2 \\ -3 \end{bmatrix}.$$

Find the value of a and the position vector of the point of intersection.

3 Show that the lines $\mathbf{r} = \begin{bmatrix} 2 \\ -1 \\ 1 \end{bmatrix} + \lambda \begin{bmatrix} 1 \\ -2 \\ 2 \end{bmatrix}$ and $\mathbf{r} = \begin{bmatrix} 1 \\ -3 \\ 4 \end{bmatrix} + \mu \begin{bmatrix} 2 \\ 3 \\ -6 \end{bmatrix}$ are skew.

9.8 The scalar product

This section involves an operation on two vectors and the angle between them. This operation is called a product but, because it involves vectors, it is not related to the product of real numbers.

The definition of the scalar product

The scalar product of two vectors **a** and **b** is denoted by $\mathbf{a} \cdot \mathbf{b}$ and defined as

$$\mathbf{a} \cdot \mathbf{b} = ab \cos \theta$$

where θ is the angle between **a** and **b**.

Since $ab \cos \theta = ba \cos \theta$, $\qquad \mathbf{a} \cdot \mathbf{b} = \mathbf{b} \cdot \mathbf{a}$.

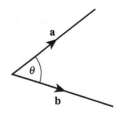

Parallel vectors

When **a** and **b** are parallel then either

$$\mathbf{a} \cdot \mathbf{b} = ab \cos 0 \qquad\qquad \text{or} \qquad\qquad \mathbf{a} \cdot \mathbf{b} = ab \cos \pi$$

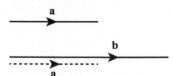

For parallel vectors in the same direction $\mathbf{a} \cdot \mathbf{b} = ab$ and for parallel vectors in the opposite direction $\mathbf{a} \cdot \mathbf{b} = -ab$.

In particular, for the unit vectors **i**, **j**, **k**,

$$\mathbf{i} \cdot \mathbf{i} = \mathbf{j} \cdot \mathbf{j} = \mathbf{k} \cdot \mathbf{k} = 1$$

In the special case when $\mathbf{a} = \mathbf{b}$

$$\mathbf{a} \cdot \mathbf{b} = \mathbf{a} \cdot \mathbf{a} = a^2$$

Perpendicular vectors

If **a** and **b** are perpendicular then $\theta = \frac{1}{2}\pi \Rightarrow \mathbf{a} \cdot \mathbf{b} = ab \cos \frac{1}{2}\pi = 0$.

For perpendicular vectors, a · b = 0.

In particular, for the unit vectors **i**, **j**, **k**,

$$\mathbf{i} \cdot \mathbf{j} = \mathbf{i} \cdot \mathbf{k} = \mathbf{j} \cdot \mathbf{k} = 0$$

Calculating a · b in Cartesian form

When $\mathbf{a} = \begin{bmatrix} x_1 \\ y_1 \\ z_1 \end{bmatrix}$ and $\mathbf{b} = \begin{bmatrix} x_2 \\ y_2 \\ z_2 \end{bmatrix}$ then $\mathbf{a} \cdot \mathbf{b} = (x_1 x_2 + y_1 y_2 + z_1 z_2)$

You can see why this is true by writing the dot product in **i**, **j**, **k** notation:

$$(x_1 \mathbf{i} + y_1 \mathbf{j} + z_1 \mathbf{k}) \cdot (x_2 \mathbf{i} + y_2 \mathbf{j} + z_2 \mathbf{k})$$

$$= x_1 x_2 \mathbf{i} \cdot \mathbf{i} + x_1 y_2 \mathbf{i} \cdot \mathbf{j} + x_1 z_2 \mathbf{i} \cdot \mathbf{k} + y_1 x_2 \mathbf{j} \cdot \mathbf{i} + y_1 y_2 \mathbf{j} \cdot \mathbf{j} + y_1 z_2 \mathbf{j} \cdot \mathbf{k} + z_1 x_2 \mathbf{k} \cdot \mathbf{i} + z_1 y_2 \mathbf{k} \cdot \mathbf{j} + z_1 z_2 \mathbf{k} \cdot \mathbf{k}$$

$$= x_1 x_2 + y_1 y_2 + z_1 z_2$$

For example, $\begin{bmatrix} 2 \\ -3 \\ 4 \end{bmatrix} \cdot \begin{bmatrix} 1 \\ 3 \\ -2 \end{bmatrix} = (2)(1) + (-3)(3) + (4)(-2) = -15$

Example 11

Find the scalar product of $\mathbf{a} = \begin{bmatrix} 2 \\ -3 \\ 5 \end{bmatrix}$ and $\mathbf{b} = \begin{bmatrix} 1 \\ -3 \\ 1 \end{bmatrix}$.

Hence find the cosine of the angle between **a** and **b**.

$$\mathbf{a} \cdot \mathbf{b} = (2)(1) + (-3)(-3) + (5)(1) = 16$$

But $\mathbf{a} \cdot \mathbf{b} = |\mathbf{a}||\mathbf{b}| \cos \theta$

$$|\mathbf{a}| = \sqrt{4 + 9 + 25} = \sqrt{38} \text{ and } |\mathbf{b}| = \sqrt{1 + 9 + 1} = \sqrt{11}$$

Hence $\cos \theta = \dfrac{\mathbf{a} \cdot \mathbf{b}}{|\mathbf{a}||\mathbf{b}|} = \dfrac{16}{\sqrt{11}\sqrt{38}} = \dfrac{16}{\sqrt{418}}$.

Example 12

Given $\mathbf{a} = \begin{bmatrix} 10 \\ -3 \\ 5 \end{bmatrix}$, $\mathbf{b} = \begin{bmatrix} 2 \\ 6 \\ -3 \end{bmatrix}$ and $\mathbf{c} = \begin{bmatrix} 1 \\ 10 \\ -2 \end{bmatrix}$, show that $\mathbf{a} \cdot \mathbf{b} + \mathbf{a} \cdot \mathbf{c} = \mathbf{a} \cdot (\mathbf{b} + \mathbf{c})$.

$$\mathbf{a} \cdot \mathbf{b} = (10)(2) + (-3)(6) + (5)(-3) = -13$$

$$\mathbf{a} \cdot \mathbf{c} = (10)(1) + (-3)(10) + (5)(-2) = -30$$

$$\mathbf{b} + \mathbf{c} = \begin{bmatrix} 3 \\ 16 \\ -5 \end{bmatrix}$$

Therefore $\mathbf{a} \cdot (\mathbf{b} + \mathbf{c}) = (10)(3) + (-3)(16) + (5)(-5) = -43$

and $\quad \mathbf{a} \cdot \mathbf{b} + \mathbf{a} \cdot \mathbf{c} = -13 - 30 = -43$

Therefore $\mathbf{a} \cdot \mathbf{b} + \mathbf{a} \cdot \mathbf{c} = \mathbf{a} \cdot (\mathbf{b} + \mathbf{c})$.

Example 13

Find the acute angle between the lines $\mathbf{r} = \begin{bmatrix} 2 \\ -1 \\ 4 \end{bmatrix} + \lambda \begin{bmatrix} 1 \\ 1 \\ -2 \end{bmatrix}$ and $\mathbf{r} = \begin{bmatrix} 3 \\ 4 \\ -7 \end{bmatrix} + \mu \begin{bmatrix} -2 \\ -5 \\ 13 \end{bmatrix}$.

The angle between the lines is the angle between their directions,

that is the angle between the vectors $\begin{bmatrix} 1 \\ 1 \\ -2 \end{bmatrix}$ and $\begin{bmatrix} -2 \\ -5 \\ 13 \end{bmatrix}$ which can be found

using the scalar product.

If θ is the angle between the lines, then

$$\begin{bmatrix} 1 \\ 1 \\ -2 \end{bmatrix} \cdot \begin{bmatrix} -2 \\ -5 \\ 13 \end{bmatrix} = \left(\sqrt{1^2 + 1^2 + (-2)^2} \times \sqrt{(-2)^2 + (-5)^2 + 13^2} \right) \cos \theta$$

$$\Rightarrow \quad -33 = \left(\sqrt{6} \right) \left(\sqrt{198} \right) \cos \theta$$

$$\Rightarrow \quad \cos \theta = -\frac{33}{6\sqrt{33}}, \text{ so } \theta = \cos^{-1} -\frac{\sqrt{33}}{6} = 162.2°$$

so the acute angle between the lines is $16.8°$ to 1 decimal place.

9.9 The coordinates of the foot of the perpendicular from a point to a line

To find the coordinates of D, the foot of the perpendicular from a point Q to a line $\mathbf{r} = \mathbf{a} + t\mathbf{b}$, we can use the fact that the angle between the vector from Q to D is 90°. The scalar product of a vector from Q to a general point P on the line is $\overrightarrow{QP} \cdot \mathbf{b}$, so D is the point where $\overrightarrow{QP} \cdot \mathbf{b} = 0$.

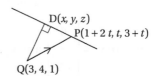

For example, an equation of a line is $\mathbf{r} = \begin{bmatrix} 1 \\ 0 \\ 3 \end{bmatrix} + t \begin{bmatrix} 2 \\ 1 \\ 1 \end{bmatrix} = \begin{bmatrix} 1+2t \\ t \\ 3+t \end{bmatrix}$

The coordinates of a general point P on the line are $(1 + 2t, t, 3 + t)$.

The point $Q(3, 4, 1)$ is not on the line and $D(x, y, z)$ is the foot of the perpendicular from Q to the line.

$$\overrightarrow{QP} = \begin{bmatrix} 3-(2t+1) \\ 4-t \\ 1-(3+t) \end{bmatrix} = \begin{bmatrix} 2-2t \\ 4-t \\ -2-t \end{bmatrix} \text{ and } \begin{bmatrix} 2 \\ 1 \\ 1 \end{bmatrix} \text{ is parallel to the line.}$$

Therefore $\overrightarrow{QP} \cdot \begin{bmatrix} 2 \\ 1 \\ 1 \end{bmatrix} = \begin{bmatrix} 2-2t \\ 4-t \\ -2-t \end{bmatrix} \cdot \begin{bmatrix} 2 \\ 1 \\ 1 \end{bmatrix} = 2(2-2t)+(4-t)+(-2-t)=6-6t.$

When the scalar product is zero, $6-6t=0$, so $t=1$.

Therefore D is the point on the line where $t=1$

So the coordinates of D are (3, 1, 4).

The distance of a point from a line

To find the distance between the foot $D(x, y, z)$ of the perpendicular from a point $Q(a, b, c)$ to a line, first find the coordinates of D.

Then the distance between Q and D is $\sqrt{(a-x)^2+(b-y)^2+(b-z)^2}$

Exercise 6

1 Calculate $\mathbf{m} \cdot \mathbf{n}$ when

a $\mathbf{m} = \begin{bmatrix} 2 \\ -4 \\ 5 \end{bmatrix}$ and $\mathbf{n} = \begin{bmatrix} 1 \\ 3 \\ 8 \end{bmatrix}$ **b** $\mathbf{m} = \begin{bmatrix} 3 \\ -7 \\ 2 \end{bmatrix}$ and $\mathbf{n} = \begin{bmatrix} 5 \\ 1 \\ -4 \end{bmatrix}$

c $\mathbf{m} = \begin{bmatrix} 2 \\ -3 \\ 6 \end{bmatrix}$ and $\mathbf{n} = \begin{bmatrix} 1 \\ 1 \\ 0 \end{bmatrix}$.

2 Find $\mathbf{p} \cdot \mathbf{q}$ and the cosine of the angle between $\mathbf{p}$ and $\mathbf{q}$ when

a $\mathbf{p} = \begin{bmatrix} 2 \\ 4 \\ 1 \end{bmatrix}$ and $\mathbf{q} = \begin{bmatrix} 1 \\ 1 \\ 1 \end{bmatrix}$ **b** $\mathbf{p} = \begin{bmatrix} -1 \\ 3 \\ -2 \end{bmatrix}$ and $\mathbf{q} = \begin{bmatrix} 1 \\ 1 \\ -6 \end{bmatrix}$

c $\mathbf{p} = \begin{bmatrix} -2 \\ 5 \\ 0 \end{bmatrix}$ and $\mathbf{q} = \begin{bmatrix} 1 \\ 1 \\ 0 \end{bmatrix}$ **d** $\mathbf{p} = \begin{bmatrix} 2 \\ 1 \\ 0 \end{bmatrix}$ and $\mathbf{q} = \begin{bmatrix} 0 \\ 1 \\ -2 \end{bmatrix}$.

3 The cosine of the angle between two vectors $\mathbf{v}_1$ and $\mathbf{v}_2$ is $\dfrac{4}{21}$

where $\mathbf{v}_1 = \begin{bmatrix} 6 \\ 3 \\ -2 \end{bmatrix}$ and $\mathbf{v}_2 = \begin{bmatrix} -2 \\ \lambda \\ -4 \end{bmatrix}$.

Find the positive value of λ.

4 In a triangle ABC, $\overrightarrow{AB} = \begin{bmatrix} 1 \\ 2 \\ 3 \end{bmatrix}$ and $\overrightarrow{BC} = \begin{bmatrix} -1 \\ 4 \\ 0 \end{bmatrix}$.

a Find the cosine of angle ABC.

b Find the vector $\overrightarrow{AC}$ and use it to calculate the angle BAC.

5 Show that $\begin{bmatrix} 1 \\ 7 \\ 3 \end{bmatrix}$ is perpendicular to both $\begin{bmatrix} 1 \\ -1 \\ 2 \end{bmatrix}$ and $\begin{bmatrix} 2 \\ 1 \\ -3 \end{bmatrix}$.

6 Show that $\begin{bmatrix} 13 \\ 23 \\ 7 \end{bmatrix}$ is perpendicular to both $\begin{bmatrix} 2 \\ 1 \\ -7 \end{bmatrix}$ and $\begin{bmatrix} 3 \\ -2 \\ 1 \end{bmatrix}$.

7 The magnitudes of two vectors **p** and **q** are 5 and 4 units respectively. The angle between **p** and **q** is 30°.

   a Find $\mathbf{p} \cdot \mathbf{q}$

   b Find the magnitude of the vector $\mathbf{p} - \mathbf{q}$.

8 Calculate the angle between the vectors $\begin{bmatrix} 2 \\ -1 \\ 3 \end{bmatrix}$ and $\begin{bmatrix} 0 \\ -1 \\ -1 \end{bmatrix}$.

9 The diagram shows a cube where the length of each edge is 4 cm.

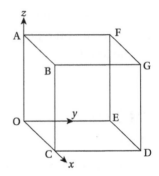

   a Express in the form $\begin{bmatrix} a \\ b \\ c \end{bmatrix}$ the vectors

     i $\overrightarrow{AE}$ **ii** $\overrightarrow{AG}$.

   b H is the midpoint of AB.

     Find the angles of the triangle AEH, giving your answers to the nearest degree.

10 Is it true to say that if $\mathbf{a} \cdot \mathbf{b} = 0$ then either $\mathbf{a} = \mathbf{0}$ or $\mathbf{b} = \mathbf{0}$?

   Explain your answer.

11 In triangle OAB, O is the origin, $\overrightarrow{OA} = \begin{bmatrix} 4 \\ -3 \\ 4 \end{bmatrix}$ and $\overrightarrow{OB} = \begin{bmatrix} 1 \\ 6 \\ -2 \end{bmatrix}$.

   a Show that triangle OAB is isosceles.

   b Find angle AOB correct to the nearest degree.

   c Hence or otherwise find the area of triangle OAB.

12 Find the acute angle between the lines

   a $\mathbf{r} = \begin{bmatrix} 1 \\ -1 \\ 1 \end{bmatrix} + \lambda \begin{bmatrix} 3 \\ -4 \\ 1 \end{bmatrix}$ and $\mathbf{r} = \mu \begin{bmatrix} -9 \\ 12 \\ -3 \end{bmatrix}$

   b $\mathbf{r} = \begin{bmatrix} 4-t \\ 8-2t \\ 3-t \end{bmatrix}$ and $\mathbf{r} = \begin{bmatrix} 7+6s \\ 6+4s \\ 5+5s \end{bmatrix}$

   c $\mathbf{r} = \begin{bmatrix} 1 \\ 0 \\ 3 \end{bmatrix} + \lambda \begin{bmatrix} 2 \\ 1 \\ 1 \end{bmatrix}$ and $\mathbf{r} = \begin{bmatrix} 2 \\ -1 \\ 1 \end{bmatrix} + \mu \begin{bmatrix} 1 \\ -2 \\ 0 \end{bmatrix}$.

13 The line l has equation $\mathbf{r} = \begin{bmatrix} 1 \\ 0 \\ 3 \end{bmatrix} + \lambda \begin{bmatrix} 2 \\ 1 \\ 1 \end{bmatrix}$. The coordinates of a point A are (2, 1, 5).

   a Show that A does not lie on the line l.

   b The point B on the line l is the foot of the perpendicular from A to l. Find the coordinates of B.

   c Hence find the perpendicular distance of the point A from the line l.

14 The line l has equation $\mathbf{r} = \begin{bmatrix} 1 \\ -1 \\ 1 \end{bmatrix} + \lambda \begin{bmatrix} 3 \\ -4 \\ 1 \end{bmatrix}$. The coordinates of a point A are (−2, 1, 0).

   a Show that A does not lie on the line l.

   b The point B on the line l is the foot of the perpendicular from A to l. Find the coordinates of B.

   c Hence find the perpendicular distance of the point A from the line l.

Summary

Vectors

A vector is a quantity with both magnitude and direction. It can be represented by a straight line segment whose length represents the magnitude and whose direction represents the direction of the vector.

When lines representing several vectors are drawn 'head to tail' in order; then the line (in the opposite sense) which completes a closed polygon, represents the sum of the vectors and is called the resultant vector.

A position vector has a fixed location in space.

Cartesian vectors

▶ The coordinates of a point P in three dimensions are (x, y, z).

▶ The position vector of P is given by $\overrightarrow{OP} = \begin{bmatrix} x \\ y \\ z \end{bmatrix}$.

▶ For vectors $\mathbf{v}_1 = \begin{bmatrix} a \\ b \\ c \end{bmatrix}$ and $\mathbf{v}_2 = \begin{bmatrix} p \\ q \\ r \end{bmatrix}$

$$\mathbf{v}_1 + \mathbf{v}_2 = \begin{bmatrix} a + p \\ b + q \\ c + r \end{bmatrix}$$

the magnitude of is $\mathbf{v}_1$ written as $|\mathbf{v}_1|$ and is equal to $\sqrt{a^2 + b^2 + c^2}$

$\mathbf{v}_1$ and $\mathbf{v}_2$ are parallel when $\mathbf{v}_1 = \lambda \mathbf{v}_2$

$\mathbf{v}_1$ and $\mathbf{v}_2$ are equal when $a = p$, $b = q$ and $c = r$.

Equation of a line

▶ A line passing through a point with position vector **a** and in the direction of the vector **b** has vector equation $\mathbf{r} = \mathbf{a} + t\mathbf{b}$, where **r** is the position vector of a point on the line and t is a parameter.

▶ Two lines with equations $\mathbf{r}_1 = \mathbf{a}_1 + \lambda\mathbf{b}_1$ and $\mathbf{r}_2 = \mathbf{a}_2 + \mu\mathbf{b}_2$

   are parallel if $\mathbf{b}_1$ is a multiple of $\mathbf{b}_2$

   intersect if there are values of λ and μ for which $\mathbf{r}_1 = \mathbf{r}_2$

   are skew (do not intersect) in all other cases.

The scalar product of two vectors

▶ If θ is the angle between two vectors $\mathbf{a} = \begin{bmatrix} x_1 \\ y_1 \\ z_1 \end{bmatrix}$ and $\mathbf{b} = \begin{bmatrix} x_2 \\ y_2 \\ z_2 \end{bmatrix}$ then

   $\mathbf{a} \cdot \mathbf{b} = |\mathbf{a}|\,|\mathbf{b}| \cos \theta$

   If **a** and **b** are perpendicular then $\mathbf{a} \cdot \mathbf{b} = 0$.

   $\mathbf{a} \cdot \mathbf{b} = x_1 x_2 + y_1 y_2 + z_1 z_2$

Review

1 ABCD is a square. Write down the single vector equivalent to

   a $\overrightarrow{AB} + \overrightarrow{BC}$ **b** $\overrightarrow{AB} + \overrightarrow{BD}$

2 OABC are four points such that $\overrightarrow{OA} = \mathbf{a}$, $\overrightarrow{OB} = \mathbf{b}$ and $\overrightarrow{OC} = \mathbf{c}$.

   Express in terms of **a**, **b** and **c**

   a $\overrightarrow{AB}$ **b** $\overrightarrow{BC}$ **c** $\overrightarrow{AC}$.

3 The vector **r** is the position vector of the point P where $\mathbf{r} = \begin{bmatrix} 2 \\ 1 \\ 0 \end{bmatrix}$.

   Write down the coordinates of P.

4 Given $\overrightarrow{OA} = \begin{bmatrix} 3 \\ 1 \\ -1 \end{bmatrix}$ and $\overrightarrow{OB} = \begin{bmatrix} 5 \\ -1 \\ 0 \end{bmatrix}$ find

   a $\left|\overrightarrow{AB}\right|$

   b a vector that is parallel to $\overrightarrow{OA} - \overrightarrow{OB}$

   c the position vector of the midpoint of AB.

5 A line passes through the point with position vector $\begin{bmatrix} 2 \\ 1 \\ 0 \end{bmatrix}$ and is parallel to the vector $\begin{bmatrix} -1 \\ 1 \\ -2 \end{bmatrix}$.

   Find a vector equation of the line.

6 A line l passes through the points A and B whose coordinates are $(3, 2, -1)$ and $(0, -2, 4)$ respectively.

  a Find a vector equation of the line l.

  b Find the coordinates of the point where l crosses the xz-plane.

7 Two points A and B are on the line l_1 where $\overrightarrow{OA} = \begin{bmatrix} 3 \\ -1 \\ 2 \end{bmatrix}$ and $\overrightarrow{OB} = \begin{bmatrix} -1 \\ 1 \\ 9 \end{bmatrix}$.

The equation of a line l_2 is $\mathbf{r} = \begin{bmatrix} 8 \\ 1 \\ -6 \end{bmatrix} + t \begin{bmatrix} 1 \\ -2 \\ -2 \end{bmatrix}$.

  a Show that l_1 and l_2 intersect.

  b Find the coordinates of the point of intersection.

8 Calculate $\mathbf{a} \cdot \mathbf{b}$ where $\mathbf{a} = \begin{bmatrix} -2 \\ 4 \\ -1 \end{bmatrix}$ and $\mathbf{b} = \begin{bmatrix} 3 \\ 1 \\ -2 \end{bmatrix}$.

9 Calculate the angle between the vectors $\mathbf{a}$ and $\mathbf{b}$ where

$\mathbf{a} = \begin{bmatrix} 5 \\ 2 \\ -1 \end{bmatrix} \qquad \mathbf{b} = \begin{bmatrix} -1 \\ 2 \\ -2 \end{bmatrix}$.

10 Calculate the acute angle between the lines whose vector equations are

$\mathbf{r} = \begin{bmatrix} 6 \\ -3 \\ 2 \end{bmatrix} + t \begin{bmatrix} 1 \\ 1 \\ -5 \end{bmatrix} \qquad \mathbf{r} = \begin{bmatrix} -2 \\ 7 \\ 7 \end{bmatrix} + s \begin{bmatrix} 4 \\ 2 \\ -5 \end{bmatrix}$.

11 The line l has equation $\mathbf{r} = \begin{bmatrix} 1 \\ 3 \\ -1 \end{bmatrix} + t \begin{bmatrix} 1 \\ 3 \\ -4 \end{bmatrix}$. The coordinates of a point A are $(1, 1, 4)$.

  a The point B on the line l is the foot of the perpendicular from A to l. Find the coordinates of B.

  b Hence find the perpendicular distance of the point A from the line l.

Assessment

1 The diagram shows a cube where the length of each edge is 2 cm.

  a Express in the form $\begin{bmatrix} a \\ b \\ c \end{bmatrix}$ the vectors

    i $\overrightarrow{AD}$

    ii $\overrightarrow{CE}$.

  b Find the angle between $\overrightarrow{AD}$ and $\overrightarrow{CE}$.

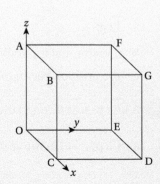

2 The points A, B and C have coordinates (2, 1, 0), (4, 0, 3) and (1, 1, 5) respectively.

   a Find a vector equation of the line through A and B.

   b Find a vector equation of the line through A and C.

   c Find the acute angle between the lines AB and AC.

3 The position vectors of the points A and B are $\begin{bmatrix} 13 \\ -4 \\ 2 \end{bmatrix}$ and $\begin{bmatrix} 18 \\ -4 \\ 3 \end{bmatrix}$ respectively.

   a Find a vector equation of the line through A and B.

   b Show that the vector $\begin{bmatrix} 1 \\ 6 \\ -5 \end{bmatrix}$ is perpendicular to the line through A and B.

4 The line l has equation $\mathbf{r} = \begin{bmatrix} 1+3s \\ 1+2s \\ -s \end{bmatrix}$. The coordinates of a point A are $(-2, 2, 1)$.

   Find the perpendicular distance of the point A from the line l.

5 The triangle ABC is such that the coordinates A and B are (1, 0, 2) and (2, 1, −1) respectively.

   The coordinates of C are (a, b, c) and angle CBA is a right angle. Find a relationship between a, b and c.

6 The points A and B have coordinates (3, 2, 10) and (5, −2, 4) respectively.

   The line l passes through A and has equation $\mathbf{r} = \begin{bmatrix} 3 \\ 2 \\ 10 \end{bmatrix} + \lambda \begin{bmatrix} 3 \\ 1 \\ -2 \end{bmatrix}$.

   a Find the acute angle between l and the line AB.

   b The point C lies on l such that angle ABC is 90°.

      Find the coordinates of C.

   c The point D is such that BD is parallel to AC and angle BCD is 90°. The point E lies on the line through B and D and is such that the length of DE is half that of AC.

      Find the coordinates of the two possible positions of E.

                      AQA MPC4 June 2015

7 The points A and B have coordinates (5, 1, −2) and (4, −1, 3) respectively.

   The line l has equation $\mathbf{r} = \begin{bmatrix} -8 \\ 5 \\ -6 \end{bmatrix} + \mu \begin{bmatrix} 5 \\ 0 \\ -2 \end{bmatrix}$.

   a Find a vector equation of the line that passes through A and B.

   b **i** Show that the line that passes through A and B intersects the line l, and find the coordinates of the point of intersection, P.

      ii The point C lies on l such that triangle PBC has a right angle at B. Find the coordinates of C.

                      AQA MPC4 June 2011

10 The Poisson Distribution

Introduction

The Poisson distribution is a special discrete probability distribution named after a French mathematician, Simeon Poisson (1781–1840). He was a professor of mathematics and wrote many papers on mathematical physics and statistics.

The Poisson distribution was published in 1837. It was derived as a limiting form of the binomial distribution.

Simeon Poisson is reputed to have said "Life is good for only two things, discovering mathematics and teaching mathematics".

Objectives

By the end of this chapter you should know how to...

▶ Recognise the conditions for a Poisson distribution.

▶ Calculate Poisson probabilities
  – using a formula
  – using cumulative Poisson probability tables.

▶ Find the mean, variance and standard deviation of a Poisson variable.

▶ Use the Poisson distribution as a limiting form of the binomial distribution.

▶ Find and use the distribution of the sum of independent Poisson random variables.

Recap

You will need to remember...

Discrete random variables   AS Textbook Chapter 14

▶ $\sum x P(X = x) = 1$

For two independent random variables, X and Y,

▶ $E(X + Y) = E(X) + E(Y)$

▶ $\text{Var}(X + Y) = \text{Var}(X) + \text{Var}(Y)$

Binomial distribution   AS Textbook Chapter 15

If $X \sim B(n, p)$, where X is the number of successes in n independent trials, each with probability p of success,

▶ $P(X = x) = \binom{n}{x} p^x q^{n-x}$ for $x = 0, 1, 2, ..., n$ where $q = 1 - p$

▶ $E(X) = np$

▶ $\text{Var}(X) = npq = np(1 - p)$

Exponential function e^x   Section 4.2

Locate the key for e^x on your calculator; it is usually the second function on the $\boxed{\ln}$ key. The value of the constant e can be found by calculating $e^1 = 2.71828...$

You will need to calculate terms with a negative index, such as:

$$e^{-0.6} = 0.5488..., \quad e^{-1} = 0.36787..., \quad e^{-3.2} = 0.0407...$$

Try these on your calculator to check that you are using the correct keys.

10.1 The Poisson distribution

The discrete random variable X, the *number of occurrences* of a certain event *in a given interval of time or space*, is said to follow a **Poisson distribution** if the events occur:

▶ at random
▶ independently
▶ singly
▶ at a constant average (mean) rate

The value of λ is all that is needed to define the Poisson distribution completely. If all the conditions are satisfied, these variables could be modelled by a Poisson distribution:

– The number of emergency calls received by an ambulance control centre in an hour.

– The number of bacteria in 10 ml of pond water.

– The number of flaws in 20 m² of material.

– The number of goals scored in a football match.

If the average rate is λ, you can write

$$X \sim Po(\lambda),$$

that is, X has a Poisson distribution with parameter λ.

> **Note**
>
> λ is the Greek letter lambda.

Poisson probability function P($X = x$)

If $X \sim Po(\lambda)$, the **probability function** of X is

$$P(X=x) = e^{-\lambda}\frac{\lambda^x}{x!}, \quad x = 0, 1, 2, 3, \dots \text{ to infinity}$$

> **Note**
>
> $P(X = x)$, $E(X)$ and $Var(X)$ are given in the examination formulae booklet for reference.

Mean, variance and standard deviation

Since λ is the average rate of occurrences, the mean, $E(X)$, is λ.

It can be shown that the variance, $Var(X)$, is also λ.

If $X \sim Po(\lambda)$, then
  mean $= E(X) = \lambda$
  variance $= Var(X) = \lambda$
  standard deviation $= \sqrt{\lambda}$

> **Note**
>
> Mean = variance = λ

Calculating Poisson probabilities

Since $P(X=x) = e^{-\lambda}\dfrac{\lambda^x}{x!}$,

$$P(X=0) = e^{-\lambda}\frac{\lambda^0}{0!} = e^{-\lambda} \qquad \lambda^0 = 1 \text{ and } 0! = 1$$

$$P(X=1) = e^{-\lambda}\frac{\lambda^1}{1!} = e^{-\lambda}\lambda \qquad \lambda^1 = \lambda \text{ and } 1! = 1$$

$$P(X=2) = e^{-\lambda}\frac{\lambda^2}{2!}$$

$$P(X=3)=e^{-\lambda}\frac{\lambda^3}{3!}$$

and so on for $x = 4, 5, 6, \ldots$ There is no upper limit for x.

The **probability distribution** can be written

x	0	1	2	3	...
P$(X=x)$	$e^{-\lambda}$	$e^{-\lambda}\lambda$	$e^{-\lambda}\dfrac{\lambda^2}{2!}$	$e^{-\lambda}\dfrac{\lambda^3}{3!}$	and so on

Example 1

Question

If $X \sim \text{Po}(4.3)$, find

  a $P(X=5)$ **b** $P(X=0)$

  c $P(X \le 2)$ **d** $P(3 < X \le 5)$

Answer

Use the Poisson probability formula with $\lambda = 4.3$:

$$P(X=x)=e^{-4.3}\frac{4.3^x}{x!}$$

a   Substitute $x = 5$ into the formula.

$$P(X = 5) = e^{-4.3} \times \frac{4.3^5}{5!}$$
$$= 0.1662\ldots = 0.166 \ (3 \text{ sf})$$

b $P(X=0)= e^{-4.3} \times \dfrac{4.3^0}{0!}$

        $= e^{-4.3}$

        $= 0.01356\ldots = 0.0136 \ (3 \text{ sf})$

It is useful to remember that
$P(X=0) = e^{-\lambda}$

c $P(X \le 2)=e^{-4.3}+e^{-4.3} \times 4.3+e^{-4.3} \times \dfrac{4.3^2}{2!}$

       $=e^{-4.3}\left(1+4.3+\dfrac{4.3^2}{2!}\right)$

       $= 0.1973\ldots = 0.197 \ (3 \text{ sf})$

Take out a factor of $e^{-4.3}$

d $P(3 < X \le 5) = P(X=4) + P(X=5)$

       $=e^{-4.3}\left(\dfrac{4.3^4}{4!}+\dfrac{4.3^5}{5!}\right)$

       $= 0.3595\ldots = 0.360 \ (3 \text{ sf})$

Include 5 but do not include 3.

Example 2

Question

The discrete random variable X follows a Poisson distribution with mean 5.4.
Calculate

  a $P(X > 3)$

  b $P(X \ge 1)$

You are given that the mean is 5.4, so $\lambda = 5.4$ and $X \sim \text{Po}(5.4)$.

a $P(X > 3) = 1 - P(X \le 3)$

$$= 1 - e^{-5.4}\left(1 + 5.4 + \frac{5.4^2}{2!} + \frac{5.4^3}{3!}\right)$$

$$= 0.7867... = 0.787 \ (3 \text{ sf})$$

b $P(X \ge 1) = 1 - P(X = 0)$

$$= 1 - e^{-5.4}$$

$$= 0.9954... = 0.995 \ (3 \text{ sf})$$

There is no upper limit for X, so use the fact that the sum of all the probabilities is 1.

Using the Poisson distribution as a model

Suppose that, on average, 3.5 vehicles per minute arrive at a fuel station and you want to know the probability that exactly 4 vehicles arrive during a particular minute. For a Poisson distribution to be a suitable model, the vehicles must arrive randomly, independently and one at a time. This would be the case if the traffic is free flowing, with no queues, but not if there are traffic lights nearby causing the vehicles to arrive in groups. You also need to ensure that the average rate per minute is constant during the time interval being considered. It is unlikely, for example, that the average rate during the night is the same as during the day.

If the conditions are met, then X, the number of cars arriving during a 1-minute interval, has a Poisson distribution with mean 3.5, that is $X \sim \text{Po}(3.5)$.

The probability that *exactly* 4 cars arrive during a particular minute is given by

$$P(X = 4) = e^{-3.5} \times \frac{3.5^4}{4!} = 0.1888... = 0.189 \ (3 \text{ sf})$$

When considering whether a Poisson model may be suitable, you must check whether *all* the conditions for a Poisson distribution are satisfied.

For example, if X is the number of people entering a shop in a 5-minute interval, the average rate at which people enter must be constant. In addition, people must enter randomly, one at a time and independently of anyone else, which would not be the case if friends or families enter together.

Example 3

The number of elephants seen per day during a safari is denoted by the random variable X, with mean 0.8.

a **i** State two conditions for X to have a Poisson distribution.

   ii For one of these conditions, explain whether you think it likely that it will be satisfied.

Now assume that the conditions are satisfied.

b Marie is on safari for 5 days. Find the probability that she sees exactly one elephant on each day.

a **i** Any two of these would be acceptable:

 - If one elephant is seen, this does not imply that another is more likely to be seen than otherwise, that is the events are independent.

 - Elephants are seen at random.

 - The mean number of elephants seen per day is constant.

ii Elephants may go around in pairs or groups, so if one elephant is seen it might be more likely that another is seen.

b If X is the number of elephants Marie sees in a day, then $X \sim Po(0.8)$.

So, $P(X = 1) = e^{-0.8} \times 0.8$ $\leftarrow$ Do not evaluate yet

P(she sees exactly one elephant on each of 5 days)

$$= (0.8e^{-0.8})^5$$

$$= 0.006001\ldots = 0.00600 \text{ (3 sf)}$$

Example 4

At a large factory, the number of accidents per day may be modelled by a Poisson distribution with parameter 2.25.

a **i** State the mean and standard deviation of the number of accidents per day.

ii Find the probability that the number of accidents in a day is within one standard deviation of the mean.

b Explain why the Poisson distribution may not be a good model for the number of people *injured* in accidents per day.

Let X be the number of accidents in a day, where $X \sim Po(\lambda)$, with $\lambda = 2.25$.

a **i** mean $= \lambda = 2.25$, standard deviation $= \sqrt{\lambda} = \sqrt{2.25} = 1.5$

ii $X \sim Po(2.25)$

$\mu - \sigma = 2.25 - 1.5 = 0.75$, $\mu + \sigma = 2.25 + 1.5 = 3.75$

$P(\mu - \sigma < X < \mu + \sigma) = P(0.75 < X < 3.75)$

$$= P(X = 1) + P(X = 2) + P(X = 3)$$ *X* takes integer values only.

$$= e^{-2.25}\left(2.25 + \frac{2.25^2}{2!} + \frac{2.25^3}{3!}\right)$$

$$= 0.7040\ldots = 0.704 \text{ (3 sf)}$$

b The Poisson distribution may not be appropriate for the number of people *injured* in accidents, as several people may be injured in the same accident. In this case, the injuries do not occur independently.

Example 5

Weak spots occur in the manufacture of a certain cable, independently and randomly, at an average rate of 2.5 weak spots per 100 m length of cable.

a Find the probability that there will be no weak spots in a 100 m length of cable.

Paul cuts 8 lengths of cable, each 100 m long.

b Find the probability that exactly 2 of them have no weak spots.

Let X be the number of weak spots in 100 m of cable, where $X \sim \text{Po}(2.5)$.

a $P(X=0) = e^{-2.5} = 0.08208 = 0.821$ (3 sf)

b Let Y be the number of lengths, in a sample of 8, that have no weak spots.

Then $Y \sim B(n, p)$ with $n = 8$, $p = e^{-2.5}$ (from part **a**) and $q = 1 - p = 1 - e^{-2.5}$

So $P(Y=2) = \begin{pmatrix} 8 \\ 2 \end{pmatrix} \times (e^{-2.5})^2 \times (1 - e^{-2.5})^6 = 0.1128... = 0.113$ (3 sf)

See AS Textbook Section 15.2

8 choose 2   2 successes   6 failures

Unit interval

Care must be taken to specify the **unit interval** being considered.

In Example 5, the mean number of weak spots in 100 m is 2.5, so the number of weak spots in 100 m is distributed Po(2.5).

Now suppose that you want a probability relating to 400 m of cable. Since weak spots occur at random, you would *expect* the mean number of weak spots in 400 m to be $4 \times 2.5 = 10$.

Since the unit interval is multiplied by 4, the mean is multiplied by 4.

So, the number of weak spots in 400 m is distributed Po(10).

Similarly, the number of weak spots in 20 m is distributed Po(0.5).

Since the unit interval is divided by 5, the mean is divided by 5.

Example 6

The number of particles emitted per second by a radioactive substance has a Poisson distribution with mean 0.8.

a Find the probability that the number of particles emitted is

  i   exactly 6 in a 5-second interval,

  ii at most 2 in a 10-second interval.

b Determine the probability that at least one particle will be emitted in each of four consecutive 1-second intervals.

a i   In 1 second, mean number of particles emitted = 0.8.

    So, in 5 seconds, mean number of particles emitted = $0.8 \times 5 = 4$.

    If X is the number of particles emitted in 5 seconds, then $X \sim \text{Po}(4)$

    $P(X=6) = e^{-4} \times \dfrac{4^6}{6!} = 0.1041... = 0.104$ (3 sf)

  Note

  Define the variable, stating the unit interval.

  ii In 10 seconds, mean number of particles emitted = $0.8 \times 10 = 8$.

    So, if Y is the number of particle emitted in 10 seconds, then $Y \sim \text{Po}(8)$

    $P(Y \leq 2) = P(Y=0) + P(Y=1) + P(Y=2)$

    $\qquad = e^{-8}\left(1 + 8 + \dfrac{8^2}{2!}\right)$

    $\qquad = 0.01375... = 0.0138$ (3 sf)

  Note

  Define a new variable and state its distribution.

(continued)

b If W is the number of particles emitted in 1 second, then $W \sim \text{Po}(0.8)$

P(at least one particle emitted in 1 second)

$$= \text{P}(W \geq 1)$$

$$= 1 - \text{P}(W = 0)$$

$$= 1 - e^{-0.8}$$

P(at least one particle emitted in each of 4 consecutive 1-second intervals)

$$= (1 - e^{-0.8})^4$$

$$= 0.09195\ldots = 0.0920 \text{ (3 sf)}$$

Note

Define another new variable.

Exercise 1

1 If $X \sim \text{Po}(1.3)$, find

   a $\text{P}(X = 2)$ **b** $\text{P}(X = 5)$ **c** $\text{P}(X \leq 2)$ **d** $\text{P}(X \geq 3)$

2 If $X \sim \text{Po}(3.5)$, find

   a $\text{P}(X = 0)$ **b** $\text{P}(X \geq 1)$ **c** $\text{P}(X < 3)$ **d** $\text{P}(X > 3)$

3 X has a Poisson distribution with mean 5.2. Find

   a $\text{P}(1 \leq X \leq 3)$ **b** $\text{P}(5 < X < 8)$ **c** $\text{P}(3 < X \leq 6)$ **d** $\text{P}(0 \leq X < 2)$

4 The discrete random variable Y has a Poisson distribution with parameter 4.3. Find

   a $\text{P}(Y = 2)$ **b** $\text{P}(Y < 3)$ **c** $\text{P}(Y \geq 2)$

5 Vehicles arrive at a large service station at an average rate of 1.8 per minute. Assuming that the number arriving may be modelled by a Poisson distribution, find the probability that, in a particular 1-minute interval, the number arriving is

   a exactly 2   **b** at least 3   **c** fewer than 5.

6 The number of births announced in the personal column of a local weekly newspaper has a Poisson distribution with mean 3.2 per week. Find the probability that, in a particular week, the number of births announced is

   a at most 3   **b** exactly 4   **c** more than 2.

7 The number of goals scored by the local team in a soccer match is represented by a random variable X with mean 2.1.

   a State two conditions for X to be modelled by a Poisson distribution.

   Assume now that $X \sim \text{Po}(2.1)$.

   b Find $\text{P}(3 < X < 7)$.

   c The manager promises the team a special party if they score at least one goal in each of the next 12 matches. Find the probability that the team has the party.

8 A car hire firm has 3 cars for hire. The number of cars requested each day may be modelled by a Poisson distribution with mean 2.2.

   a Find the probability that, on any given day,

     i exactly 2 cars are requested

     ii all the cars are in use

     iii the firm will not be able to meet the demand.

   b Find the probability that at least one car is requested each day over a period of 6 days.

9 The number of letters advertising various products that Fahed receives per day may be modelled by a Poisson distribution with parameter 0.5. Find the probability that Fahed receives

  a exactly 2 letters on a particular day

  b no letters on 3 consecutive days

  c more than 4 letters in a 5-day period.

10 At a certain factory the number of accidents per week may be modelled by a Poisson distribution with parameter 0.5.

  a Find the probability that there are at least 2 accidents in a week.

  b Find the probability that there are exactly 3 accidents in 2 weeks.

  c In a period of 50 consecutive months (each consisting of 4 weeks), how many months would you expect to be accident-free?

11 The number of bacterial colonies on a Petri dish can be modelled by a Poisson distribution with average number 2.5 per cm^2. Find the probability that there are

  a no bacterial colonies in 1 cm^2

  b more than 4 bacterial colonies in 2 cm^2

  c exactly 6 bacterial colonies in 4 cm^2.

12 A shop sells printers. The number of printers sold per week may be modelled by a Poisson distribution with standard deviation 2.

  Find

  a the mean number of printers sold in a week

  b the probability that fewer than the mean number of printers are sold in a week

  c the probability that exactly 8 printers are sold in a fortnight.

13 The number of telephone calls made to a large office between 9.30 am and 10.30 am on a weekday follows a Poisson distribution with mean 6.

  Find the probability that

  a there will be at least 2 calls between 9.30 am and 10.30 am on Tuesday

  b there will be exactly 2 calls between 9.30 am and 9.40 am on Wednesday

  c during a period of 5 days there will be exactly 3 days when there are no calls between 10.00 am and 10.10 am.

14 $X \sim Po(\lambda)$. Find the value of λ if

  a $P(X=0) = 2P(X=1)$ b $P(X=5) = P(X=6)$

15 X follows a Poisson distribution with standard deviation 2.   See AS Textbook Chapter 14

  Find

  a $E(X)$ b $Var(X)$ c $E(3X+2)$ d $Var(5X-1)$

  e the standard deviation of $3X$.

10.2 Cumulative Poisson probability tables

Using the Poisson formula can be time-consuming, especially when you have several probabilities to calculate. As with the binomial distribution studied in AS, a cumulative probability table is available that enables you to find probabilities quickly.

The table gives cumulative Poisson probabilities $P(X \leq x)$, where
$$P(X \leq x) = P(X=0) + P(X=1) + P(X=2) + \ldots + P(X=x)$$

In the examination you will have access to Table 2 in the booklet *Formulae and Statistical Tables*.

An extract is printed below. Notice that:

▶ The values of λ are written across the top.
▶ x-values are written down the side.
▶ The cumulative probabilities are given to 4 decimal places.
▶ For any value of λ, the values of x increase indefinitely. However, for a certain value of x and all subsequent values, $P(X \leq x) = 1.0000$ to 4 decimal places. Once this x-value has been reached, the table is left blank for higher values of x.

The following is an extract from **Table 2: Cumulative Poisson distribution function**
The tabulated value is $P(X \leq x)$, where X has a Poisson distribution with mean λ.

λ	0.10	0.20	0.30	0.40	0.50	0.60	0.70	0.80	0.90	1.0	1.2	1.4	1.6	1.8	λ
x															x
0	0.9048	0.8187	0.7408	0.6703	**0.6065**	0.5488	0.4966	0.4493	0.4066	0.3679	0.3012	0.2466	0.2019	0.1653	0
1	0.9953	0.9825	0.9631	0.9384	0.9098	0.8781	0.8442	0.8088	0.7725	0.7358	0.6626	0.5918	0.5249	0.4628	1
2	0.9998	0.9989	0.9964	0.9921	0.9856	0.9769	0.9659	**0.9526**	0.9371	0.9197	0.8795	0.8335	0.7834	0.7306	2
3	1.0000	0.9999	0.9997	0.9992	0.9982	0.9966	0.9942	0.9909	0.9865	0.9810	0.9662	0.9463	0.9212	**0.8913**	3
4		1.0000	1.0000	0.9999	0.9998	0.9996	0.9992	0.9986	0.9977	0.9963	0.9923	0.9857	0.9763	0.9636	4
5				1.0000	1.0000	1.0000	0.9999	0.9998	0.9997	0.9994	0.9985	0.9968	0.9940	0.9896	5
6							1.0000	1.0000	1.0000	0.9999	0.9997	0.9994	0.9987	0.9974	6
7										1.0000	1.0000	0.9999	0.9997	0.9994	7
8												1.0000	1.0000	0.9999	8
9														1.0000	9

For example,

▶ when $\lambda = 0.80$, $P(X \leq 2) = 0.9526$
▶ when $\lambda = 1.8$, $P(X \leq 3) = 0.8913$
▶ when $\lambda = 0.50$, $P(X \leq 0) = P(X = 0) = 0.6065$

> **Note**
>
> $P(X \leq 0) = P(X = 0)$

Example 7

X, the number of insurance claims per week made by a large engineering company, is modelled by a Poisson distribution with mean 1.8.

a Find the probability that, in a particular week, the number of claims is

   i   at most 5

   ii at least 4

   iii fewer than 7

   iv at least 1 but at most 5

   v more than 2 but fewer than 7

   vi exactly 4.

b The director of the engineering firm is writing a report and wants to state that there is less than a 1% chance that the firm will make more than a certain number of claims in a week.

   What is the least value he can use in his report for the number of claims?

x	$\lambda = 1.8$	
0	0.1653	$\longleftarrow$ P($X=0$)
1	0.4628	$\longleftarrow$ P($X \le 1$)
2	0.7306	$\longleftarrow$ P($X \le 2$)
3	0.8913	$\longleftarrow$ P($X \le 3$)
4	0.9636	$\longleftarrow$ P($X \le 4$)
5	0.9896	$\longleftarrow$ P($X \le 5$)
6	0.9974	$\longleftarrow$ P($X \le 6$)
7	0.9994	$\longleftarrow$ P($X \le 7$)
8	0.9999	$\longleftarrow$ P($X \le 8$)
9	1.0000	$\longleftarrow$ P($X \le 9$)

X is the number of claims in a week, where $X \sim$ Po(1.8).

Possible values of *X*

a i P(at most 5)

     $= $ P($X \le 5$)   **0 1 2 3 4 5 6 7 …**

     $= 0.9896$

  ii P(at least 4)

     $= $ P($X \ge 4$)   **0 1 2 3 4 5 6 7 …**

     $= 1 - $ P($X \le 3$)

     $= 1 - 0.8913$

     $= 0.1087$

  iii P(fewer than 7)

     $= $ P($X < 7$)   **0 1 2 3 4 5 6 7 …**

     $= $ P($X \le 6$)

     $= 0.9974$

  iv P(at least 1 but at most 5)

     $= $ P($1 \le X \le 5$)   **0 1 2 3 4 5 6 7 …**

     $= $ P($X \le 5$) $- $ P($X = 0$)

     $= 0.9896 - 0.1653$

     $= 0.8243$

  v P(more than 2 but fewer than 7)

     $= $ P($2 < X < 7$)   **0 1 2 3 4 5 6 7 8 …**

     $= $ P($X \le 6$) $- $ P($X \le 2$)

     $= 0.9974 - 0.7306$

     $= 0.2668$

(continued)

(continued)

vi P(exactly 4)

$$= P(X = 4)$$

0 1 2 3 4 5 6 7...

$$= P(X \le 4) - P(X \le 3)$$

$$= 0.9636 - 0.8913$$

$$= 0.0723$$

b You need to find the least value of x for which $P(X > x) < 0.01$

Now $P(X > x) = 1 - P(X \le x)$

so you want $1 - P(X \le x) < 0.01$

Rearranging:

$$1 - 0.01 < P(X \le x)$$

Take care when rearranging.

so $$0.99 < P(X \le x)$$

$$P(X \le x) > 0.99$$

Look down the table until you find a probability greater than 0.99.

Now $P(X \le 5) = 0.9896 < 0.99$

$P(X \le 6) = 0.9974 > 0.99$

So the least value of x is 6 and the director can say that there is less than a 1% chance that the firm will make more 6 claims in a week.

Exercise 2

Use Table 2 from the *Formulae and Statistical Tables* booklet where appropriate.

1 If $X \sim \text{Po}(3.8)$, find

    a $P(X \le 2)$ **b** $P(X > 3)$ **c** $P(X = 2)$ **d** $P(1 \le X \le 10)$

2 If $X \sim \text{Po}(10.0)$, find

    a $P(X < 9)$ **b** $P(5 < X < 10)$ **c** $P(X > 18)$ **d** $P(11 \le X < 17)$

3 The discrete random variable Y follows a Poisson distribution with parameter 7.5. Find

    a $P(Y \le 6)$ **b** $P(Y > 9)$ **c** $P(6 < Y < 13)$

4 Call-outs at a fire station in a town occur at an average rate of 4.5 per night.

    a State what needs to be assumed in order to justify a Poisson model.

    Now assume that the number of call-outs follows a Poisson distribution.

    b Find the probability that, on a particular night, the number of call-outs is

        i 3 or more   **ii** at most 5   **iii** at least 1

    c Find the probability that the number of call-outs is within one standard deviation of the mean.

5 A grain store containing wheat has been contaminated by weed seeds. The farmer finds that there are, on average, 4 weed seeds per kilogram of wheat. Assuming that the number of weed seeds may be modelled by a Poisson distribution, find the probability that 3 kg of wheat contains

    a at most 9 weed seeds   **b** at least 16 weed seeds.

6 In spring, the number of tadpoles in a particular pond follows a Poisson distribution with a mean of 1.4 tadpoles per 100 ml of pond water. Gershon collects 1 litre of pond water in his bucket and examines it for tadpoles. Find the probability that the bucket will contain

  a more than 16 tadpoles

  b fewer than 14 tadpoles

  c at least 11 but at most 23 tadpoles.

7 Emails arrive randomly on Oliver's computer at an average rate of 6.5 per hour. Stating a necessary assumption, find the probability that Oliver receives

  a exactly 6 emails in an hour

  b no more than 12 emails in 2 hours

  c at least 2 emails in 30 minutes.

8 During each afternoon, helicopters land on a holiday island one at a time and at a constant average rate of one every 20 minutes. Find the probability that, on a particular afternoon

  a at most 2 helicopters land between 2.30 pm and 3.30 pm

  b at least 2 helicopters land between 3.40 pm and 4.20 pm

  c no helicopters land between 2.00 pm and 2.15 pm.

9 The number of calls per hour received by an office switchboard follows a Poisson distribution with parameter 30. Find the probability that there are

  a at least 7 calls in 6 minutes

  b fewer than 8 calls in 20 minutes

  c more than 4 but fewer than 13 calls in 15 minutes.

10 The number of family groups arriving at a theme park may be modelled by a Poisson distribution with mean 1.5 per minute.

  a Find the probability that during a particular 2-minute interval 3 family groups arrive at the theme park.

  b Find the probability that during a particular 5-minute interval more than 4 family groups arrive at the theme park.

  c Would the Poisson distribution provide a good model for the number of visitors to the theme park in a 1-minute interval? Give a reason for your answer.

11 In a factory, the number of breakdowns per week of a bottling machine has a Poisson distribution with mean 2.

  a Find the probability that, in a period of two weeks, there are more than 5 breakdowns.

  b Find the probability that, in a period of 4 weeks, the number of breakdowns is at most 16 but more than the mean number of breakdowns in 4 weeks.

12 The random variable X follows a Poisson distribution.

Given that $\text{Var}(X) = 4.5$, find

a $\text{P}(X < \text{E}(X))$ **b** the standard deviation of X.

13 In the manufacture of a certain commercial carpet, small faults occur in the carpet independently and at random at an average rate of 1.0 per 20 square metres.

a Find the probability that in a randomly selected 20 square metres of the carpet there are

     i no faults   **ii** no more than 2 faults.

A new office block has 10 rooms. Each room has a floor area of 80 square metres and is carpeted using this commercial carpet.

b For any one of these rooms, find the probability that the carpet in the room contains

     i at least 2 faults   **ii** exactly 3 faults   **iii** at most 5 faults.

14 The random variable X is distributed $\text{Po}(1.4t)$.

a If $t = 2$, find $\text{P}(X = 4)$.

b If $t = 4$, find $\text{P}(X < 2)$.

c If $t = 10$, find $\text{P}(X > \text{E}(X))$.

15 The random variable X has a Poisson distribution with mean 10. Find the least value of n such that

a $\text{P}(X \leq n) > 0.95$ **b** $\text{P}(X > n) < 0.1$.

16 Gamma-ray bursts (GRBs) are pulses of gamma rays lasting a few seconds, which are produced by explosions in distant galaxies. They are detected by satellites in orbit around earth. One particular satellite detects GRBs at a constant average rate of 3.5 per week (7 days).

You may assume that the detection of GRBs by this satellite may be modelled by a Poisson distribution.

a Find the probability that the satellite detects:

     i exactly 4 GRBs during one particular week;

     ii at least 2 GRBs on one particular day;

     iii more than 10 GRBs but fewer than 20 GRBs during the 28 days of February 2013.

b Give one reason, apart from the constant average rate, why it is likely that the detection of GRBs by this satellite may be modelled by a Poisson distribution.   AQA MS2B June 2013

10.3 The Poisson distribution as a limiting case of the binomial distribution

In certain circumstances, binomial probabilities can be calculated approximately by using the Poisson distribution.

For a Poisson distribution

If $X \sim \text{Po}(\lambda)$, then $\text{E}(X) = \lambda$ and $\text{Var}(X) = \lambda$,

so mean = variance = λ

See Section 10.1

For a binomial distribution

If $X \sim \text{B}(n, p)$, then $\text{E}(X) = np$ and $\text{Var}(X) = np(1-p)$

When n is large and p is small,

See AS Textbook Section 15.2

$$1 - p \to 1$$

so that $\text{E}(X) = np$ and $\text{Var}(X) \to np$

so mean = variance = np

This satisfies the conditions for a Poisson distribution with mean np.

> If n is large and p is small,
> $X \sim \text{B}(n, p)$ can be approximated by $X \sim \text{Po}(np)$

Note

$\lambda = np$

Although there is no specific rule of how large n should be or how small p should be, $n \geq 50$ and $p \leq 0.1$ works well.

The larger the value of n and the smaller the value of p, the better the approximation.

Example 8

In a certain city the probability that a person has blood type AB is 0.0375.

A random sample of 120 people is selected from the city. By using a suitable approximation, find the probability that

a   exactly 5 people in the group have blood type AB

b   at least 2 people in the group have blood type AB.

Let X be the number of people in 120 with blood type AB.

$X \sim \text{B}(120, 0.0375)$ $\qquad n = 120, p = 0.0375$

$np = 120 \times 0.0375 = 4.5$

Since n is large and p is small, use the Poisson approximation, where $X \sim \text{Po}(4.5)$.

Note

Define the variable as binomial, then justify the use of the Poisson approximation.

Using Table 2 with $\lambda = 4.5$:

a $P(X = 5) = P(X \leq 5) - P(X \leq 4)$

$\qquad = 0.7029 - 0.5321$

$\qquad = 0.1708 = 0.171 \ (3 \text{ sf})$

b $P(X \geq 2) = 1 - P(X \leq 1)$

$\qquad = 1 - 0.0611$

$\qquad = 0.9389 = 0.939 \ (3 \text{ sf})$

Using the binomial distribution directly:

$P(X = 5) = \begin{pmatrix} 120 \\ 5 \end{pmatrix} \times 0.0375^5 \times 0.9625^{115}$

$\qquad = 0.174 \ (3 \text{ sf})$

$P(X \geq 2) = 1 - P(X \leq 1)$

$\qquad = 1 - (0.9625^{120} + 120 \times 0.0375 \times 0.9625^{119})$

$\qquad = 0.942 \ (3 \text{ sf})$

Notice that in Example 8 the answers obtained using the Poisson approximation are comparable to 2 sf with those obtained directly using the binomial distribution.

Example 9

The probability that a particular brand of cell phone works properly when it is first used is 0.97.

During a busy weekend, a store sells 200 of this brand of cell phone. Using a suitable approximation, find the probability that more than 196 of the cell phones work properly when they are first used.

Let X be the number of cell phones in 200 that work properly when they are first used, where $X \sim B(200, 0.97)$.

Now n is large, but $p = 0.97$ is not small, so the conditions for the Poisson approximation are not satisfied.

Consider instead the variable Y, where Y is the number of cell phones that *do not work properly*.

$Y \sim B(200, 0.03)$ $\qquad n = 200, p = 0.03$

$np = 200 \times 0.03 = 6$

Now, since n is large and p is small, use the Poisson approximation, where

$Y \sim Po(6)$

$P(X > 196) = P(X = 197) + P(X = 198) + P(X = 199) + P(X = 200)$

$\qquad\qquad = P(Y = 3) + P(Y = 2) + P(Y = 1) + P(Y = 0)$

$\qquad\qquad = P(Y \le 3) \qquad \leftarrow$ Use Table 2 with $\lambda = 6.0$

$\qquad\qquad = 0.1512... = 0.151 \ (3 \text{ sf})$

> **Note**
>
> This is a useful method when n is large and p is close to 1.

Exercise 3

Use a suitable approximation where appropriate.

1 The random variable X follows a binomial distribution, where $X \sim B(100, 0.03)$. Find

  a $P(X = 0)$ **b** $P(X = 2)$ **c** $P(X = 4)$

  using **i** a binomial distribution   **ii** a suitable Poisson approximation.

2 Calculate these probabilities, using a suitable approximation where appropriate.

  a $P(X > 3)$ when $X \sim B(350, 0.004)$

  b $P(X \le 2)$ when $X \sim B(40, 0.15)$

3 The probability that a bolt is defective is 0.2%.

  Bolts are packed in boxes of 500.

  a Find the probability that in a randomly chosen box

    i there are 2 defective bolts

    ii there are more than 3 defective bolts.

b Two boxes are picked at random from the production line. Find the probability that one has 2 defective bolts and the other has no defective bolts.

c Three boxes are selected at random from the production line. Find the probability that they contain no defective bolts.

4 On average 1 in 200 cars develops a particular fault within a year of purchase.

a Find the probability that none of a randomly chosen sample of 250 cars develops the fault within a year of purchase.

b Find the probability that, in a randomly chosen sample of 300 cars, exactly 2 develop the fault within a year of purchase.

5 Two fair cubical dice are thrown.

a When both dice show 6, it is called a 'double six'. What is the probability of throwing a double six?

Two fair cubical dice are thrown a total of 90 times.

b Find the probability that at least 2 double sixes are thrown.

6 It is known that the probability that a person has blood type X is $\frac{1}{80}$. Using a suitable approximation, find the probability that there are at least 5 people with blood type X in a random sample of 240 people.

7 The probability that I dial a wrong number when making a telephone call is 0.018.

In a typical week I make 50 telephone calls.

a Using a Poisson approximation to a binomial model, find the probability that in a given week

i I dial no wrong numbers

ii I dial more than 2 wrong numbers.

b Comment on the suitability of the binomial model and of the Poisson approximation.

8 A manufacturer produces jelly beans in a variety of colours. On average 2% of the jelly beans are purple. The jelly beans are randomly placed into cartons, each containing 150 jelly beans.

a A carton is selected at random. Find the probability that the carton contains at least 4 purple jelly beans.

b Two cartons are selected at random. Find the probability that both cartons contain at least 1 purple jelly bean.

9 A nurseryman found that, on average, 97% of sunflower seeds germinate. Using a suitable approximation, find the probability that in a pack containing 150 sunflower seeds

a more than 4 do not germinate

b at least 145 germinate.

10 An aircraft has 240 seats. Past records show that, on average, 2% of people who have booked tickets for a flight do not turn up. The airline sells 244 tickets for a particular flight.

Using a suitable approximation, find the probability that

a exactly 4 people do not turn up for the flight

b there are empty seats on the flight

c there are not enough seats for everyone who turns up for the flight.

10.4 The sum of independent Poisson variables

Consider two independent Poisson variables, $X \sim Po(\lambda_1)$ and $Y \sim Po(\lambda_2)$.

It can be shown that $X + Y$, the sum of the two variables, is also a Poisson variable.

Now $E(X + Y) = E(X) + E(Y) = \lambda_1 + \lambda_2$

See AS Textbook Section 15.4

$Var(X + Y) = Var(X) + Var(Y) = \lambda_1 + \lambda_2$

So, for *independent* Poisson variables X and Y, such that $X \sim Po(\lambda_1)$ and $Y \sim Po(\lambda_2)$,

$X + Y \sim Po(\lambda_1 + \lambda_2)$

This result can be extended to the sum of n independent Poisson variables.

Note

Mean $= \lambda_1 + \lambda_2$
Variance $= \lambda_1 + \lambda_2$

Example 10

Question

X and Y are independent Poisson variables such that $X \sim Po(3.5)$ and $Y \sim Po(6.5)$. Find

a $P(X + Y = 8)$

b $P(X + Y < 15)$

c $P(7 < X + Y < 13)$

d $P(Y = 0 \mid X < 5)$

Answer

$X \sim Po(3.5)$, $Y \sim Po(6.5)$

If $T = X + Y$, then $T \sim Po(3.5 + 6.5)$, so $T \sim Po(10)$.

Use Table 2 with $\lambda = 10.0$:

a $P(X + Y = 8)$ $= P(T = 8)$

$= P(T \leq 8) - P(T \leq 7)$ ... 6 7 **8** 9 10 11 12 13 ...

$= 0.3328 - 0.2202$

$= 0.1126 = 0.113$ (3 sf)

b $P(X + Y < 15) = P(T < 15)$ ... 10 11 12 13 **14** 15 16 ...

$= P(T \leq 14)$

$= 0.9165 = 0.917$ (3 sf)

(continued)

(continued)

c $P(7 < X + Y < 13)$

$\quad = P(7 < T < 13)$ ... 6 **7 8 9 10 11 12** 13 ...

$\quad = P(T \le 12) - P(T \le 7)$

$\quad = 0.7916 - 0.2202$

$\quad = 0.5714 = 0.571 \,(3\,sf)$

d Since X and Y are independent,

$\quad P(Y = 0 \mid X < 5) = P(Y = 0)$

$\qquad\qquad = e^{-6.5}$

$\qquad\qquad = 0.0015034... = 0.00150 \,(3\,sf)$

See AS Textbook Section 13.3

> **Note**
>
> $P(A \mid B) = P(A)$ when A and B are independent.

Example 11

During a weekday, heavy lorries pass an electronic counter on a street independently and at random times. The mean rate for the northbound lorries is 6 in any hour and, independently, the mean rate for southbound lorries is 3 in any hour.

Find the probability that

a exactly 4 lorries pass the counter in a given 40-minute interval

b at least 1 lorry from each direction will pass the counter in a given 40-minute interval.

a Let N be the number of northbound lorries in a 40-minute interval.
If the mean number in 60 minutes is 6, the mean number in 40 minutes is 4,
so $N \sim Po(4)$.

Let S be the number of southbound lorries in a 40-minute interval.
If the mean number in 60 minutes is 3, the mean number in 40 minutes is 2,
so $S \sim Po(2)$.

Let T be the total number of lorries in a 40-minute interval, where $T = N + S$.
Now $T \sim Po(4 + 2)$, so $T \sim Po(6)$

$P(T = 4) = P(T \le 4) - P(T \le 3)$ ← Use Table 2 with $\lambda = 6.0$

$\qquad = 0.2851 - 0.1512$

$\qquad = 0.1339 = 0.134 \,(3\,sf)$

b Since $N \sim Po(4)$,

$P(N \ge 1) = 1 - P(N = 0)$

$\qquad = 1 - e^{-4}$ Do not evaluate yet

Since $S \sim Po(2)$,

$P(S \ge 1) = 1 - P(S = 0)$

$\qquad = 1 - e^{-2}$ Do not evaluate yet

P(at least 1 lorry from each direction in a 40-minute interval)

$\qquad = P(N \ge 1) \times P(S \ge 1)$

$\qquad = (1 - e^{-4}) \times (1 - e^{-2})$

$\qquad = 0.8488...$

$\qquad = 0.849 \,(3\,sf)$

Exercise 4

1 X, Y and W are independent random variables, where

$X \sim \text{Po}(2)$, $Y \sim \text{Po}(3.5)$ and $W \sim \text{Po}(4)$

Find

a $P(X + Y = 6)$ **b** $P(W + Y < 3)$

c $P(3 \leq X + W \leq 9)$ **d** $P(X + Y + W > 16)$

2 A restaurant kitchen has two microwave ovens, A and B.
The number of times that A breaks down has a Poisson distribution with mean 0.4 per week. Independently, the number of times that B breaks down has a Poisson distribution with mean 0.1 per week. Find the probability that, in the next 3 weeks,

a A will not break down and B will break down twice

b there will be a total of 3 breakdowns.

3 Two identical racing cars are being tested independently on a circuit. For each car, the number of mechanical breakdowns may be modelled by a Poisson distribution with a mean of 1 breakdown in 100 laps. If a car breaks down, the service team attends to it and the car then continues on the circuit. The first car completes 20 laps and the second car completes 40 laps. Find the probability that, during the test drives, the service team is called out to attend to breakdowns

a once   **b** more than twice.

4 The number of houses sold by an estate agent may be modelled by a Poisson distribution with mean 2.75 per week and, independently, the number of semi-detached houses sold may be modelled by a Poisson distribution with mean 3.25 per week.

a Find the mean and variance of the total number of houses sold in a week.

b Find the probability that fewer than 5 houses are sold in a week.

5 In hockey matches, the school team scores goals independently and at random times. In home games they score an average of 3 goals per match and in away games they score an average of 2.5 goals per match.

a Find the probability that the team will score fewer than 8 goals in their next 2 home games.

The team captain sets a target of a total of more than 7 goals in the next 2 games consisting of a home game followed by an away game.

b Find the probability that the team will meet the target.

6 The two centre pages of the *Weekly Press* consist of a page of film reviews and a page of classified advertisements. The number of misprints on the film-reviews page may be modelled by a Poisson distribution with mean 1.4 and, independently, the number of misprints on the classified advertisement page may be modelled by a Poisson distribution with mean 2.2. Find the probability that there will be more than 4 but fewer than 9 misprints on the two centre pages of the *Weekly Press*.

7 During the evening, vehicles pass a marker point on a motorway independently, at random times, at a constant average rate of 2 vehicles every 3 minutes.

  a Find the probability that fewer than 4 vehicles pass the marker point in a period of 3 minutes.

  b Find the probability that a total number of 10 vehicles pass the marker point in two separate periods of 3 minutes and 5 minutes.

8 The number of A-grades, X, achieved in total by students at Lowkey School in their Mathematics examinations each year can be modelled by a Poisson distribution with mean 3.

  a Determine the probability that, during a 5-year period, students at Lowkey School achieve a total of more than 18 A-grades in their Mathematics examinations.

  b The number of A-grades, Y, achieved in total by students at Lowkey School in their English examinations each year can be modelled by a Poisson distribution with a mean of 7.

    i Determine the probability that, during a year, students at Lowkey School achieve a total of fewer than 15 A-grades in their Mathematics and English examinations.

    ii What assumption did you make in answering **b i**?

AQA MS2B June 2006

9 The administration department at Hitech College uses two photocopiers, A and B, which operate independently.

The number of breakdowns each week, X, for photocopier A can be modelled by a Poisson distribution with a mean of 0.5.

The number of breakdowns each week, Y, for photocopier B can be modelled by a Poisson distribution with a mean of 2.5.

  a Calculate:

    i $P(X \le 1)$;   **ii** $P(Y = 2)$.

  b **i** In a given week, show that the probability of a total of exactly 2 photocopier breakdowns is 0.224, correct to three decimal places.

    ii Hence find the probability that there will be exactly 2 photocopier breakdowns in each of four consecutive weeks.

  c The total number of photocopier breakdowns, T, in a 4-week period can be modelled by a Poisson distribution with mean μ.

    i Find the value of μ.

    ii Hence find the probability that, in a given 4-week period, there will be a total of at least 18 photocopier breakdowns.

AQA MS2A January 2006

10. The independent Poisson variables X and Y have means 3 and 4 respectively.

   a   Find the mean and variance of the random variable S, such that
   $$S = X_1 + X_2 + Y_1 + Y_2 + Y_3$$
   where X_1 and X_2 are two independent observations of X and Y_1, Y_2 and Y_3 are three independent observations of Y.

   b   Find the mean and variance of the random variable D, such that
   $$D = 2X + 3Y$$

   c   i   Is S a Poisson variable? Give a reason for your answer.
   ii   Is D a Poisson variable? Give a reason for your answer.

10.5 Further applications

Example 12

Customers enter a large health food shop either alone or in groups.

a   The number of customers entering alone between 10.00 am and 11.00 am may be modelled by a Poisson distribution with mean 0.7 per minute.

   Find the probability that, during a particular minute between 10.00 am and 11.00 am, the number of customers entering the shop alone is

   i   3 or fewer.   ii   exactly 3.

b   The number of groups of customers entering the shop between 10.00 am and 11.00 am may be modelled by a Poisson distribution with mean 0.2 per minute.

   i   Find the probability that, during a particular minute between 10.00 am and 11.00 am, more than one group of customers enters the shop.

   ii   Find the probability that 4 or more groups of customers enter the shop between 10.15 am and 10.25 am.

c   State whether the Poisson distribution is likely to provide a suitable model for the **number of customers** entering the shop in groups during each minute between 10.00 am and 11.00 am. Explain your answer.

d   The shop is open from 8.00 am until 6.00 pm. State whether the Poisson distribution is likely to provide a suitable model for the number of customers entering the shop alone during each minute when the shop is open. Explain your answer.

AQA SS02 January 2006

a   Let X be the number of customers entering alone in 1 minute, during the time period from 10.00 am to 11.00 am, so $X \sim \text{Po}(0.7)$.

   i $P(X \leq 3) = 0.9942 = 0.994$ (3 sf)   ← Use Table 2 with $\lambda = 0.70$

   ii $P(X = 3) = P(X \leq 3) - P(X \leq 2)$
   $$= 0.9942 - 0.9659$$
   $$= 0.0283 \text{ (3 sf)}$$

(continued)

(continued)

Answer

b Let Y be the number of groups entering in 1 minute, during the time period from 10.00 am to 11.00 am, so $Y \sim \text{Po}(0.2)$.

  i $P(Y > 1) = 1 - P(Y \leq 1)$ ← Use Table 2 with $\lambda = 0.20$

              $= 1 - 0.9825$

              $= 0.0175 \, (3 \text{ sf})$

  ii Let W be the number of groups entering between 10.15 am and 10.25 am, that is in 10 minutes, so $W \sim \text{Po}(10 \times 0.2)$, that is $W \sim \text{Po}(2)$.

     $P(W \geq 4) = 1 - P(W \leq 3)$ ← Use Table 2 with $\lambda = 2.0$

                $= 1 - 0.8571$

                $= 0.1429 = 0.143 \, (3 \text{ sf})$

c The Poisson distribution would not provide a suitable model for the number of *customers* entering in groups as the customers who enter in a group do not enter independently of each other.

d The mean is unlikely to be constant throughout the day as some times will be busier than others, such as during the lunch break. So the Poisson distribution would not provide a suitable model for the number of customers entering the shop alone during each minute that the shop is open.

Example 13

Question

Joe owns two garages, Acefit and Bestjob, each specialising in the fitting of the latest satellite navigation device.

The daily demand, X, for the device at Acefit garage may be modelled by a Poisson distribution with mean 3.6.

The daily demand, Y, for the device at Bestjob garage may be modelled by a Poisson distribution with mean 4.4.

a Calculate:

  i $P(X \leq 3)$;   **ii** $P(Y = 5)$.

b The total demand for the device at Joe's two garages is denoted by T.

  i Write down the distribution of T, stating any assumptions you make.

  ii Determine $P(6 < T < 12)$.

  iii Calculate the probability that the total demand for the device will exceed 14 on each of two consecutive days. Give your answer to one significant figure.

  iv Determine the minimum number of devices that Joe should have in stock if he is to meet his total demand on at least 99% of days.

AQA MS2B June 2009

a i $X \sim \text{Po}(3.6)$, so $P(X \le 3) = 0.5152 = 0.515$ (3 sf) ← Use Table 2 with $\lambda = 3.6$

ii $Y \sim \text{Po}(4.4)$, so $P(Y = 5) = e^{-4.4} \times \dfrac{4.4^5}{5!} = 0.1687\ldots = 0.169$ (3 sf)

Since $\lambda = 4.4$ is not in the tables, you need to use the formula.

b i T is the total demand, where $T = X + Y$.

Assume that the daily demand for the device at Acefit is independent of the daily demand for the device at Bestjob.

Since the sum of two independent Poisson variables is also Poisson,
$T \sim \text{Po}(3.6 + 4.4)$, so $T \sim \text{Po}(8)$

ii $P(6 < T < 12) = P(T \le 11) - P(T \le 6)$ ... 6 **7 8 9 10 11** 12 13 ...

$\qquad\qquad\quad = 0.8881 - 0.3134$

$\qquad\qquad\quad = 0.5747 = 0.575$ (3 sf)

iii $P(T > 14) = 1 - P(T \le 14)$ ... 13 14 **15 16 17** ...

$\qquad\qquad\quad = 1 - 0.9827$

$\qquad\qquad\quad = 0.0173$

$P(T$ will exceed 14 on each of two consecutive days)

$\qquad = (0.0173)^2 = 0.000299\ldots = 0.0003$ (1 sf)

iv You need to find the least value of t such that $P(T \le t) \ge 0.99$.

You can see from Table 2, with $\lambda = 8.0$, that

$\qquad P(T \le 14) = 0.9827 < 0.99$

$\qquad P(T \le 15) = 0.9918 > 0.99$

so the least value of t is 15.

The minimum number of devices he should have in stock is 15.

Example 14

The number of customers entering a certain branch of a bank may be modelled by a Poisson distribution with a mean of 1.8 per minute.

a Find the probability that at least 5 customers enter the branch during a particular 2-minute interval.

The probability is 0.005 that a customer who enters the branch intends to open a new account. During a particular morning 400 customers enter the bank.

b Use a suitable approximation to find

i the probability that at most 3 of these 400 customers intend to open a new account

ii the least value of a such that the probability that more than a customers in the 400 intend to open an account is less than 0.001.

a Let X be the number of customers entering the bank in a 2-minute interval.

In 2 minutes, mean $= 2 \times 1.8 = 3.6$, so $X \sim \text{Po}(3.6)$.

$P(X \ge 5) = 1 - P(X \le 4)$ ← Use Table 2 with $\lambda = 3.6$

$\qquad\qquad = 1 - 0.7064$

$\qquad\qquad = 0.2936 = 0.294$ (3 sf) *(continued)*

(continued)

b Let Y be the number in 400 who intend to open a new account.

Assuming that whether a customer opens an account is independent of whether any other customer opens an account, then

$Y \sim \text{B}(n, p)$ with $n = 400$, $p = 0.005$

$np = 400 \times 0.005 = 2$

Since n is large and p is small, use the Poisson approximation, where

$Y \sim \text{Po}(2)$.

i Using Table 2 with $\lambda = 2.0$:

P(at most 3 customers intend to open an account)

$= \text{P}(Y \leq 3) = 0.8571 = 0.857$ (3 sf)

ii You require the least a such that $\text{P}(Y > a) < 0.001$,

so $\text{P}(Y \leq a) > 0.999$

From Table 2, with $\lambda = 2.0$:

$\text{P}(Y \leq 7) = 0.9989 < 0.999$

$\text{P}(Y \leq 8) = 0.9998 > 0.999$

So the least value of a is 8.

Note

Define the distribution as binomial before using the Poisson approximation.

Note

Justify the use of the Poisson approximation.

Summary

Conditions for a Poisson distribution

The random variable X is the number of occurrences of a certain event in a given interval of space or time.

$X \sim \text{Po}(\lambda)$ when events occur:

► at random
► independently
► singly
► at a constant average (mean) rate.

Poisson probability distribution

When $X \sim \text{Po}(\lambda)$, the probability of x occurrences in a given interval is given by

$$\text{P}(X = x) = e^{-\lambda} \frac{\lambda^x}{x!}, \quad x = 0, 1, 2, 3, \ldots \text{ to infinity}$$

Mean, variance and standard deviation

Mean $= \text{E}(X) = \lambda$

Variance $= \text{Var}(X) = \lambda$ and standard deviation $= \sqrt{\lambda}$

Note

Mean $=$ variance $= \lambda$

Poisson approximation to the binomial distribution

When n is large ($n \geq 50$ and $p \leq 0.1$, say),

$X \sim \text{B}(n, p)$

can be approximated by

$X \sim \text{Po}(np)$

Note

Mean $=$ variance $= np$

Sum of independent Poisson variables

If X and Y are independent Poisson variables, where $X \sim \text{Po}(\lambda_1)$ and $Y \sim \text{Po}(\lambda_2)$,

then $\quad X + Y \sim \text{Po}(\lambda_1 + \lambda_2)$

This result can be extended to the sum of n independent Poisson variables.

> **Note**
>
> Mean = variance = $\lambda_1 + \lambda_2$

Review

1 The random variable X has the distribution Po(6). Find

a $P(X < 5)$ **b** $P(6 < X \leq 9)$

c $P(X = 9)$ **d** the standard deviation of X.

2 A radioactive disintegration gives counts that follow a Poisson distribution with a mean count of 15 per second. Find the probability that, in a 1-second interval, the count is:

a greater than 16;   **b** at least 12 but not more than 20.

3 The number of cars, X, passing along a road each minute can be modelled by a Poisson distribution with a mean of 2.6.

a Calculate $P(X = 2)$.

b **i** Write down the distribution of Y, the number of cars passing along this road in a 5-minute interval.

    ii Hence calculate the probability that at least 15 cars pass along this road in each of four successive 5-minute intervals.

AQA MS2B June 2005

4 The number of faults per square metre of a particular type of cloth is denoted by X.

a State conditions that must be satisfied for X to follow a Poisson distribution.

Now assume that X has the distribution Po(0.85).

b Find the probability that there are

    i no faults in 2 m^2 of cloth

    ii at least 2 faults in 3 m^2 of cloth.

5 Along a stretch of motorway, vehicle breakdowns occur at an average rate of 2.4 per day. Stating any necessary assumptions

a find the probability that there will be more than 4 breakdowns in a given day

b find the least integer n such that the probability of more than n breakdowns in a day is less than 0.03.

6 The number of telephone calls made to a health centre may be modelled by a Poisson distribution.

The probability that there are exactly 2 calls during a 5-minute interval is the same as the probability that there are exactly 3 calls during a 5-minute interval.

a Find the mean number of calls in a 5-minute interval.

b Find the standard deviation of the number of calls in a 10-minute interval.

7 On a factory production line, broken biscuits occur randomly and 1.5% of biscuits are broken.

A check is carried out in which 300 biscuits are randomly selected from the production line and inspected to see whether they are broken.

Use a suitable approximation to find the probability that

a fewer than 3 are broken

b at least 5 are broken.

8 A study undertaken by Goodhealth Hospital found that the number of patients each month, X, contracting a particular superbug may be modelled by a Poisson distribution with mean 1.5.

a i Calculate $P(X = 2)$.

ii Hence determine the probability that exactly 2 patients will contract this superbug in each of three consecutive months.

b i Write down the distribution of Y, the number of patients contracting this superbug in a given 6-month period.

ii Find the probability that at least 12 patients will contract this superbug during a given 6-month period.

c State **two** assumptions implied by the use of the Poisson model for the number of patients contracting this superbug.

AQA MS2B January 2006

9 The number of telephone calls per day, X, received by Candice may be modelled by a Poisson distribution with mean 3.5.

The number of e-mails per day, Y, received by Candice may be modelled by a Poisson distribution with mean 6.0.

a For any particular day, find:

i $P(X = 3)$; **ii** $P(Y \geq 5)$.

b i Write down the distribution of T, the total number of telephone calls and e-mails per day received by Candice.

ii Determine $P(7 \leq T \leq 10)$.

iii Hence calculate the probability that, on each of three consecutive days, Candice will receive a total of at least 7 but at most 10 telephone calls and e-mails.

AQA MS2B June 2007

10 A large office block is busy during five weekdays, Monday to Friday, and less busy during the two weekend days, Saturday and Sunday. The block is illuminated by fluorescent light tubes which frequently fail and must be replaced with new tubes by John, the caretaker.

The number of fluorescent tubes that fail on a particular weekday can be modelled by a Poisson distribution with mean 1.5.

The number of fluorescent tubes that fail in a particular weekend day can be modelled by a Poisson distribution with mean 0.5.

a Find the probability that:

   i on a particular Monday, exactly 3 fluorescent light tubes fail;

   ii during the two days of a weekend, more than 1 fluorescent light tube fails;

   iii during a complete seven-day week, fewer than 10 fluorescent light tubes fail.

b John keeps a supply of new fluorescent light tubes. More new tubes are delivered every Monday to replace those he has used during the previous week. John wants the probability that he runs out of new tubes before the next Monday morning to be less than 1 per cent. Find the minimum number of new tubes that he should have available on a Monday morning.

c Give a reason why a Poisson distribution with mean 0.375 is unlikely to provide a satisfactory model for the number of fluorescent light tubes that fail between 1 am and 7 am on a weekday.

AQA MS2B January 13

Assessment

1 A charity shop opens for 8 hours each day. The daily number of customers requesting a refund for an earlier purchase may be modelled by a Poisson distribution with mean 0.6.

a Find the probability that on a particular day there are:

   i no requests for refunds;

   ii more than 2 requests for refunds.   [3]

b On a particular day, the shop manager is absent for 4 hours and leaves an assistant in charge of the shop. Find the probability that, during the manger's 4-hour absence, there are:

   i one or more requests for refunds;

   ii exactly 2 requests for refunds.   [5]

AQA SS02 June 2009

2 It is known that a particular allergy affects 1.5% of children under 10 years old.

The random variable X is the number of children who suffer from the allergy in a random sample of 300 children under 10 years old.

a Write down the probability distribution of X.   [1]

b Using a suitable approximation, find

   i $P(X = 3)$ **ii** $P(X \geq 5)$ [5]

3. The number of accidents, X, occurring during one week at Joanne's place of work can be modelled by a Poisson distribution with a mean of 0.7.

The number of accidents, Y, occurring at Pete's place of work can be modelled by a Poisson distribution with a mean of 1.3.

a i Determine $P(X < 3)$. [1]

   ii Calculate $P(Y = 2)$. [2]

b Find the probability that, during a particular week, there are at least 4 accidents in total at these two places of work. [3]

AQA MS2A January 2007

4. **a** The number of genuine telephone calls, X, made to the emergency services each hour may be modelled by a Poisson distribution with mean 6.5.

   Determine the probability that, in a given hour, there will be fewer than 6 genuine telephone calls made to the emergency services. [2]

b The number of bogus telephone calls, Y, made to the emergency services each hour may be modelled by a Poisson distribution with mean 1.5.

   Calculate the probability that, in a given hour, there will be at least 1 bogus telephone call made to the emergency services. [3]

c Assuming that X and Y are independent Poisson variables, find:

   i $P(X + Y = 6)$; [3]

   ii $P(Y \geq 1 \mid X < 6)$. [1]

AQA MS2A June 2008

5. **a** The number of telephone calls, X, received per hour for Dr Able may be modelled by a Poisson distribution with mean 6.

   Determine $P(X = 8)$. [2]

b The number of telephone calls, Y, received per hour by Dr Bracken may be modelled by a Poisson distribution with mean λ and standard deviation 3.

   i Write down the value of λ. [1]

   ii Determine $P(Y > \lambda)$. [2]

c i Assuming that X and Y are independent Poisson variables, write down the distribution of the total number of telephone calls received per hour for Dr Able and Dr Bracken. [1]

   ii Determine the probability that a total of at most 20 telephone calls will be received during any one-hour period. [1]

   iii The total number of telephone calls received during each of 6 one-hour periods is recorded. Calculate the probability that a total of at least 21 calls will be received during exactly 4 of these one-hour periods. [3]

AQA MS2B June 2008

11 Continuous Random Variables

Introduction

There are two types of random variable: discrete and continuous. Earlier in the course you studied discrete random variables. These usually arise from *counting* the number of occurrences of a particular event, such as in the Bernoulli, binomial and Poisson distributions.

In contrast, a continuous random variable arises from *measuring* a characteristic such as time, height, length, area, volume or weight.

Recap

You will need to remember...

Pure mathematics

▶ Shapes of graphs, especially linear, quadratic and cubic functions.
▶ Integration methods especially $\int_a^b x^n \, dx, n \neq -1$
and $\int (ax+b)^n \, dx, n \neq -1$

Discrete random variables   AS Textbook Chapter 14

▶ A discrete random variable X takes individual values, each with a given probability. It is defined by its probability distribution $P(X=x)$ and is illustrated by a vertical line graph.

For example

x	1	2	3	4
$P(X=x)$	0.1	0.3	0.5	0.1

$P(X=3) = 0.5$
$P(X<3) = 0.1 + 0.3 = 0.4$
$P(X \leq 3) = 0.1 + 0.3 + 0.5 = 0.9$

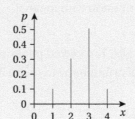

▶ Expectation and variance
$\sum P(X=x) = \sum p_i = 1 \quad$ for $i = 1, 2, \ldots, n$
$E(X) = \mu = \sum x_i p_i$ and $E(g(X)) = \sum g(x_i) p_i$
$Var(X) = \sigma^2 = E[(X-\mu)^2] = E(X^2) - \mu^2$
$\qquad = \sum (x-\mu)^2 p_i = \sum x^2 p_i - \mu^2$

Objectives

By the end of this chapter you should know how to...

▶ Recognise the difference between discrete and continuous random variables.
▶ Calculate probabilities
  – using the probability density function f
  – using the cumulative distribution function F.
▶ Obtain F(x) from f(x) and obtain f(x) from F(x).
▶ Calculate the mean, variance and standard deviation of X and of a simple function of X.
▶ Calculate the mean and variance of
  – a sum or difference of two independent continuous random variables
  – a sum of independent continuous random variables.

11.1 Continuous random variables

A **continuous random variable** arises from *measuring* a characteristic. It cannot take exact values, being defined only within a specified interval. For example, when a boy's height is given as 126 cm, measured to the nearest cm, this means that the height could be anywhere in the interval 125.5 cm ≤ height < 126.5 cm.

Examples of continuous random variables are:

- The mass, in kg, of sacks of flour.
- The time, in minutes, to perform a particular task.
- The length, in cm, of leaves on a particular type of bush.
- The time, in seconds, between vehicles passing a checkpoint.
- The error, in grams, made by scales when weighing potatoes.

Probability density function

A continuous random variable X is defined by a *function* of x. This is known as the **probability density function** f and is defined by f(x), which is specified for the set of values for which it is valid. It can be illustrated on a graph.

For example, a teacher suggests that the time, in hours, spent by students on a certain task may be modelled by the continuous random variable X with probability density function

$$f(x) = \begin{cases} \dfrac{1}{2} + \dfrac{1}{3}x & 1 \le x \le 2 \\ 0 & \text{otherwise} \end{cases}$$

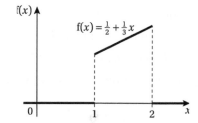

Note that a thicker line is drawn on the x-axis when $x < 1$ and when $x > 2$. This shows that f$(x) = 0$ outside the interval $1 \le x \le 2$, indicating that no student spends less than 1 hour or more than 2 hours on the task.

> In general, when X is continuous,
> ▶ the function f cannot be negative, so f$(x) \ge 0$ for all values of x
> ▶ the graph of $y = $ f(x) never goes below the x-axis

Finding probabilities

The probability that x lies in a particular interval is given by the *area* under the graph of f.

> For a continuous random variable X with probability density function f, defined for $a \le x \le b$,
> $$P(x_1 \le X \le x_2) = \int_{x_1}^{x_2} f(x)\,dx$$

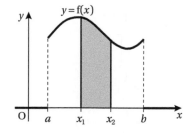

> Since $P(a \le X \le b) = 1$, the total area under the curve is 1, so
> $$\int_{\text{all } x} f(x)\,dx = \int_a^b f(x)\,dx = 1$$

Note about inequalities

When X is discrete, the probability that X is exact, such as $P(X = 3)$, has a definite value which usually is not zero, as in the Recap on page **176**.

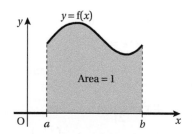

However, when X is continuous, the probability that X is exact is *always* zero. For example, a reaction time given as 3 seconds, correct to the nearest tenth of a second, could be anywhere in the interval 2.95 seconds $\le X < 3.05$ seconds. This interval becomes narrower and narrower as you try to approach the instant of time of 3 seconds and the probability becomes smaller and smaller such that the probability that the reaction time is *exactly* 3 seconds is zero, so $P(X = 3) = 0$.

When X is continuous, the probability that X takes an *exact* value is zero,

so $P(X=x) = 0$, for *any* value of x

Since it is only possible to find the probability that X lies in a *particular interval*, it does not matter whether the inequality is $<$ or $\leq$, or whether it is $>$ or $\geq$.

So, for example, it is not possible to distinguish between these probabilities:

▸ $P(X < 3)$ and $P(X \leq 3)$

▸ $P(X > 3)$ and $P(X \geq 3)$

▸ $P(2 < X < 5)$, $P(2 \leq X < 5)$, $P(2 < X \leq 5)$ and $P(2 \leq X \leq 5)$

Example 1

The continuous random variable X has probability density function f defined by

$$f(x) = \begin{cases} kx^2 & 1 \leq x \leq 4 \\ 0 & \text{otherwise} \end{cases}$$

where k is a constant.

a   Sketch the graph of f.   **b**   Find the value of k.

c   Calculate $P(2.5 \leq X \leq 3.5)$.

d   Determine   **i** $P(X = 2)$ **ii** $P(X \neq 2)$

a   The graph of $y = f(x)$ is part of a quadratic curve.

When $x = 1$, $y = k$

When $x = 4$, $y = 16k$

Remember to draw a thicker line on the x-axis to show that $f(x) = 0$ for all x-values outside the interval $1 \leq x \leq 4$.

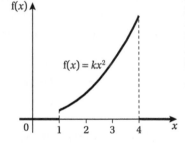

$f(x) = kx^2$

b   Total area under curve

$$= \int_1^4 kx^2 \, dx$$

$$= k\left[\frac{x^3}{3}\right]_1^4$$

$$= \frac{k}{3}[x^3]_1^4$$

$$= \frac{k}{3}(4^3 - 1^3)$$

$$= 21k$$

Since total area $= 1$

$$21k = 1$$

$$k = \frac{1}{21} \quad \leftarrow \text{Leave this as a fraction.}$$

c $f(x) = \dfrac{1}{21}x^2$

$$P(2.5 \leq X \leq 3.5) = \int_{2.5}^{3.5} \frac{1}{21}x^2 \, dx$$

$$= \frac{1}{21}\left[\frac{x^3}{3}\right]_{2.5}^{3.5}$$

$$= \frac{1}{63}[x^3]_{2.5}^{3.5}$$

$P(2.5 \leq X \leq 3.5)$ is given by this area

(continued)

(continued)

$$= \frac{1}{63}(3.5^3 - 2.5^3)$$

$$= 0.4325... = 0.433 \,(3\,\text{sf})$$

d i It is *impossible* for X to take an exact value, so $P(X = 2) = 0$

ii It is *certain* that X does not take an exact value, so $P(X \neq 2) = 1$

Example 2

The random variable W denotes the mass, in kg, of a particular chemical produced in an industrial process. The probability density function of W is defined by

$$f(w) = \begin{cases} \dfrac{1}{36}w(6 - w) & 0 \leq w \leq 6 \\ 0 & \text{otherwise} \end{cases}$$

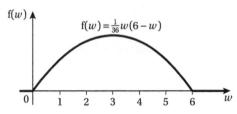

Determine the probability that more than 5 kg of the chemical is produced in the industrial process.

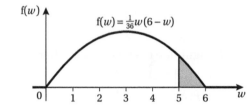

$$P(W > 5) = \int_5^6 \frac{1}{36} w(6 - w)\,dw$$

$$= \frac{1}{36}\left[3w^2 - \frac{w^3}{3} \right]_5^6$$

$$= \frac{1}{36}\left(\left(3 \times 6^2 - \frac{6^3}{3} \right) - \left(3 \times 5^2 - \frac{5^3}{3} \right) \right)$$

$$= \frac{2}{27}$$

Note that it is usually easier to substitute a limit of 0 so, in this example, this alternative method may be quicker:

$$P(W > 5) = 1 - P(W \leq 5)$$

$$= 1 - \frac{1}{36}\left[3w^2 - \frac{w^3}{3} \right]_0^5$$

$$= 1 - \frac{1}{36}\left(\left(3 \times 5^2 - \frac{5^3}{3} \right) - 0 \right)$$

$$= 1 - \frac{25}{27}$$

$$= \frac{2}{27}$$

Example 3

The error, in grams, made by a set of weighing scales may be modelled by the random variable X with probability density function defined by

$$f(x) = \begin{cases} k & -3 \leq x \leq 7 \\ 0 & \text{otherwise} \end{cases}$$

where k is a constant.

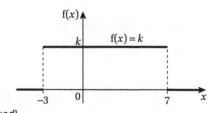

a Find the value of k.

(continued)

Continuous Random Variables

Question

b Find the probability that an error is positive.

c Given that an error is positive, find the probability that the error is less than 4 g.

d Find the probability that the **magnitude** of an error is less than 2 g.

e Find the probability that the scales are weighing exactly 1 gram underweight.

> **Note**
>
> This is a **uniform** or **rectangular distribution**.

Answer

a Total area = $10k$ ← Area of rectangle

$$10k = 1$$
$$k = 0.1$$

So $f(x) = 0.1$

b P(error is positive) = P($X > 0$)

$$= 0.1 \times 7$$
$$= 0.7$$

c P($X < 4 \mid X > 0$) = $\dfrac{\text{P}(0 < X < 4)}{\text{P}(X > 0)}$ See AS Textbook Section 13.3

$$= \frac{0.4}{0.7}$$

$$= \frac{4}{7}$$

d P(magnitude of error is more than 2 g)

$$= \text{P}(|X| > 2)$$
$$= \text{P}(X < -2) + \text{P}(X > 2)$$
$$= 0.1 \times 1 + 0.1 \times 5$$
$$= 0.6$$

e Since P($X = x$) = 0 for all values of x,

P(exactly 1 gram underweight) = 0

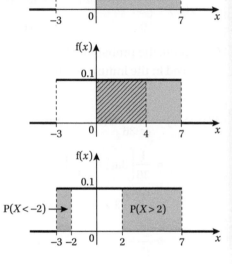

The probability density function may be given in more than one part.

Example 4

Question

The continuous random variable T has probability density function defined by

$$f(t) = \begin{cases} kt & 0 \le t < 1 \\ kt(2 - t) & 1 \le t \le 2 \\ 0 & \text{otherwise} \end{cases}$$

where k is a positive constant.

a Find the value of k.

b Draw a sketch of f.

c Find P($T > 1$).

Answer

a Since the total area is 1,

$$1 = \int_0^1 kt \, dt + \int_1^2 kt(2 - t) dt$$

$$= k \int_0^1 t \, dt + k \int_1^2 (2t - t^2) dt$$

(continued)

(continued)

$$= k\left[\frac{t^2}{2}\right]_0^1 + k\left[t^2 - \frac{t^3}{3}\right]_1^2$$

$$= k\left(\frac{1^2}{2} - 0\right) + k\left(\left(2^2 - \frac{2^3}{3}\right) - \left(1^2 - \frac{1^3}{3}\right)\right)$$

$$= \frac{7}{6}k$$

So $k = \frac{6}{7}$ ← Leave this as a fraction.

b The first section, $f(t) = \frac{6}{7}t$, is a straight line.

When $t = 1$, $f(1) = \frac{6}{7} \times 1 = \frac{6}{7}$

The second section, $f(t) = \frac{6}{7}t(2 - t)$, is part of a parabola.

When $t = 1$, $f(1) = \frac{6}{7} \times 1 \times (2 - 1) = \frac{6}{7}$

When $t = 2$, $f(2) = 0$

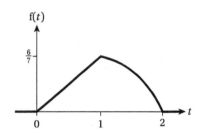

c To find $P(T > 1)$, you could evaluate $\int_1^2 \frac{6}{7}t(2 - t)\,dt$.

However, it is quicker to use the fact that the total area is 1.

$P(T > 1) = 1 - P(T < 1)$

$$= 1 - \frac{1}{2} \times 1 \times \frac{6}{7}$$

$$= \frac{4}{7}$$

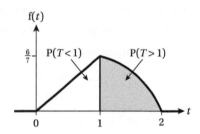

Example 5

The continuous random variable X has probability density function defined by

$$f(x) = \begin{cases} \dfrac{1}{8}(x + 2)^2 & -2 \le x < 0 \\ \dfrac{1}{2} & 0 \le x \le 1\dfrac{1}{3} \\ 0 & \text{otherwise} \end{cases}$$

Find

a $P(X > 1)$

b $P(-1 \le X \le 1)$

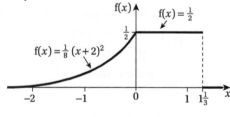

a $P(X > 1)$ can be calculated using geometry.

From the diagram,

$P(X > 1)$ = area of shaded rectangle

$$= \frac{1}{3} \times \frac{1}{2} = \frac{1}{6}$$

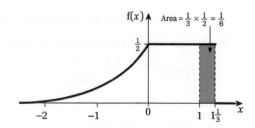

(continued)

(continued)

b $P(-1 \le X \le 1)$ is given by the shaded area.

It must be found in two stages.

$$P(-1 \le X \le 0) = \frac{1}{8}\int_{-1}^{0}(x+2)^2\,dx$$

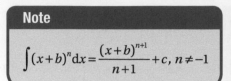

Note

$$\int(x+b)^n\,dx = \frac{(x+b)^{n+1}}{n+1}+c,\; n \ne -1$$

$$= \frac{1}{8}\left[\frac{(x+2)^3}{3}\right]_{-1}^{0}$$

$$= \frac{1}{8}\left(\frac{2^3}{3}-\frac{1^3}{3}\right)$$

$$= \frac{7}{24}$$

$P(0 \le X \le 1)$ = area of rectangle

$$= 1 \times \frac{1}{2} = \frac{1}{2}$$

So, $P(-1 \le X \le 1) = \frac{7}{24}+\frac{1}{2}=\frac{19}{24}$ ← Add the two probabilities

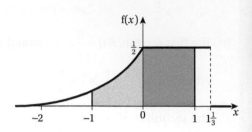

Exercise 1

If you are not given a sketch of the probability density function, it may be helpful to draw one even when it is not requested.

1 The continuous random variable X has probability density function defined by

$$f(x)=\begin{cases} kx^2 & 0 \le x \le 2 \\ 0 & \text{otherwise} \end{cases}$$

where k is a constant.

Determine

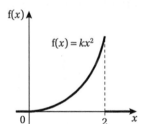

a the value of k **b** $P(X \ge 1)$

c $P(0.5 \le X \le 1.5)$ **d** $P(X = 1.3)$.

2 The continuous random variable X has probability density function defined by

$$f(x)=\begin{cases} k & -2 \le x \le 3 \\ 0 & \text{otherwise} \end{cases}$$

where k is a constant.

a Find

   i the value of k **ii** $P(-1.6 \le X \le 2.1)$

   iii $P(-2.5 < X < 2.5)$, i.e. $P(|X| < 2.5)$.

b **i** Find $P(X > 1)$.

   ii Given than X is greater than 1, find the probability that X is less than 1.5.

3 The continuous random variable X has probability density function defined by

$$f(x)=\begin{cases} kx & 0 \le x \le 4 \\ 0 & \text{otherwise} \end{cases}$$

where k is a constant.

a Find the value of k.   **b** Find $P(1 \le X \le 3)$.

c State the value of $P(X \ne 2.4)$.

4 The delay, in hours, of a flight from Chicago can be modelled by the continuous random variable T with probability density function defined by

$$f(t) = \begin{cases} k(10 - t) & 0 \le t \le 10 \\ 0 & \text{otherwise} \end{cases}$$

a Find the value of k.

b Find the probability that the delay will be

   i less than 4 hours

   ii between 2 hours and 6 hours

   iii exactly 3 hours.

5 The continuous random variable X has probability density function defined by

$$f(x) = \begin{cases} k(x+2)^2 & 0 \le x \le 2 \\ 0 & \text{otherwise} \end{cases}$$

where k is a constant.

a Find the value of k.

b i Find $P(0 \le X \le 1)$.   **ii** Hence find $P(X > 1)$.

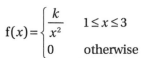

6 The continuous random variable Y has probability density function defined by

$$f(y) = \begin{cases} k\sqrt{y} & 1 \le y \le 9 \\ 0 & \text{otherwise} \end{cases}$$

where k is a constant.

a Find the value of k.   **b** Find $P(4 \le Y \le 9)$.

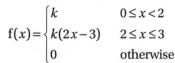

7 The continuous random variable X has probability density function defined by

$$f(x) = \begin{cases} \dfrac{k}{x^2} & 1 \le x \le 3 \\ 0 & \text{otherwise} \end{cases}$$

where k is a constant.

a Find the value of k.   **b** Find $P(X \le 2)$.

8 The continuous random variable x has probability density function defined by

$$f(x) = \begin{cases} k & 0 \le x < 2 \\ k(2x - 3) & 2 \le x \le 3 \\ 0 & \text{otherwise} \end{cases}$$

where k is a positive constant.

a Find the value of k.   **b** Find $P(X < 1)$.

c Find $P(X > 2.5)$.   **d** Find $P(1 \le X \le 2.3)$.

9 The continuous random variable X has probability density function defined by

$$f(x) = \begin{cases} \dfrac{6}{5}(x+1) & -1 \le x < 0 \\ \dfrac{6}{5}(x-1)^2 & 0 \le x \le 1 \\ 0 & \text{otherwise} \end{cases}$$

Find

a $P(X < 0)$ **b** $P(X > 0.3)$ **c** $P(X = 1)$

d $P(-0.5 \le X \le 1)$ **e** $P(X \ne 0)$

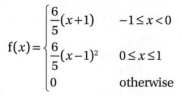

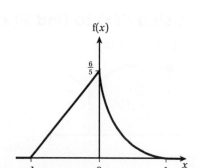

Continuous Random Variables 183

10 Explain, with a reason, whether each of these functions could be the probability density function of a continuous random variable X.

a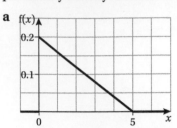

b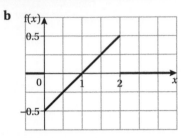

c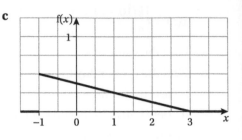

11.2 Cumulative distribution function

When finding several probabilities relating to a particular probability density function, the process of integrating $f(x)$ is the same each time, but with different limits. This can be made quicker by setting up a function which gives the area under the graph as far as a general x-value. You then need only to substitute the particular value required each time.

This function is called the **cumulative distribution function**, or sometimes just the **distribution function**, and it is denoted by F, where

$$F(x) = P(X \le x) = P(X < x)$$

$F(x)$ is obtained from $f(x)$ by integrating, using a lower limit of $-\infty$ and an upper limit of x. However, to avoid confusion between x (the upper limit) and x (the variable of the integration), the variable in the definite integral is changed from x to another letter, say t. This does not affect the evaluation of the integral and will give F as a function of x.

> **When X is continuous, the cumulative distribution function F is given by**
> $$F(x) = \int_{-\infty}^{x} f(t)\,dt$$

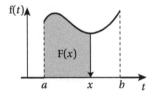

In practice, if the probability density function f is defined for $a \le x \le b$,

▶ the lower limit is taken to be a, where
$$F(a) = P(X \le a) = 0$$

▶ the upper limit is taken to be b and, since the total area is 1,
$$F(b) = P(X \le b) = 1$$

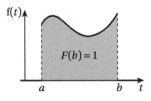

Using F(x) to find P(x₁ ≤ X ≤ x₂)

$P(X \le x_2) = F(x_2)$

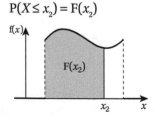

$P(X \le x_1) = F(x_1)$

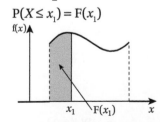

$P(x_1 \le X \le x_2) = F(x_2) - F(x_1)$

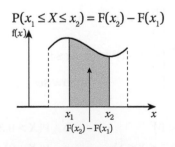

Using F(x) to find the median and quartiles

For a continuous random variable X:

▶ The median, m, is the value which divides the area under the graph of the probability density function $f(x)$ in half, so

$$P(X \leq m) = F(m) = 0.5$$

▶ The quartiles are the values which, together with the median, divide the area under the graph of $f(x)$ into quarters.

Denoting the lower quartile by q_1 and the upper quartile by q_3

$$P(X \leq q_1) = F(q_1) = 0.25$$

$$P(X \leq q_3) = F(q_3) = 0.75$$

The interquartile range is found from the quartiles, where

Interquartile range = $q_3 - q_1$

See AS Textbook Section 14.1

Example 6

The continuous random variable X has cumulative distribution function F defined by

$$F(x) = \begin{cases} 0 & x \leq 1 \\ \frac{1}{6}(x+4)(x-1) & 1 \leq x \leq 2 \\ 1 & x \geq 2 \end{cases}$$

a Find

  i $P(X < 1.6)$

  ii $P(X > 1.8)$

  iii $P(1.6 < X < 1.8)$

b **i** Show that the median, m, satisfies the equation $m^2 + 3m - 7 = 0$.

  ii Hence find the median.

a **i** Substitute $x = 1.6$ into $F(x)$:

$$P(X < 1.6) = F(1.6) = \frac{1}{6}(1.6 + 4)(1.6 - 1) = 0.56$$

  ii $P(X > 1.8) = 1 - F(1.8)$

$$= 1 - \frac{1}{6}(1.8 + 4)(1.8 - 1)$$

$$= 1 - 0.7733...$$

$$= 0.2266... = 0.227 \ (3 \text{ sf})$$

  iii $P(1.6 < X < 1.8) = F(1.8) - F(1.6)$

$$= 0.7733... - 0.56$$

$$= 0.2133... = 0.213 \ (3 \text{ sf})$$

b **i** Since $F(m) = 0.5$,

$$\frac{1}{6}(m + 4)(m - 1) = 0.5$$

Multiply throughout by 6

$$(m + 4)(m - 1) = 3$$

$$m^2 + 3m - 4 = 3$$

$$m^2 + 3m - 7 = 0$$

(continued)

Continuous Random Variables 185

ii Using the quadratic formula:

$$m = \frac{-3 \pm \sqrt{3^3 - 4 \times 1 \times (-7)}}{2} = \frac{-3 \pm \sqrt{37}}{2}$$

so $m = \frac{-3 + \sqrt{37}}{2} = 1.541...$ or $m = \frac{-3 - \sqrt{37}}{2} = -4.541...$ ← Reject this value, since m lies between 1 and 2.

So the median of X is 1.54 (3 sf)

Example 7

The continuous random variable X has probability density function given by

$$f(x) = \begin{cases} \dfrac{1}{8}x & 0 \le x \le 4 \\ 0 & \text{otherwise} \end{cases}$$

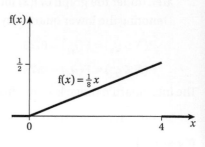

a Find the cumulative distribution function F(x) and illustrate it on a sketch.

b Use F(x) to determine

  i P($X < 2$)   **ii** P($X > 3$)   **iii** P($0.3 \le X \le 1.8$)

c Find the interquartile range.

a Since f(x) is defined for $0 \le x \le 4$:

When $x < 0$, F(x) = 0 and when $x > 4$, F(x) = 1

When $0 \le x \le 4$,

$$F(x) = \int_0^x f(t)\,dt \qquad \leftarrow \text{ Use } f(t) = \frac{1}{8}t$$

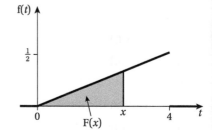

$$= \int_0^x \frac{1}{8}t\,dt$$

$$= \frac{1}{8}\left[\frac{t^2}{2}\right]_0^x$$

$$= \frac{1}{16}(x^2 - 0)$$

$$= \frac{1}{16}x^2$$

So F(x) = $\begin{cases} 0 & x < 0 \\ \dfrac{1}{16}x^2 & 0 \le x \le 4 \\ 1 & x > 4 \end{cases}$

Check: $F(4) = \dfrac{1}{16} \times 4^2 = 1$, as expected.

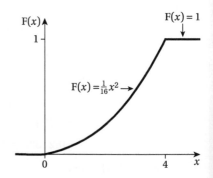

b **i** P($X < 2$) = F(2)

$$= \frac{1}{16} \times 2^2$$

$$= 0.25$$

ii P($X > 3$) = 1 − F(3)

$$= 1 - \frac{1}{16} \times 3^2$$

$$= 0.4375$$

(continued)

(continued)

iii $P(0.3 \leq X \leq 1.8)$

$$= F(1.8) - F(0.3)$$

$$= \frac{1}{16} \times 1.8^2 - \frac{1}{16} \times 0.3^2$$

$$= 0.1968... = 0.197 \text{ (3 sf)}$$

c Let the lower quartile be q_1 and the upper quartile q_3.

By definition, $F(q_1) = 0.25$

From part **b i**, $F(2) = 0.25$, so $q_1 = 2$

By definition, $F(q_3) = 0.75$

so $\quad \dfrac{1}{16} q_3^2 = 0.75$

$$q_3^2 = 12$$

$$q_3 = \sqrt{12}$$

Interquartile range $= q_3 - q_1$

$$= \sqrt{12} - 2$$

$$= 1.464... = 1.46 \text{ (3 sf)}$$

Example 8

The continuous random variable T has probability density function given by

$$f(t) = \begin{cases} \dfrac{4}{15}t^3 & 1 \leq t \leq 2 \\ 0 & \text{otherwise} \end{cases}$$

Find the cumulative distribution function $F(t)$.

Since $f(t)$ is defined for $1 \leq t \leq 2$:

When $t < 1$, $F(t) = 0$ and when $t > 2$, $F(t) = 1$

When $1 \leq t \leq 2$,

$$F(t) = \int_1^t f(y)\,dy \qquad \leftarrow \text{Use } f(y) = \frac{4}{15}y^3$$

$$= \int_1^t \frac{4}{15} y^3 \, dy$$

$$= \frac{4}{15}\left[\frac{y^4}{4}\right]_1^t$$

$$= \frac{1}{15}(t^4 - 1^4)$$

$$= \frac{1}{15}(t^4 - 1)$$

Note

Since the variable is T, you need to use a different letter in the integration when finding $F(t)$.

Now summarise fully

$$F(t) = \begin{cases} 0 & t < 1 \\ \dfrac{1}{15}(t^4 - 1) & 1 \leq t \leq 2 \\ 1 & t > 2 \end{cases} \qquad \leftarrow \text{Check: } F(2) = \frac{1}{15}(2^4 - 1) = 1, \text{ as expected.}$$

Continuous Random Variables

Take special care when f(x) is in more than one part.

Example 9

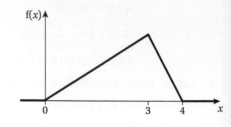

The continuous random variable X has probability density function defined by

$$f(x) = \begin{cases} \dfrac{1}{6}x & 0 \le x \le 3 \\ \dfrac{1}{2}(4-x) & 3 \le x \le 4 \\ 0 & \text{otherwise} \end{cases}$$

a Find the cumulative distribution function F(x) and illustrate it in a sketch.

b Use F(x) to find $P\left(\dfrac{1}{2} < X < 3\dfrac{1}{2}\right)$.

a Since f(x) is defined for values between 0 and 4,

F(x) = 0 when $x \le 0$ and F(x) = 1 when $x \ge 4$

For the rest of the function, you need to find F(x) in stages.

First consider x where $0 \le x \le 3$.

$$F(x) = \int_0^x \frac{1}{6}t \, dt \qquad \leftarrow \text{Use } f(t) = \frac{1}{6}t$$

$$= \frac{1}{6}\left[\frac{t^2}{2}\right]_0^x$$

$$= \frac{1}{12}\left[t^2\right]_0^x$$

$$= \frac{1}{12}(x^2 - 0)$$

$$= \frac{1}{12}x^2$$

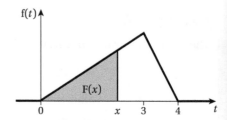

Note that $F(3) = \dfrac{1}{12} \times 3^2 = \dfrac{3}{4}$

Now consider x where $3 \le x \le 4$.

It is very important to include F(3) here.

$$F(x) = F(3) + \int_3^x \frac{1}{2}(4-t)\,dt \qquad \leftarrow \text{Use } f(t) = \frac{1}{2}(4-t)$$

$$= \frac{3}{4} + \frac{1}{2}\left[4t - \frac{t^2}{2}\right]_3^x$$

$$= \frac{3}{4} + \frac{1}{2}\left(\left(4x - \frac{1}{2}x^2\right) - \left(4 \times 3 - \frac{3^2}{2}\right)\right)$$

$$= \frac{3}{4} + 2x - \frac{1}{4}x^2 - 6 + \frac{9}{4}$$

$$= -\frac{1}{4}x^2 + 2x - 3$$

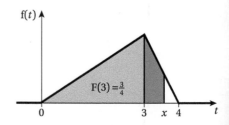

Make sure that F(4) = 1: $F(4) = -\frac{1}{4} \times 4^2 + 2 \times 4 - 3 = 1$, as expected

(continued)

(continued)

The cumulative distribution function $F(x)$ is summarised as follows:

$$F(x) = \begin{cases} 0 & x \leq 0 \\ \dfrac{1}{12}x^2 & 0 \leq x \leq 3 \\ -\dfrac{1}{4}x^2 + 2x - 3 & 3 \leq x \leq 4 \\ 1 & x \geq 4 \end{cases}$$

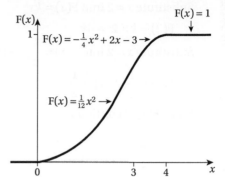

b To find $F\left(\dfrac{1}{2}\right)$, since $\dfrac{1}{2}$ is in the interval $0 \leq x \leq 3$, use $F(x) = \dfrac{1}{12}x^2$

$$F\left(\frac{1}{2}\right) = \frac{1}{12} \times \left(\frac{1}{2}\right)^2 = \frac{1}{48}$$

To find $F\left(3\dfrac{1}{2}\right)$, since $3\dfrac{1}{2}$ is in the interval $3 \leq x \leq 4$, use $F(x) = -\dfrac{1}{4}x^2 + 2x - 3$

$$F\left(3\frac{1}{2}\right) = -\frac{1}{4} \times \left(3\frac{1}{2}\right)^2 + 2 \times 3\frac{1}{2} - 3 = \frac{15}{16}$$

$$P\left(\frac{1}{2} < X < 3\frac{1}{2}\right) = F\left(3\frac{1}{2}\right) - F\left(\frac{1}{2}\right)$$

$$= \frac{15}{16} - \frac{1}{48}$$

$$= \frac{11}{12}$$

The graph of the cumulative distribution function F

You can see from the sketches of $F(x)$ in Example 7 and Example 9 that the shape of the graph of the cumulative distribution function is similar to that of a cumulative frequency curve which you may have studied in earlier work.

> In general, when $f(x)$ is defined for $a \leq x \leq b$, the cumulative distribution function F is a continuous, increasing function.

Note that when $x \geq b$, the graph is the horizontal line $F(x) = 1$.

Example 10

The continuous random variable X has cumulative distribution function defined by

$$F(x) = \begin{cases} 0 & x < 0 \\ kx^2 & 0 \leq x \leq 2 \\ \dfrac{1}{2}(x-1) & 2 \leq x \leq 3 \\ 1 & x > 3 \end{cases}$$

a Find the value of the constant k.

b Find

   i $P(X < 1.2)$ **ii** $P(X < 2.4)$ **iii** $P(1.2 < X < 2.4)$

c Show that the median is 2.

a To find k, consider F(2).

Substitute $x = 2$ into $F(x) = kx^2$

$$F(2) = k \times 2^2 = 4k$$

Substitute $x = 2$ into $F(x) = \frac{1}{2}(x-1)$

$$F(2) = \frac{1}{2}(2-1) = \frac{1}{2}$$

Since $F(x)$ is continuous,

$$4k = \frac{1}{2}$$

$$k = \frac{1}{8}$$

Therefore, when $0 \le x \le 2$, $F(x) = \frac{1}{8} x^2$

Note

You will get the same value for F(2) using $F(x) = kx^2$ or $F(x) = \frac{1}{2}(x-1)$.

b i Since 1.2 is in the interval $0 \le x \le 2$, use $F(x) = \frac{1}{8} x^2$

$$P(X < 1.2) = F(1.2)$$

$$= \frac{1}{8} \times 1.2^2$$

$$= 0.18$$

ii Since 2.4 is in the interval $2 \le x \le 3$, use $F(x) = \frac{1}{2}(x-1)$

$$P(X < 2.4) = F(2.4)$$

$$= \frac{1}{2}(2.4-1)$$

$$= 0.7$$

iii $P(1.2 < X < 2.4) = F(2.4) - F(1.2)$

$$= 0.7 - 0.18$$

$$= 0.52$$

c If m is the median, then $F(m) = \frac{1}{2}$.

You found in part **a** that $F(2) = \frac{1}{2}$, so the median, m is 2.

Relationship between F(x) and f(x)

Since

$$F(x) = \int_{-\infty}^{x} f(t)\, dt$$

it follows that

$$f(x) = \frac{d}{dx} F(x) = F'(x)$$

Note that $f(x)$ is given by the **gradient** of the $F(x)$ graph.

In particular, if $F(x) = k$, where k is a constant, then $f(x) = 0$.

Example 11

The continuous random variable X has cumulative distribution function $F(x)$ where

$$F(x) = \begin{cases} 0 & x < 0 \\ \dfrac{1}{27}x^3 & 0 \le x \le 3 \\ 1 & x > 3 \end{cases}$$

Find the probability density function of X.

When $x < 0$ and $x > 3$,

F(x) is constant, so $f(x) = 0$.

When $0 \le x \le 3$,

$$f(x) = \frac{d}{dx}F(x)$$

$$= \frac{d}{dx}\left(\frac{1}{27}x^3\right)$$

$$= \frac{1}{9}x^2$$

The probability density function of X is

$$f(x) = \begin{cases} \dfrac{1}{9}x^2 & 0 \le x \le 3 \\ 0 & \text{otherwise} \end{cases}$$

Example 12

The continuous random variable X has cumulative distribution function $F(x)$ where

$$F(x) = \begin{cases} 0 & x < -2 \\ \dfrac{1}{12}(2+x) & -2 \le x \le 0 \\ \dfrac{1}{6}(1+x) & 0 \le x \le 4 \\ \dfrac{1}{12}(6+x) & 4 \le x \le 6 \\ 1 & x > 6 \end{cases}$$

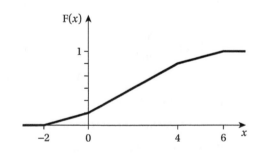

Find and sketch f, the probability distribution function of X.

When $x < -2$ and when $x > 6$,

$f(x) = 0$

When $-2 \le x < 0$,

$$f(x) = \frac{d}{dx}\left(\frac{1}{12}(2+x)\right) = \frac{1}{12}$$

When $0 \le x < 4$,

$$f(x) = \frac{d}{dx}\left(\frac{1}{6}(1+x)\right) = \frac{1}{6}$$

(continued)

Continuous Random Variables

(continued)

When $4 \leq x < 6$,

$$f(x) = \frac{d}{dx}\left(\frac{1}{12}(6 + x)\right) = \frac{1}{12}$$

So, the probability density function of X is

$$f(x) = \begin{cases} \dfrac{1}{12} & -2 \leq x < 0 \\[2mm] \dfrac{1}{6} & 0 \leq x < 4 \\[2mm] \dfrac{1}{12} & 4 \leq x < 6 \\[2mm] 0 & \text{otherwise} \end{cases}$$

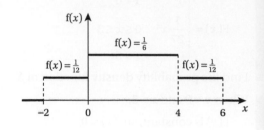

Exercise 2

Remember to check whether you are given $f(x)$ or $F(x)$.

1 The continuous random variable X has cumulative distribution function given by

$$F(x) = \begin{cases} 0 & x < 0 \\[2mm] \dfrac{1}{125}x^3 & 0 \leq x \leq 5 \\[2mm] 1 & x > 5 \end{cases}$$

Find

a $F(4)$

b $P(X > 4)$

c $P(1 < X < 3)$

d the median of X

e the probability density function $f(x)$.

2 The continuous random variable X has cumulative distribution function given by

$$F(x) = \begin{cases} 0 & x < -3 \\[2mm] \dfrac{1}{36}(3 + x)^2 & -3 \leq x \leq 3 \\[2mm] 1 & x > 3 \end{cases}$$

a Find $P(0 < X < 2)$

b Find $P(X > 1)$

c Find the lower quartile of X.

3 The continuous random variable T has cumulative distribution function given by

$$F(t) = \begin{cases} 0 & t < 0 \\ t(2 - t) & 0 \leq t \leq 1 \\ 1 & t > 1 \end{cases}$$

a Find

i $P(T < 0.2)$

ii $P(T > 0.4)$

iii $P(0.1 < T < 0.8)$

iv $P(T < 2)$

b **i** If the upper quartile is denoted by q, show that $4q^2 - 8q + 3 = 0$.

ii Hence find the upper quartile.

c Find the probability density function, $f(t)$.

4 The continuous random variable X has cumulative distribution function $F(x)$ where

$$F(x) = \begin{cases} 0 & x < 0 \\ \dfrac{2}{3}x & 0 \le x \le 1 \\ \dfrac{1}{3}(x + c) & 1 \le x \le 2 \\ 1 & x > 2 \end{cases}$$

a Find the value of c.

b Find $P(0.6 < X < 1.2)$.

c Find the median and interquartile range.

d Find the probability density function $f(x)$.

5 The continuous random variable X has probability density function given by

$$f(x) = \begin{cases} \dfrac{3}{8}x^2 & 0 \le x \le 2 \\ 0 & \text{otherwise} \end{cases}$$

a Find the cumulative distribution function $F(x)$.

b Using $F(x)$, find

  i $P(X < 1)$

  ii $P(1 < X < 1.3)$

  iii the value of c if $P(X \le c) = 0.7$.

6 The continuous random variable X has probability density function given by

$$f(x) = \begin{cases} \dfrac{1}{4} & 1 \le x \le 5 \\ 0 & \text{otherwise} \end{cases}$$

a Sketch the graph of f.

b Find the cumulative distribution function $F(x)$.

c Sketch the graph of F.

d Find $P(2 < X < 4)$.

7 The continuous random variable T has probability density function given by

$$f(t) = \begin{cases} 1 - \dfrac{1}{4}t & 1 \le t \le 3 \\ 0 & \text{otherwise} \end{cases}$$

a Find the cumulative distribution function $F(t)$.

b Find $P(1.5 \le T \le 2)$.

c Find the median.

d Given that T is greater than the median, find the probability that T is less than the upper quartile.

8 The continuous random variable Y has probability density function given by

$$f(y) = \begin{cases} \dfrac{32}{3y^3} & 2 \le y \le 4 \\ 0 & \text{otherwise} \end{cases}$$

a Find the cumulative distribution function F(y).

b Find the median of Y.

9 The continuous random variable T has probability density function given by

$$f(t) = \begin{cases} \dfrac{5}{6}(t^4 + 1) & 0 \le t \le 1 \\ 0 & \text{otherwise} \end{cases}$$

a Find the cumulative distribution function F(t).

b Find P($0.2 < T < 0.7$).

10 The continuous random variable X has probability density function defined by

$$f(x) = \begin{cases} \dfrac{1}{3}x & 0 \le x < 2 \\ -\dfrac{2}{3}x + 2 & 2 \le x \le 3 \\ 0 & \text{otherwise} \end{cases}$$

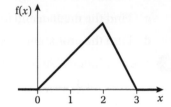

a Find the cumulative distribution function F(x).

b Sketch the graph of F.

c Find

   i P($X < 1$)   **ii** P($0.3 < X < 0.9$)

   iii P($X > 2.4$)   **iv** P($1 < X < 2.4$)

   v the median.

11 The continuous random variable X has probability density function defined by

$$f(x) = \begin{cases} \dfrac{1}{4} & 0 \le x \le 2 \\ \dfrac{1}{4}(2x - 3) & 2 \le x \le 3 \\ 0 & \text{otherwise} \end{cases}$$

a Find the cumulative distribution function F(x).

b Find

   i P($X < 1$)   **ii** P($1 < X < 2.7$)

   iii the median   **iv** P($X = 2$).

12 The continuous random variable X has cumulative distribution function given by

$$F(x) = \begin{cases} 0 & x < 0 \\ x(1 - kx) & 0 \le x \le 2 \\ 1 & x \ge 2 \end{cases}$$

a **i** Find the value of k.

   ii Sketch the graph of F.

b **i** Find the probability density function f(x).

   ii Sketch the graph of f.

13 The continuous random variable X has cumulative distribution function $F(x)$ where

$$F(x) = \begin{cases} 0 & x < 0 \\ \dfrac{1}{2}x^2 & 0 \le x \le 1 \\ 2x - \dfrac{1}{2}x^2 - a & 1 \le x \le 2 \\ 1 & x \ge 2 \end{cases}$$

a Find the value of a.

b Find the probability density function $f(x)$ and draw a sketch of f.

c State the median of X.

14 The continuous random variable X has probability density function defined by

$$f(x) = \begin{cases} \dfrac{1}{5}(2x + 1) & 0 \le x \le 1 \\ \dfrac{1}{15}(4 - x)^2 & 1 \le x \le 4 \\ 0 & \text{otherwise} \end{cases}$$

a Sketch the graph of f.

b **i** Show that the cumulative distribution function, $F(x)$, for $0 \le x \le 1$ is

$$F(x) = \frac{1}{5}x(x + 1)$$

ii Hence write down the value of $P(X \le 1)$.

iii Find the value of x for which $P(X \ge x) = \dfrac{17}{20}$.

iv Find the lower quartile of the distribution.

AQA MS2B June 2006

11.3 Expectation of *X*

The expected value of X, known as the expectation or mean, is written $E(X)$ and denoted by μ.

> For a continuous random variable X with probability density function $f(x)$, the *expectation*, or mean, of X is given by
> $$E(X) = \mu = \int_{\text{all } x} x f(x) \, dx$$

Note about symmetry

If the graph of f has a line of symmetry, you can state the value of μ straight away.

This is the case with the random variable W defined in Example 2, where

$$f(w) = \begin{cases} \dfrac{1}{36} w(6 - w) & 0 \le w \le 6 \\ 0 & \text{otherwise} \end{cases}$$

Since the graph is part of a quadratic curve, it is symmetrical about $w = 3$, so you can see straight away that the mean is 3.

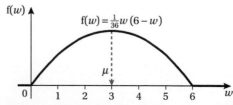

Example 13

Question

The random variable X has probability density function defined by

$$f(x) = \begin{cases} \dfrac{1}{18}(6 - x) & 0 \le x \le 6 \\ 0 & \text{otherwise} \end{cases}$$

Find $E(X)$.

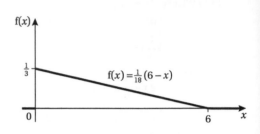

Answer

$$E(X) = \int_{\text{all } x} x\, f(x)\, dx$$

$$= \int_0^6 x \times \frac{1}{18}(6 - x)\, dx. \qquad \leftarrow \text{Multiply } x \text{ by } f(x) \text{ before integrating.}$$

$$= \frac{1}{18} \int_0^6 (6x - x^2)\, dx$$

$$= \frac{1}{18}\left[3x^2 - \frac{x^3}{3} \right]_0^6$$

$$= \frac{1}{18}\left(\left(3 \times 6^2 - \frac{6^3}{3} \right) - 0 \right)$$

$$= 2$$

Example 14

Question

The random variable X has probability density function given by

$$f(x) = \begin{cases} \dfrac{1}{9}x^2 & 0 \le x \le 3 \\ 0 & \text{otherwise} \end{cases}$$

The mean of X is μ and the median of X is m.

a Find μ.

b Find $P(X \le \mu)$.

c Find the probability that X lies between μ and m.

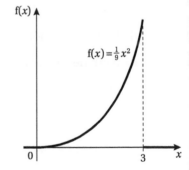

Answer

a $$\mu = \int_0^3 x \times \frac{1}{9}x^2\, dx$$

$$= \frac{1}{9} \int_0^3 x^3\, dx$$

$$= \frac{1}{9}\left[\frac{x^4}{4} \right]_0^3$$

$$= \frac{1}{36}(3^4 - 0)$$

$$= 2.25$$

(continued)

(continued)

b $P(X \le \mu) = P(X \le 2.25)$

$$= \int_0^{2.25} \frac{1}{9}x^2 \, dx$$

$$= \frac{1}{9}\left[\frac{x^3}{3}\right]_0^{2.25}$$

$$= \frac{1}{27}(2.25^3 - 0)$$

$$= 0.421875 = 0.422 \text{ (3 sf)}$$

c $P(X \le m) = 0.5$ and $P(X \le \mu) = 0.421875$, so $\mu < m$.

$P(\mu \le X \le m)$

$\quad = 0.5 - 0.421875$

$\quad = 0.078125 = 0.0781 \text{ (3 sf)}$

When $f(x)$ is given in more than one part, the integration is done in sections also.

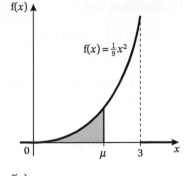

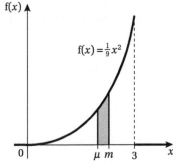

Example 15

Find the mean of the continuous random variable T with probability density function given by

$$f(t) = \begin{cases} \dfrac{1}{4} & 0 \le t \le 2 \\[2mm] \dfrac{1}{4}(2t - 3) & 2 \le t \le 3 \\[2mm] 0 & \text{otherwise} \end{cases}$$

$$E(T) = \int_0^2 \left(t \times \frac{1}{4}\right) dt + \int_2^3 t \times \frac{1}{4}(2t - 3) \, dt$$

$$= \frac{1}{4}\int_0^2 t \, dt + \frac{1}{4}\int_2^3 (2t^2 - 3t) \, dt$$

$$= \frac{1}{4}\left[\frac{t^2}{2}\right]_0^2 + \frac{1}{4}\left[\frac{2t^3}{3} - \frac{3t^2}{2}\right]_2^3$$

$$= \frac{1}{4}\left(\frac{2^2}{2} - 0\right) + \frac{1}{4}\left(\left(\frac{2 \times 3^3}{3} - \frac{3 \times 3^2}{2}\right) - \left(\frac{2 \times 2^3}{3} - \frac{3 \times 2^2}{2}\right)\right)$$

$$= 1\frac{19}{24}$$

Expectation of g(X)

If $g(x)$ is any function of the continuous random variable X having probability density function $f(x)$, then

$$E(g(X)) = \int_{\text{all } x} g(x)f(x) \, dx$$

In particular, if $g(x) = x^2$,

$$E(X^2) = \int_{all\ x} x^2 f(x)\,dx$$

As with discrete random variables, the following results hold when X is continuous and a and b are constants:

$E(a) = a$	For example	$E(4) = 4$
$E(aX) = aE(X)$		$E(3X) = 3E(X)$
$E(aX + b) = aE(X) + b$		$E(10X + 4) = 10E(X) + 4$

These results can be extended to other simple functions of X, such as

$$E(3X^2 + 6X - 2) = 3E(X^2) + 6E(X) - 2$$

$$E(4X^{-2} + 1) = 4E(X^{-2}) + 1 = 4E\left(\frac{1}{X^2}\right) + 1$$

Example 16

Question

The continuous random variable X has probability density function given by

$$f(x) = \begin{cases} \dfrac{3}{8}x^2 & 0 \le x \le 2 \\ 0 & \text{otherwise} \end{cases}$$

Find

a i $E(X^2)$ **ii** $E(6X^2 + 2)$

b i $E\left(\dfrac{1}{X}\right)$ **ii** $E\left(\dfrac{5}{X} - 3\right)$

Answer

a i $E(X^2) = \displaystyle\int_{all\ x} x^2 f(x)\,dx$

$= \displaystyle\int_0^2 \left(x^2 \times \frac{3}{8}x^2\right) dx$

$= \dfrac{3}{8}\displaystyle\int_0^2 x^4\,dx$

$= \dfrac{3}{8}\left[\dfrac{x^5}{5}\right]_0^2$

$= \dfrac{3}{40}(2^5 - 0)$

$= 2.4$

b i $E\left(\dfrac{1}{X}\right) = \displaystyle\int_{all\ x} \frac{1}{x}f(x)\,dx$

$= \displaystyle\int_0^2 \left(\frac{1}{x} \times \frac{3}{8}x^2\right) dx$

$= \dfrac{3}{8}\displaystyle\int_0^2 x\,dx$

$= \dfrac{3}{8}\left[\dfrac{x^2}{2}\right]_0^2$

$= \dfrac{3}{16}(2^2 - 0)$

$= 0.75$

ii $E(6X^2 + 2) = 6E(X^2) + 2$

$= 6 \times 2.4 + 2$

$= 16.4$

ii $E\left(\dfrac{5}{X} - 3\right) = 5\,E\left(\dfrac{1}{X}\right) - 3$

$= 5 \times 0.75 - 3$

$= 0.75$

Exercise 3

1 The continuous random variable X has probability density function given by

$$f(x) = \begin{cases} \dfrac{1}{16}x & 2 \le x \le 6 \\ 0 & \text{otherwise} \end{cases}$$

Find

a $E(X)$ **b** $E(3X - 2)$

2 The continuous random variable X has probability density function given by

$$f(x) = \begin{cases} \dfrac{1}{20}(x + 3) & 0 \le x \le 4 \\ 0 & \text{otherwise} \end{cases}$$

Find

a $E(X)$ **b** $E(X^2)$ **c** $E(3X^2 - 5X + 2)$

3 The continuous random variable T has probability density function given by

$$f(t) = \begin{cases} \dfrac{3}{4}(t^2 + 1) & 0 \le t \le 1 \\ 0 & \text{otherwise} \end{cases}$$

Find the mean of T.

4 The continuous random variable X has probability density function given by

$$f(x) = \begin{cases} \dfrac{3}{8}(1 + x^2) & -1 \le x \le 1 \\ 0 & \text{otherwise} \end{cases}$$

a Sketch the probability density function.

b *State* the value of the mean.

c Find $E(2X + 5)$.

5 The continuous random variable Y has probability density function given by

$$f(y) = \begin{cases} \dfrac{3}{14}\sqrt{y} & 1 \le y \le 4 \\ 0 & \text{otherwise} \end{cases}$$

Find the value of $E(Y)$.

6 The continuous random variable X has probability density function given by

$$f(x) = \begin{cases} \dfrac{3}{4}x(2 - x) & 0 \le x \le 2 \\ 0 & \text{otherwise} \end{cases}$$

a Sketch the probability density function of X.

b Find the mean of X.

7 The continuous random variable X has mean μ and probability density function given by

$$f(x) = \begin{cases} \dfrac{1}{4}x^3 & 0 \le x \le 2 \\ 0 & \text{otherwise} \end{cases}$$

a Find μ.   **b** Find $P(X < \mu)$.

c Is the mean of X less than or greater than the median of X? Justify your answer.

8 The random variable T denotes the lifetime, in years, of a particular type of light bulb. The mean lifetime is μ.

The probability density function of T is defined by

$$f(t) = \begin{cases} kt(5-t) & 0 \le t \le 5 \\ 0 & \text{otherwise} \end{cases}$$

a Show that $k = \dfrac{6}{125}$.   **b** Find μ.

c Find $P(T < \mu)$.

d Two light bulbs are selected at random. Find the probability that both light bulbs last longer than the mean lifetime of this type of light bulb.

9 The continuous random variable X has probability density function given by

$$f(x) = \begin{cases} \dfrac{5}{32}x^4 & 0 \le x \le 2 \\ 0 & \text{otherwise} \end{cases}$$

a Find $E\left(\dfrac{1}{X}\right)$.   **b** Find $E\left(10 - \dfrac{8}{X}\right)$.

10 The continuous random variable X has probability density function given by

$$f(x) = \begin{cases} k & 1 \le x \le 9 \\ 0 & \text{otherwise} \end{cases}$$

where k is a constant.

a Find k.   **b** Find $E(X)$.   **c** Find $E(X^{-2})$.

11 The continuous random variable X has probability density function given by

$$f(x) = \begin{cases} \dfrac{6}{17}x & 0 \le x \le 1 \\ \dfrac{6}{17}x^2 & 1 \le x \le 2 \\ 0 & \text{otherwise} \end{cases}$$

Find

a $E(X)$ **b** $E(34X - 40)$

12 The continuous random variable X has mean μ and probability density function given by

$$f(x) = \begin{cases} \dfrac{3}{32}x^2 & 0 \le x \le 2 \\ \dfrac{3}{32}(6-x) & 2 \le x \le 6 \\ 0 & \text{otherwise} \end{cases}$$

Find μ.

13 The continuous random variable X has probability density function given by

$$f(x) = \begin{cases} p - qx & 0 \le x \le 2 \\ 0 & \text{otherwise} \end{cases}$$

where p and q are constants.

a Using the fact that the total area under the curve is 1, show that $2p - 2q = 1$.

b Given that the mean of X is $\dfrac{2}{3}$,

    i form a second equation in p and q

    ii find the value of p and the value of q.

11.4 Variance of X

A very important application of $\mathrm{E}(\mathrm{g}(X))$ is in finding the variance of X, denoted by σ^2.

For a continuous random variable X,

$$\mathbf{Var}(X) = \sigma^2 = \mathrm{E}((X-\mu)^2) \quad \text{where } \mu = \mathrm{E}(X)$$

    See AS Textbook Chapter 14

As in the discrete case, the formula can be written

$$\mathbf{Var}(X) = \sigma^2 = \mathrm{E}(X^2) - (\mathrm{E}(X))^2$$

$$= \mathrm{E}(X^2) - \mu^2 \quad \leftarrow \text{This is 'the mean of the squares minus the square of the mean'.}$$

So, when X is continuous,

$$\mathbf{Var}(X) = \int_{\text{all } x} (x-\mu)^2\, \mathrm{f}(x)\,\mathrm{d}x \;\text{ where }\; \mu = \int_{\text{all } x} x\mathrm{f}(x)\,\mathrm{d}x$$

Alternatively,

$$\mathbf{Var}(X) = \int_{\text{all } x} x^2 \mathrm{f}(x)\,\mathrm{d}x - \mu^2$$

Note that the standard deviation of X, denoted by σ, is the square root of the variance.

    i.e. $\quad \sigma = \sqrt{\mathbf{Var}(X)}$

Example 17

The random variable X has mean μ, where $\mu = 0.75$. The probability density function of X is defined by

$$\mathrm{f}(x) = \begin{cases} 3x^2 & 0 \le x \le 1 \\ 0 & \text{otherwise} \end{cases}$$

a Show that the standard deviation, σ, is 0.1936, correct to 4 significant figures.

b Find $(\mu - \sigma \le X \le \mu + \sigma)$.

a Use $\sigma^2 = \mathrm{E}(X^2) - \mu^2$, first finding $\mathrm{E}(X^2)$.

$$\mathrm{E}(X^2) = \int_{\text{all } x} x^2 \mathrm{f}(x)\,\mathrm{d}x$$

$$= \int_0^1 (x^2 \times 3x^2)\,\mathrm{d}x$$

$$= 3\int_0^1 x^4\,\mathrm{d}x$$

$$= 3\left[\frac{x^5}{5}\right]_0^1$$

(continued)

(continued)

$$= 3\left(\frac{1^5}{5} - 0\right)$$

$$= 0.6$$

$$\sigma^2 = E(X^2) - \mu^2$$

$$= 0.6 - 0.75^2$$

$$= 0.0375$$

$$\sigma = \sqrt{0.0375} = 0.19364... = 0.1936 \,(4\,\text{sf})$$

b $\mu + \sigma = 0.75 + \sqrt{0.0375} = 0.9436...$

$\mu - \sigma = 0.75 - \sqrt{0.0375} = 0.5563...$

$$P(\mu - \sigma \le X \le \mu + \sigma) = \int_{\mu - \sigma}^{\mu + \sigma} 3x^2 \, dx$$

$$= \left[x^3\right]_{\mu - \sigma}^{\mu + \sigma}$$

$$= (0.9436...)^3 - (0.5563...)^3$$

$$= 0.6680... = 0.668 \,(3\,\text{sf})$$

> **Note**
>
> If you are not given μ, calculate $E(X)$ first.

> **Note**
>
> This is the probability that X lies within one standard deviation of the mean.

If the function is given in more than one part, the integrations are carried out in sections, as with the mean.

Example 18

The continuous random variable X has probability density function defined by

$$f(x) = \begin{cases} \dfrac{1}{4}x & 0 \le x \le 2 \\[2mm] 1 - \dfrac{1}{4}x & 2 \le x \le 4 \\[2mm] 0 & \text{otherwise} \end{cases}$$

Find the standard deviation of X.

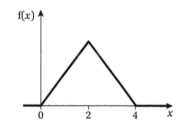

By symmetry, $\mu = 2$.

$$E(X^2) = \int_{\text{all } x} x^2 f(x) \, dx$$

$$= \int_0^2 \left(x^2 \times \frac{1}{4}x\right) dx + \int_2^4 x^2 \times \left(1 - \frac{1}{4}x\right) dx$$

$$= \frac{1}{4}\int_0^2 x^3 dx + \int_2^4 \left(x^2 - \frac{1}{4}x^3\right) dx$$

$$= \frac{1}{4}\left[\frac{x^4}{4}\right]_0^2 + \left[\frac{x^3}{3} - \frac{x^4}{16}\right]_2^4$$

$$= \frac{1}{4}\left(\frac{2^4}{4} - 0\right) + \left(\left(\frac{4^3}{3} - \frac{4^4}{16}\right) - \left(\frac{2^3}{3} - \frac{2^4}{16}\right)\right)$$

$$= \frac{14}{3}$$

$$\text{Var}(X) = \sigma^2 = E(X^2) - \mu^2$$

$$= \frac{14}{3} - 2^2 = \frac{2}{3}$$

$$\sigma = \sqrt{\frac{2}{3}} = 0.8164... = 0.816 \,(3\,\text{sf})$$

> **Note**
>
> Remember to look for symmetry when finding the mean.

Variance of a simple function of X

As with discrete random variables, the following results hold when X is continuous, where a and b are constants:

$\text{Var}(a) = 0$	For example	$\text{Var}(2) = 0$	
$\text{Var}(aX) = a^2\text{Var}(X)$		$\text{Var}(4X) = 16\text{Var}(X)$	
$\text{Var}(aX+b) = a^2\text{Var}(X)$		$\text{Var}(4X+2) = 16\text{Var}(X)$	
$\text{Var}(aX-b) = a^2\text{Var}(X)$		$\text{Var}(4X-1) = 16\text{Var}(X)$	

These results can be extended to other simple functions of X, such as

$$\text{Var}(4X^3 + 2) = 16\text{Var}(X^3)$$

$$\text{Var}(3X^{-2} + 5) = 9\text{Var}(X^{-2}) = 9\text{Var}\left(\frac{1}{X^2}\right)$$

Example 19

Question

The continuous random variable X has mean μ and standard deviation σ.

Find, in terms of μ and/or σ

a $E(X^2)$

b $E(X^2 + 3X)$

c $\text{Var}\left(\frac{1}{2}X - 5\right)$

Answer

a $E(X) = \mu$, $\text{Var}(X) = \sigma^2$

You know that $\quad \text{Var}(X) = E(X^2) - \mu^2$

that is $\qquad\qquad \sigma^2 = E(X^2) - \mu^2$

so $\qquad\qquad E(X^2) = \sigma^2 + \mu^2$

b $E(X^2 + 3X) = E(X^2) + 3E(X) = \sigma^2 + \mu^2 + 3\mu$

c $\text{Var}\left(\frac{1}{2}X - 5\right) = \left(\frac{1}{2}\right)^2 \text{Var}(X) = \frac{1}{4}\sigma^2$

Example 20

Question

The continuous random variable X has the probability density function given by

$$f(x) = \begin{cases} \dfrac{1}{9}x^2 & 0 \le x \le 3 \\ 0 & \text{otherwise} \end{cases}$$

a Find

  i $E\left(\dfrac{1}{X}\right)$

  ii $E\left(\dfrac{1}{X^2}\right)$

  iii $\text{Var}\left(\dfrac{1}{X}\right)$

b Find the mean and variance of $\left(\dfrac{3X+4}{X}\right)$.

a i $E\left(\dfrac{1}{X}\right) = \displaystyle\int_{\text{all } x}\left(\dfrac{1}{x}\right)f(x)\,dx$

$= \displaystyle\int_0^3\left(\dfrac{1}{x}\times\dfrac{1}{9}x^2\right)dx$

$= \dfrac{1}{9}\displaystyle\int_0^3 x\,dx$

$= \dfrac{1}{9}\left[\dfrac{x^2}{2}\right]_0^3$

$= \dfrac{1}{9}\left(\dfrac{3^2}{2}-0\right)$

$= \dfrac{1}{2}$

ii $E\left(\dfrac{1}{X^2}\right) = \displaystyle\int_{\text{all } x}\left(\dfrac{1}{x^2}\right)f(x)\,dx$

$= \displaystyle\int_0^3\left(\dfrac{1}{x^2}\times\dfrac{1}{9}x^2\right)dx$

$= \dfrac{1}{9}\displaystyle\int_0^3 1\,dx$

$= \dfrac{1}{9}\left[x\right]_0^3$

$= \dfrac{1}{3}$

iii $\text{Var}\left(\dfrac{1}{X}\right) = E\left(\dfrac{1}{X^2}\right) - \left[E\left(\dfrac{1}{X}\right)\right]^2$

$= \dfrac{1}{3} - \left(\dfrac{1}{2}\right)^2$

$= \dfrac{1}{12}$

b $E\left(\dfrac{3X+4}{X}\right) = E\left(3+\dfrac{4}{X}\right)$

$= 3 + 4E\left(\dfrac{1}{X}\right)$

$= 3 + 4\times\dfrac{1}{2}$

$= 5$

$\text{Var}\left(\dfrac{3X+4}{X}\right) = \text{Var}\left(3+\dfrac{4}{X}\right)$

$= 4^2\,\text{Var}\left(\dfrac{1}{X}\right)$

$= 16\times\dfrac{1}{12}$

$= \dfrac{4}{3}$

11.5 Independent continuous random variables

As with discrete random variables, the following results hold for **independent** continuous random variables.

AS Textbook Chapter 14

For two independent continuous random variables X and Y

$$E(aX + bY) = aE(X) + bE(Y)$$
$$E(aX - bY) = aE(X) - bE(Y)$$
$$Var(aX + bY) = a^2Var(X) + b^2Var(Y)$$
$$Var(aX - bY) = a^2Var(X) + b^2Var(Y) \quad \leftarrow \text{Note the } + \text{ sign here.}$$

For n independent continuous random variables $X_1, X_2, \cdots, X_n$

$$E(X_1 + X_2 + \cdots + X_n) = E(X_1) + E(X_2) + \cdots + E(X_n)$$
so $\quad E(\Sigma X_i) = \Sigma E(X_i)$
$$Var(X_1 + X_2 + \cdots + X_n) = Var(X_1) + Var(X_2) + \cdots + Var(X_n)$$
so $\quad Var(\Sigma X_i) = \Sigma Var(X_i)$

In particular, for n observations of the same variable X

$$E(X_1 + X_2 + \cdots + X_n) = nE(X)$$

$$Var(X_1 + X_2 + \cdots + X_n) = nVar(X)$$

> **Note**
>
> The mean of a sum is equal to the sum of the means. The variance of a sum is equal to the sum of the variances.

Example 21

Children are taking part in a competition in which they complete a puzzle then answer some general knowledge questions.

X minutes is the time taken to complete the puzzle. The mean of X is 10 and the standard deviation is 3. Y minutes is the time taken to answer the general knowledge questions. The mean of Y is 15 and the standard deviation is 4.

X and Y are independent.

a The random variable T minutes is the total time taken by a child to complete the two tasks. Find the mean and standard deviation of T.

b The random variable D minutes is the amount by which the time to answer the general knowledge questions exceeds the time to complete the puzzle. Find $E(D)$ and $Var(D)$.

c Find

  i $E(3X + 4Y)$ ii $Var(10X - 5Y)$

a $T = X + Y$
  $E(T) = E(X) + E(Y) = 10 + 15 = 25$
  $Var(T) = Var(X) + Var(Y) = 3^2 + 4^2 = 25$
  standard deviation of $T = \sqrt{25} = 5$

The mean is 25 minutes and the standard deviation is 5 minutes.

(continued)

(continued)

b $D = Y - X$

$E(D) = E(Y - X) = E(Y) - E(X) = 15 - 10 = 5$

$Var(D) = Var(Y - X)$

$\qquad = Var(Y) + Var(X) \qquad \leftarrow$ Remember + here.

$\qquad = 3^2 + 4^2$

$\qquad = 25$

c i $E(3X + 4Y) = 3E(X) + 4E(Y) = (3 \times 10) + (4 \times 15) = 90$

ii $Var(10X - 5Y) = 10^2 Var(X) + 5^2 Var(Y) = (100 \times 3^2) + (25 \times 4^2) = 1300$

Example 22

Every week day Ishmael catches a bus from his local bus stop to his place of work. The time, X minutes, that he waits for the bus has probability density function

$$f(x) = \begin{cases} \dfrac{1}{10} & 0 \le x \le 10 \\ 0 & \text{otherwise} \end{cases}$$

The times are independent from day to day.

a What is the longest time that he waits for a bus?

b Determine $E(X)$ and $Var(X)$.

T is the total time, in minutes, that he waits for a bus in a working week of 5 days.

c Find the mean and variance of T.

d Find the probability that he waits more than 2 minutes on each of the 5 days.

a Since $0 \le x \le 10$, the longest time he waits is 10 minutes.

b By symmetry, $E(X) = \mu = 5$

$$E(X^2) = \int_0^{10} \left(x^2 \times \frac{1}{10} \right) dx$$

$$= \frac{1}{10} \left[\frac{x^3}{3} \right]_0^{10}$$

$$= \frac{1}{30}(10^3 - 0)$$

$$= 33\frac{1}{3}$$

$$Var(X) = E(X^2) - \mu^2$$

$$= 33\frac{1}{3} - 5^2$$

$$= 8\frac{1}{3}$$

c $T = X_1 + X_2 + X_3 + X_4 + X_5$

$E(T) = 5E(X) = 5 \times 5 = 25$

$Var(T) = 5Var(X) = 5 \times 8\frac{1}{3} = 41\frac{2}{3}$

> **Note**
>
> T is the sum of 5 independent observations of X.

d $P(X > 2) = 8 \times 0.1 = 0.8 \qquad \leftarrow$ area of a rectangle

$P(X > 2$ on each of 5 days)

$\qquad = 0.8^5$

$\qquad = 0.32768 = 0.328$ (3 sf)

Exercise 4

1 The continuous random variable X has probability density function given by

$$f(x) = \begin{cases} \dfrac{2}{5}x & 2 \le x \le 3 \\ 0 & \text{otherwise} \end{cases}$$

a Find $E(X)$.

b Find

    i $E(X^2)$ **ii** $Var(X)$ **iii** $Var(3X-2)$

2 The continuous random variable X has mean μ and standard deviation σ. The probability density function of X is given by

$$f(x) = \begin{cases} \dfrac{1}{32}(8 - x) & 0 \le x \le 8 \\ 0 & \text{otherwise} \end{cases}$$

a Show that $\mu = 2\dfrac{2}{3}$ and $\sigma^2 = 3\dfrac{5}{9}$.

b Find $P(X < \mu - \sigma)$.

3 The continuous random variable X has probability density function given by

$$f(x) = \begin{cases} \dfrac{3}{7}x^2 & 1 \le x \le 2 \\ 0 & \text{otherwise} \end{cases}$$

a Find

    i $E(X)$ **ii** $E(X^2)$ **iii** the standard deviation of X.

b Given that $Y = \dfrac{1}{X}$, find

    i $E(Y)$ **ii** $E(Y^2)$ **iii** the standard deviation of Y.

4 The continuous random variable X has probability density function given by

$$f(x) = \begin{cases} k & -2 \le x \le 3 \\ 0 & \text{otherwise} \end{cases}$$

a Find the value of k.

b *State* the value of $E(X)$.

c Find $Var(X)$.

d Find the standard deviation of X.

5 The continuous random variable T has probability density function given by

$$f(t) = \begin{cases} \dfrac{3}{32}(4 - t^2) & -2 \le t \le 2 \\ 0 & \text{otherwise} \end{cases}$$

a Sketch the graph of f.

b State the mean of T.

c Calculate $Var(T)$.

d The random variable S is the sum of three independent observations of T. Find $E(S)$ and $Var(S)$.

6 The continuous random variable W has probability density function given by

$$f(w) = \begin{cases} \dfrac{k}{\sqrt{w}} & 1 \le w \le 4 \\ 0 & \text{otherwise} \end{cases}$$

where k is a constant.

a Find the value of k. **b** Find the mean of W.

c Find the standard deviation of W.

7 The continuous random variable X has probability density function defined by

$$f(x) = \begin{cases} \dfrac{1}{27}x^2 & 0 \le x \le 3 \\ \dfrac{1}{3} & 3 \le x \le 5 \\ 0 & \text{otherwise} \end{cases}$$

Find

a $E(X)$ **b** $Var(X)$ **c** $Var(5X-6)$

8 Y is a continuous random variable with a mean of 20 and a standard deviation of 4.

Find

a $Var(3-Y)$ **b** $E(Y^2)$

9 The random variable X has mean μ and standard deviation σ. The probability density function of X is defined by

$$f(x) = \begin{cases} \dfrac{3}{8}(x+1)^2 & 0 \le x \le 1 \\ 0 & \text{otherwise} \end{cases}$$

Calculate μ and σ.

10 X and Y are independent continuous random variables such that $E(X) = 10$ and $Var(X) = 9$, $E(Y) = 14$ and $Var(Y) = 4$.

Find

a $E(X+Y)$ **b** $E(3Y-2X)$

c $Var\left(\dfrac{1}{2}X+4Y\right)$

d the standard deviation of $6X-4Y$.

11 X and Y are independent random variables such that

$$E(X) = 8, Var(X) = 2$$
$$E(Y) = 10, Var(Y) = 3$$

Find

a $Var(X-Y)$ **b** $E(3X+Y)$

c $Var(2X-8)$

d the standard deviation of $4X-3Y$.

12 A machine produces two types of items, A and B. The time taken to produce item A is distributed with a mean of 3 minutes and a standard deviation of 0.3 minutes. The time taken to produce item B is distributed with a mean of 5 minutes and a standard deviation of 0.9 minutes.

Find the mean and standard deviation of the time taken to produce

a 1 item of each type

b 4 items of Type A

c 4 items of Type A and 2 items of Type B.

13 X and Y are independent continuous random variables.

X has probability density function given by

$$f(x) = \begin{cases} \frac{1}{8}(4 - x) & 0 \leq x \leq 4 \\ 0 & \text{otherwise} \end{cases}$$

Y has probability density function given by

$$f(y) = \begin{cases} \frac{1}{8}y & 0 \leq y \leq 4 \\ 0 & \text{otherwise} \end{cases}$$

Find

a the mean and variance of X

b the mean and variance of Y

c i $E(X + Y)$ **ii** $\text{Var}(X + Y)$

d i $E(5X - 3Y)$ **ii** $\text{Var}(5X - 3Y)$

11.6 Further applications

Example 23

The continuous random variable X has a cumulative distribution function defined by

$$F(x) = \begin{cases} 0 & x < -5 \\ \frac{x + 5}{20} & -5 \leq x \leq 15 \\ 1 & x > 15 \end{cases}$$

a Show that, for $-5 \leq x \leq 15$, the probability density function, $f(x)$ of X is given by $f(x) = \frac{1}{20}$.

b Find:

i $P(X \geq 7)$;   **ii** $P(X \neq 7)$;

iii $E(X)$;   **iv** $E(3X^2)$.

AQA MS2B June 2012

a For $-5 \leq x \leq 15$,

$$f(x) = \frac{d}{dx}F(x)$$

$$= \frac{d}{dx}\left(\frac{x+5}{20}\right)$$

$$= \frac{d}{dx}\left(\frac{1}{20}x + \frac{5}{20}\right)$$

$$= \frac{1}{20}$$

b **i** $P(X \geq 7) = 1 - F(7)$

$$= 1 - \frac{7+5}{20}$$

$$= 0.4$$

ii Since it is certain that X cannot take an *exact* value,

$$P(X \neq 7) = 1$$

iii By symmetry, μ is halfway between -5 and 15, so

$$E(X) = \frac{1}{2}(-5 + 15) = 5$$

iv $E(3X^2) = \int_{-5}^{15}\left(3x^2 \times \frac{1}{20}\right)dx$

$$= \frac{1}{20}\left[x^3\right]_{-5}^{15}$$

$$= \frac{1}{20}(15^3 - (-5)^3)$$

$$= 175$$

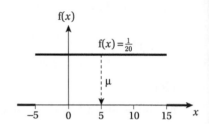

Example 24

The continuous random variable X has probability density function given by

$$f(x) = \begin{cases} \frac{1}{2}(x^2 + 1) & 0 \leq x \leq 1 \\ (x-2)^2 & 1 \leq x \leq 2 \\ 0 & \text{otherwise} \end{cases}$$

a Sketch the graph of f.

b Calculate $P(X \leq 1)$.

c Show that $E(X^2) = \frac{4}{5}$.

d **i** Given that $E(X) = \frac{19}{24}$ and $Var(X) = \frac{499}{k}$, find the numerical value of k.

ii Find $E(5X^2 + 24X - 3)$.

iii Find $Var(12X - 5)$.

AQA MS2B January 2010

a When $x < 0$ and $x > 2$, $f(x) = 0$.

When $0 \le x \le 1$, $f(x) = \dfrac{1}{2}(x^2 + 1)$

This is part of a quadratic curve, with

$f(0) = \dfrac{1}{2}(0^2 + 1) = \dfrac{1}{2}$ and $f(1) = \dfrac{1}{2}(1^2 + 1) = 1$

When $1 \le x \le 2$, $f(x) = (x - 2)^2$

This is also part of a quadratic curve, with

$f(1) = (1 - 2)^2 = 1$ and $f(2) = (2 - 2)^2 = 0$

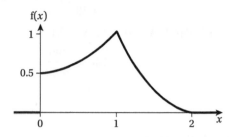

b To find $P(X \le 1)$, use $f(x) = \dfrac{1}{2}(x^2 + 1)$

$$P(X \le 1) = \int_0^1 \dfrac{1}{2}(x^2 + 1)\,dx$$

$$= \dfrac{1}{2}\int_0^1 (x^2 + 1)\,dx$$

$$= \dfrac{1}{2}\left[\dfrac{x^3}{3} + x\right]_0^1$$

$$= \dfrac{1}{2}\left(\left(\dfrac{1^3}{3} + 1\right) - 0\right)$$

$$= \dfrac{2}{3}$$

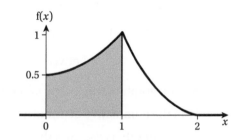

c $E(X^2) = \displaystyle\int_{\text{all } x} x^2 f(x)\,dx$

$$= \int_0^1 x^2 \times \dfrac{1}{2}(x^2 + 1)\,dx + \int_1^2 x^2 \times (x - 2)^2\,dx \qquad \leftarrow \text{The integration is done in two stages}$$

$$= \dfrac{1}{2}\int_0^1 (x^4 + x^2)\,dx + \int_1^2 x^2(x^2 - 4x + 4)\,dx$$

$$= \dfrac{1}{2}\int_0^1 (x^4 + x^2)\,dx + \int_1^2 (x^4 - 4x^3 + 4x^2)\,dx$$

$$= \dfrac{1}{2}\left[\dfrac{x^5}{5} + \dfrac{x^3}{3}\right]_0^1 + \left[\dfrac{x^5}{5} - x^4 + \dfrac{4x^3}{3}\right]_1^2$$

$$= \dfrac{1}{2}\left(\dfrac{1^5}{5} + \dfrac{1^3}{3} - 0\right) + \left(\left(\dfrac{2^5}{5} - 2^4 + 4 \times \dfrac{2^3}{3}\right) - \left(\dfrac{1^5}{5} - 1^4 + 4 \times \dfrac{1^3}{3}\right)\right)$$

$$= \dfrac{4}{5}, \text{ as required.}$$

d i $\text{Var}(X) = E(X^2) - (E(X))^2$

$$= \dfrac{4}{5} - \left(\dfrac{19}{24}\right)^2$$

$$= \dfrac{499}{2880}$$

Since $\text{Var}(X) = \dfrac{499}{k}$, $k = 2880$

(continued)

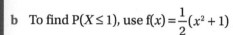

(continued)

ii $E(5X^2 + 24X - 3) = 5E(X^2) + 24E(X) - 3$

$$= 5 \times \frac{4}{5} + 24 \times \frac{19}{24} - 3$$

$$= 20$$

iii $\text{Var}(12X - 5) = 12^2 \text{Var}(X)$

$$= 144 \times \frac{499}{2880}$$

$$= 24.95$$

Example 25

A continuous random variable X has the probability density function defined by

$$f(x) = \begin{cases} x^2 & 0 \le x \le 1 \\ \dfrac{1}{3}(5 - 2x) & 1 \le x \le 2 \\ 0 & \text{otherwise} \end{cases}$$

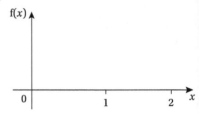

a Sketch the graph of f on a copy of the axes.

b i Find the cumulative distribution function, F, for $0 \le x \le 1$.

ii Hence, or otherwise, find the value of the lower quartile of X.

c i Show that the cumulative distribution function for $1 \le x \le 2$ is defined by

$$F(x) = \frac{1}{3}(5x - x^2 - 3)$$

ii Hence, or otherwise, find the value of the upper quartile of X.

AQA MS2B June 2013

a When $x < 0$ and $x > 2$, $f(x) = 0$.

The first section is part of a quadratic curve.

When $x = 0$, $f(0) = 0$, when $x = 1$, $f(1) = 1^2 = 1$

The second section is a straight line with a negative gradient.

When $x = 1$, $f(1) = \dfrac{1}{3}(5 - 2 \times 1) = 1$

When $x = 2$, $f(2) = \dfrac{1}{3}(5 - 2 \times 2) = \dfrac{1}{3}$

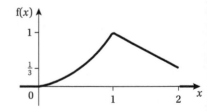

b i When $0 \le x \le 1$,

$$F(x) = \int_0^x t^2 \, dt \qquad \leftarrow \text{Use } f(t) = t^2$$

$$= \left[\frac{t^3}{3} \right]_0^x$$

$$= \frac{1}{3}(x^3 - 0)$$

$$= \frac{1}{3}x^3$$

(continued)

(continued)

ii If q_1 is the lower quartile, then $F(q_1) = \dfrac{1}{4}$

Now $F(1) = \dfrac{1}{3} \times 1^3 = \dfrac{1}{3} > \dfrac{1}{4}$, so $q_1 < 1$ and q_1 lies in the first section.

So $\dfrac{1}{3} q_1^3 = \dfrac{1}{4}$

$q_1^3 = \dfrac{3}{4}$

$q_1 = \sqrt[3]{\dfrac{3}{4}} = 0.9085\ldots = 0.909 \ (3\ \text{sf})$

c i When $1 \leq x \leq 2$,

$F(x) = F(1) + \displaystyle\int_1^x \dfrac{1}{3}(5 - 2t)\,dt$ $\leftarrow$ Use $f(t) = \dfrac{1}{3}(5 - 2t)$

$= \dfrac{1}{3} + \dfrac{1}{3}\Big[5t - t^2\Big]_1^x$

$= \dfrac{1}{3} + \dfrac{1}{3}\big((5x - x^2) - (5 \times 1 - 1^2)\big)$

$= \dfrac{1}{3} + \dfrac{1}{3}(5x - x^2 - 4)$

$= \dfrac{1}{3}(1 + 5x - x^2 - 4)$

$= \dfrac{1}{3}(5x - x^2 - 3)$

Remember to include $F(1)$, the area under the first curve.

ii For upper quartile q_3, $F(q_3) = \dfrac{3}{4}$.

Since $F(1) = \dfrac{1}{3} < \dfrac{3}{4}$, you know that q_3 is in the second section.

Writing q for q_3:

$\dfrac{1}{3}(5q - q^2 - 3) = \dfrac{3}{4}$

$5q - q^2 - 3 = \dfrac{9}{4}$

$20q - 4q^2 - 12 = 9$

$4q^2 - 20q + 21 = 0$

$(2q - 3)(2q - 7) = 0$

$q = 1.5$ or $q = 3.5$ (reject since out of range)

So, upper quartile $= 1.5$.

Summary

Probability density function f

$$P(x_1 \leq X \leq x_2) = \int_{x_1}^{x_2} f(x)\,dx$$

$$\text{Total area} = \int_{\text{all }x} f(x)\,dx = 1$$

Cumulative distribution function F

$$F(x) = P(X \leq x) = \int_{-\infty}^{x} f(t)\,dt$$

If $a \leq x \leq b$, $F(a) = 0$, $F(b) = 1$

$$P(X \geq x) = 1 - F(x)$$

$$P(x_1 \leq X \leq x_2) = F(x_2) - F(x_1)$$

Median: $F(m) = 0.5$

Quartiles: $F(q_1) = 0.25$ and $F(q_3) = 0.75$

Interquartile range $= q_3 - q_1$

Expectation and variance

$$E(X) = \mu = \int_{\text{all }x} x f(x)\,dx$$

$$E(g(X)) = \int_{\text{all }x} g(x) f(x)\,dx$$

In particular

$$E(X^2) = \int_{\text{all }x} x^2 f(x)\,dx$$

$$\text{Var}(X) = \sigma^2 = E((X-\mu)^2) = E(X^2) - \mu^2$$

So $\quad \text{Var}(X) = \sigma^2 = \int_{\text{all }x} (x-\mu)^2 f(x)\,dx = \int_{\text{all }x} x^2 f(x)\,dx - \mu^2$

Standard deviation $\sigma = \sqrt{\text{Var}(X)}$

Expectation algebra, where *a* and *b* are constants

$E(a) = a$	$\text{Var}(a) = 0$
$E(aX) = aE(X)$	$\text{Var}(aX) = a^2\text{Var}(X)$
$E(aX + b) = aE(X) + b$	$\text{Var}(aX + b) = a^2\text{Var}(X)$
$E(aX - b) = aE(X) - b$	$\text{Var}(aX - b) = a^2\text{Var}(X)$

For two independent continuous random variables X and Y

$E(aX + bY) = aE(X) + bE(Y) \quad \text{Var}(aX + bY) = a^2\text{Var}(X) + b^2\text{Var}(Y)$

$E(aX - bY) = aE(X) - bE(Y) \quad \text{Var}(aX - bY) = a^2\text{Var}(X) + b^2\text{Var}(Y) \quad \leftarrow \text{Note} + \text{sign}$

For n independent continuous random variables $X_1, X_2, ..., X_n$

$$E(\textstyle\sum X_i) = \sum E(X_i)$$

$$\text{Var}(\textstyle\sum X_i) = \sum \text{Var}(X_i)$$

In particular, for n independent observations of the same variable X

$$E(X_1 + X_2 + \cdots + X_n) = nE(X)$$
$$\text{Var}(X_1 + X_2 + \cdots + X_n) = n\text{Var}(X)$$

Review

1 A machine cuts lengths of hose for washing machines. Each length cut is at least 2 metres. The extra length, X millimetres, above 2 metres may be modelled by the following probability density function, where c is a positive constant.

$$f(x) = \begin{cases} c & 0 \le x \le 4 \\ \dfrac{c}{16}(20 - x) & 4 \le x \le 20 \\ 0 & \text{otherwise} \end{cases}$$

a Sketch the graph of f.

b By considering your sketch, or otherwise, find the exact value of c.

c Hence determine the probability that a randomly selected length of hose is less than 2.01 metres.

AQA MAS1 2005

2 The mass, in kilograms, of metal extracted from 10 g of ore from a certain mine is a continuous random variable X with probability density function

$$f(x) = \begin{cases} \dfrac{3}{4}x(2 - x)^2 & 0 \le x \le 2 \\ 0 & \text{otherwise} \end{cases}$$

a Show that the mean mass is 0.8 kg.

b Find the standard deviation of the mass of metal extracted.

c The random variable Y is the sum of 4 independent observations of X. Find $E(Y)$ and $\text{Var}(Y)$.

3 A continuous random variable X has probability density function given by

$$f(x) = \begin{cases} \dfrac{1}{c} & 0 \le x \le c \\ 0 & \text{otherwise} \end{cases}$$

a State the value of $E(X)$ in terms of c.

b Find $\text{Var}(X)$ in terms of c.

c If $c = 6$, find the standard deviation of X.

4 The continuous random variable X has the cumulative distribution function

$$F(x) = \begin{cases} 0 & x \le -4 \\ \dfrac{x + 4}{9} & -4 \le x \le 5 \\ 1 & x \ge 5 \end{cases}$$

a Determine the probability density function, f(x) of X.

b Sketch the graph of f.

c Determine $P(X > 2)$.

d Evaluate the mean and variance of X.

AQA MS2B January 2007

5 The continuous random variable X has probability density function defined by

$$f(x) = \begin{cases} \dfrac{3}{8}x^2 & 0 \le x \le \dfrac{1}{2} \\ \dfrac{3}{32} & \dfrac{1}{2} \le x \le 11 \\ 0 & \text{otherwise} \end{cases}$$

a Sketch the graph of f.

b Show that:

i $P\left(X \ge 8\dfrac{1}{3}\right) = \dfrac{1}{4}$;

ii $P(X \ge 3) = \dfrac{3}{4}$.

c Hence write down the **exact** value of

i the interquartile range of X,   **ii** the median, m, of X.

d Find the **exact** value of $P(X < m \mid X \ge 3)$.

AQA MS2B January 2011

6 The waiting times, T minutes, before being served at a local newsagents can be modelled by a continuous random variable with probability density function

$$f(t) = \begin{cases} \dfrac{3}{8}t^2 & 0 \le t \le 1 \\ \dfrac{1}{16}(t + 5) & 1 \le t \le 3 \\ 0 & \text{otherwise} \end{cases}$$

a Sketch the graph of f.

b For a customer selected at random, calculate $P(T \ge 1)$.

c **i** Show that the cumulative distribution function for $1 \le t \le 3$ is given by

$$F(t) = \dfrac{1}{32}(t^2 + 10t - 7)$$

ii Hence find the median waiting time.

AQA MS2B January 2007

7 The random variable X has probability density function defined by

$$f(x) = \begin{cases} \dfrac{1}{2} & 0 \le x \le 1 \\ \dfrac{1}{18}(x - 4)^2 & 1 \le x \le 4 \\ 0 & \text{otherwise} \end{cases}$$

a State the values for the median and lower quartile of X.

b Show that, for $1 \leq x \leq 4$, the cumulative distribution function, $F(x)$, of X is given by

$$F(x) = 1 + \frac{1}{54}(x-4)^3$$

(You may assume that $\int (x-4)^2\, dx = \frac{1}{3}(x-4)^3 + c$)

c Determine $P(2 \leq X \leq 3)$.

d i Show that q, the upper quartile of X, satisfies the equation $(q-4)^3 = -13.5$.

   ii Hence evaluate q to three decimal places.

AQA MS2B June 2010

8 X, Y and W are independent continuous random variables such that

Var $(X) = 2$, Var$(Y) = 3$, Var$(W) = 4$

Find

a Var$(X - Y)$ b Var$\left(Y - \frac{1}{4}W\right)$ c Var$(X + Y + W)$

Assessment

1 Customers at an internet café pay a flat fee which entitles them to use a terminal for up to two hours. The actual amount of time, in hours, that a customer uses a terminal, may be modelled by the random variable, T, with probability density function

$$f(t) = \begin{cases} kt & 0 < t < 2 \\ 0 & \text{otherwise} \end{cases}$$

a Show that the value of the constant k is 0.5. [3]

b Find the mean and the standard deviation of T. [6]

c Find the probability that a customer spends less than one hour using a terminal. [2]

AQA MBS4 June 2002

2 A town's library offers a booking system for free internet access. Users are allowed a maximum of one hour connection time at any one booking. The actual connection times, X minutes, of users may be modelled by the following probability density function, where k is a constant.

$$f(x) = \begin{cases} \dfrac{kx}{40} & 0 \leq x \leq 40 \\ k & 40 \leq x \leq 60 \\ 0 & \text{otherwise} \end{cases}$$

a Sketch the graph of f. [4]

b By considering your sketch, or otherwise, show that the value of k is 0.025. [3]

c Hence determine $P(X > 30)$. [3]

AQA MAS1 June 2004

3 The continuous random variable X has probability density function defined by

$$f(x) = \begin{cases} k(x^2 + 1) & 0 \le x \le 3 \\ 0 & \text{otherwise} \end{cases}$$

  a **i** Show that the value of k is $\dfrac{1}{12}$.   [3]

      ii Find the distribution function, $F(x)$, for all x.   [3]

      iii Sketch the graph of F.   [2]

      iv Find $P(X \ge 2)$.   [2]

  b Calculate the **exact** value of $E(X)$.   [3]

AQA MS2A January 2008

4 The continuous random variable X has cumulative distribution function given by

$$F(x) = \begin{cases} 0 & x \le 0 \\ x^3 & 0 \le x \le 1 \\ 1 & x \ge 1 \end{cases}$$

  a Find the median of X.   [2]

  b Find the probability density function, $f(x)$.   [2]

  c Determine

      i $E(X)$ [2]

      ii the probability that X lies between the median and the mean.   [2]

  d Determine

      i $Var(X)$ [2]

      ii the probability that X lies within one standard deviation of the mean.   [2]

5 The time, T hours, that the supporters of Bracken Football Club have to queue in order to obtain their Cup Final tickets has the following probability distribution function.

$$f(t) = \begin{cases} \dfrac{1}{5} & 0 \le t < 3 \\ \dfrac{1}{45}t(6 - t) & 3 \le t \le 6 \\ 0 & \text{otherwise} \end{cases}$$

  a Sketch the graph of f.   [3]

  b Write down the value of $P(T = 3)$.   [1]

  c Find the probability that a randomly selected supporter has to queue for at least 3 hours in order to obtain tickets.   [2]

  d Show that the median queuing time is 2.5 hours.   [2]

  e Calculate $P(\text{median} < T < \text{mean})$.   [6]

AQA MS2B June 2005

6 A continuous random variable X has probability density function defined by

$$f(x) = \begin{cases} kx^2 & 0 \le x \le 3 \\ 9k & 3 \le x \le 4 \\ 0 & \text{otherwise} \end{cases}$$

a Sketch the graph of f. [3]

b Show that the value of k is $\dfrac{1}{18}$. [4]

c **i** Write down the median value of X.

ii Calculate the value of the lower quartile of X. [4]

AQA MS2B January 2013

7 The random variable X has probability density function defined by

$$f(x) = \begin{cases} \dfrac{1}{40}(x+7) & 1 \le x \le 5 \\ 0 & \text{otherwise} \end{cases}$$

a Sketch the graph of f. [2]

b Find the **exact** value of $E(X)$. [3]

c Prove that the distribution function, F, for $1 \le x \le 5$, is defined by

$$F(x) = \frac{1}{80}(x+15)(x-1)$$ [4]

d Hence, or otherwise:

i find $P(2.5 \le X \le 4.5)$; [2]

ii show that the median, m, of X satisfies the equation

$m^2 + 14m - 55 = 0$. [3]

e Calculate the value of the median of X, giving your answer to three decimal places. [2]

AQA MS2B January 2012

8 X and Y are independent random variables such that

$E(X) = 8$, $\text{Var}(X) = 2$

$E(Y) = 10$, $\text{Var}(Y) = 3$

Find

a $E(3X + Y)$ [1]

b $\text{Var}(X + Y)$ [1]

c $\text{Var}(2X + 8Y)$ [2]

d the standard deviation of $4X - 3Y$. [3]

12 The Exponential Distribution

Introduction

The exponential distribution is a special continuous distribution with a very important link to the Poisson distribution.

Examples of continuous random variables which may follow an exponential model are:

- The time between accidents on a large building site.
- The length of material between randomly occurring faults on a roll of material.
- The time between successive particle emissions from a radioactive source.

Objectives

By the end of this chapter you should know how to...

- Recognise the probability density function $f(x)$ for an exponential distribution.
- Use the formulae for the mean, variance and standard deviation.
- Derive the cumulative distribution function $F(x)$.
- Calculate probabilities using either $f(x)$ or $F(x)$.
- Use the relationship between Poisson events and an exponential distribution.
- Use the 'no memory' property of an exponential distribution.

Recap

You will need to remember...

Continuous random variables

- Calculating probabilities by integrating the probability density function $f(x)$.
- Obtaining the cumulative distribution function $F(x)$ from $f(x)$.
- Calculating probabilities using $F(x)$.
- Obtaining $f(x)$ by differentiating $F(x)$.

Poisson distribution

If $X \sim \text{Po}(\lambda)$,

- mean = variance = λ
- $P(X = x) = e^{-\lambda} \dfrac{\lambda^x}{x!}$
- Using cumulative Poisson probabilities (Table 2)

From pure mathematics

- If $f(x) = e^{-kx}$, then $f'(x) = -ke^{-kx}$
- $\displaystyle\int e^{-kx}\,dx = -\frac{1}{k}e^{-kx} + c$
- The solution of exponential equations of the form $e^{-kx} = b$
- As $x \to \infty$, $e^{-x} \to 0$

  Try values on your calculator, with $x = 10, 50, 100, 200$, etc.

12.1 The exponential distribution

Probability density function f(x)

A continuous random variable X has an **exponential distribution** if, for some positive constant λ, it has a **probability density function** defined by

$$f(x) = \begin{cases} \lambda e^{-\lambda x} & x \geq 0 \\ 0 & \text{otherwise} \end{cases}$$

You can write

$$X \sim \text{Exp}(\lambda)$$

where λ is the **parameter** of the exponential distribution.

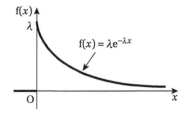

Note that

▶ x cannot be negative.

▶ There is no upper limit for x.

▶ As $x \to \infty$, $f(x) \to 0$

> **Note**
>
> $f(x)$, $E(X)$ and $Var(X)$ are given in the examination formulae booklet for reference.

Mean, variance and standard deviation

It can be shown that

$$\mathbf{mean} = E(X) = \frac{1}{\lambda}$$

$$\mathbf{variance} = \text{Var}(X) = \frac{1}{\lambda^2}$$

$$\mathbf{standard\ deviation} = \frac{1}{\lambda}$$

> **Note**
>
> Mean = standard deviation = $\dfrac{1}{\lambda}$

Example 1

The continuous random variable X has an exponential distribution with probability density function defined by

$$f(x) = \begin{cases} \dfrac{1}{4} e^{-\frac{1}{4}x} & x \geq 0 \\ 0 & \text{otherwise} \end{cases}$$

State

a the value of λ, the parameter of the distribution of X

b the mean of X

c the variance of X.

a $\lambda = \dfrac{1}{4}$

b $\text{mean} = \dfrac{1}{\lambda} = \dfrac{1}{\frac{1}{4}} = 4$

c $\text{variance} = \dfrac{1}{\lambda^2} = \dfrac{1}{\left(\frac{1}{4}\right)^2} = 16$

Finding probabilities using f(x)

Since X is a continuous random variable, probabilities are given by the area under the graph of f between specified limits.

See Section 11.1

$$P(a < X < b) = \int_a^b f(x)\,dx$$

$$= \int_a^b \lambda e^{-\lambda x}\,dx$$

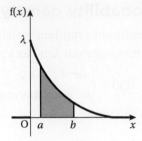

Example 2

Question

A continuous random variable has probability density function defined by

$$f(x) = \begin{cases} 2e^{-2x} & x \geq 0 \\ 0 & \text{otherwise} \end{cases}$$

Find

a $P(X < 3)$

b $P(X > 3)$

c $P(1 \leq X \leq 3)$

Answer

This is an exponential distribution with parameter 2.

a $P(X < 3) = \int_0^3 2e^{-2x}\,dx$

$= \left[-e^{-2x} \right]_0^3$

$= -e^{-6} + e^0 \qquad e^0 = 1$

$= 1 - e^{-6}$

$= 0.9975 \cdots = 0.998 \ (3\ \text{sf})$

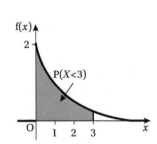

b $P(X > 3) = 1 - P(X < 3)$

$= 1 - (1 - e^{-6})$

$= e^{-6}$

$= 0.002478 \cdots = 0.00248 \ (3\ \text{sf})$

Alternatively,

$$P(X > 3) = \int_3^\infty 2e^{-2x}\,dx$$

$= \left[-e^{-2x} \right]_3^\infty$

$= 0 - (-e^{-6}) \qquad \text{As } x \to \infty,\ e^{-2x} \to 0$

$= e^{-6}$

$= 0.002478 \cdots = 0.00248 \ (3\ \text{sf})$, as before

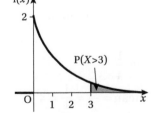

(continued)

The Exponential Distribution

Answer

(continued)

c $\quad P(1 \leq X \leq 3) = \int_1^3 f(x)\,dx$

$\qquad = \int_1^3 2e^{-2x}\,dx$

$\qquad = \left[-e^{-2x}\right]_1^3$

$\qquad = -e^{-6} - (-e^{-2}) = e^{-2} - e^{-6}$

$\qquad = 0.1328\cdots = 0.133 \,(3\,\text{sf})$

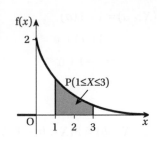

Cumulative distribution function F(x)

Rather than integrate f(x) each time to find various probabilities, it is useful to derive the *general case* by setting up the **cumulative distribution function** F, where

$$F(x) = P(X \leq x) = P(X < x) \qquad\qquad \text{See Section 11.2}$$

When $x < 0$, $F(x) = 0$

When $x \geq 0$,

$$F(x) = P(X \leq x) = \int_0^x f(t)\,dt$$

$$\qquad = \int_0^x \lambda e^{-\lambda t}\,dt$$

$$\qquad = \left[-e^{-\lambda t}\right]_0^x$$

$$\qquad = -e^{-\lambda x} - (-e^0)$$

$$\qquad = 1 - e^{-\lambda x}$$

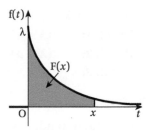

If $X \sim \text{Exp}(\lambda)$,

$$F(x) = \begin{cases} 0 & x < 0 \\ 1 - e^{-\lambda x} & x \geq 0 \end{cases}$$

Note that, as $x \to \infty$, $e^{-\lambda x} \to 0$, so $F(x) \to 1$.

Note that, since X is continuous, if you differentiate $F(x)$ you obtain f(x),

so $\quad \dfrac{d}{dx}F(x) = \dfrac{d}{dx}(1 - e^{-\lambda x}) \qquad\qquad$ See Section 11.2

$$\qquad\qquad = \lambda e^{-\lambda x}$$

$$\qquad\qquad = f(x)$$

Finding probabilities using F(x)

Important probability results

F(x) can be used to derive general results that you can use straight away in your solutions.

Result 1

$$P(X < a) = F(a)$$

$$\qquad = 1 - e^{-\lambda a}$$

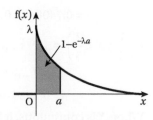

Result 2

$$P(X > a) = 1 - F(a)$$
$$= 1 - (1 - e^{-\lambda a})$$
$$= 1 - 1 + e^{-\lambda a}$$
$$= e^{-\lambda a}$$

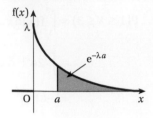

Result 3

$$P(a < X < b) = P(X < b) - P(X < a)$$
$$= F(b) - F(a)$$
$$= 1 - e^{-\lambda b} - (1 - e^{-\lambda a})$$
$$= 1 - e^{-\lambda b} - 1 + e^{-\lambda a}$$
$$= e^{-\lambda a} - e^{-\lambda b}$$

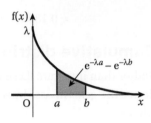

Alternatively, using Result 2 directly:

$$P(a < X < b) = P(X > a) - P(X > b)$$
$$= e^{-\lambda a} - e^{-\lambda b}$$

Summarising:

> If $X \sim \text{Exp}(\lambda)$
>
> **Result 1:** $P(X < a) = 1 - e^{-\lambda a}$
> **Result 2:** $P(X > a) = e^{-\lambda a}$
> **Result 3:** $P(a < X < b) = e^{-\lambda a} - e^{-\lambda b}$

Example 3

The continuous random variable X has probability density function defined by

$$f(x) = \begin{cases} 3e^{-3x} & x \geq 0 \\ 0 & \text{otherwise} \end{cases}$$

Determine

a $P(X < 1.4)$ **b** $P(X > 0.1)$

c $P\left(\dfrac{1}{3} < X < \dfrac{2}{3}\right)$ **d** $P(X = 0.5)$

This is an exponential distribution with $\lambda = 3$.

a $P(X < 1.4) = 1 - e^{-3 \times 1.4}$ Result 1
$$= 1 - e^{-4.2}$$
$$= 0.9850\cdots = 0.985 \text{ (3 sf)}$$

b $P(X > 0.1) = e^{-3 \times 0.1}$ Result 2
$$= e^{-0.3}$$
$$= 0.7408\cdots = 0.741 \text{ (3 sf)}$$

c $P\left(\dfrac{1}{3} < X < \dfrac{2}{3}\right) = e^{-3 \times \frac{1}{3}} - e^{-3 \times \frac{2}{3}}$ Result 3
$$= e^{-1} - e^{-2}$$
$$= 0.2325\cdots = 0.233 \text{ (3 sf)}$$

d When X is continuous, it is impossible for X to take an exact value,

so $P(X = 0.5) = 0$ See Section 11.1

Example 4

The continuous random variable X has an exponential distribution with parameter 5.

a Find the mean μ.

b Find the median m.

a mean $= \dfrac{1}{\lambda} = \dfrac{1}{5} = 0.2$

so $\mu = 0.2$

b By definition, $P(X < m) = F(m) = 0.5$ See Section 11.2

Now, since $P(X < m) = 1 - e^{-5m}$ Result 1

$$1 - e^{-5m} = 0.5$$

$$e^{-5m} = 0.5$$

$$e^{5m} = \dfrac{1}{0.5} = 2$$ See AS Textbook Chapter 12

$$5m = \ln 2$$

$$m = \dfrac{\ln 2}{5} = 0.1386\cdots = 0.139 \,(3\text{ sf})$$

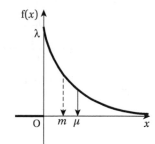

Notice that, since median $= 0.139$ and mean $= 0.2$

median $<$ mean.

In fact, this relationship is true for *any* exponential distribution.

Example 5

The continuous random variable X has an exponential distribution with mean 10.

Find the probability that X is less than the mean.

$X \sim \text{Exp}(\lambda)$

mean $= \dfrac{1}{\lambda} = 10$, so $\lambda = \dfrac{1}{10}$

$P(X < \text{mean}) = P(X < 10)$

$\qquad = 1 - e^{-\frac{1}{10} \times 10}$

$\qquad = 1 - e^{-1}$

$\qquad = 0.6321\cdots = 0.632 \,(3\text{ sf})$

> **Note**
>
> $P(X < \text{median}) = 0.5$
> $P(X < \text{mean}) = 0.632$
> This also confirms that the
> median is less than the mean.

Notice that, in general, for *any* exponential distribution with parameter λ,

$P(X < \text{mean}) = 1 - e^{-(\lambda \times \frac{1}{\lambda})}$

$\qquad = 1 - e^{-1}$

$\qquad = 0.632 \,(3\text{ sf})$

Example 6

Question

The continuous random variable T has cumulative distribution function F defined by

$$F(t) = \begin{cases} 0 & t < 0 \\ 1 - e^{-6t} & t \geq 0 \end{cases}$$

a i Find the value of c such that $P(X < c) = 0.75$.

ii Find the value of d such that $P(X > d) = 0.005$.

b i Find the probability density function $f(t)$ and state the value of the parameter λ.

ii Find the standard deviation of T.

a i $P(X < c) = F(c)$

$$= 1 - e^{-6c}$$

So, $1 - e^{-6c} = 0.75$

$$e^{-6c} = 0.25$$

$$e^{6c} = \frac{1}{0.25} = 4$$

$$6c = \ln 4$$

$$c = \frac{\ln 4}{6} = 0.2310\cdots = 0.231 \ (3 \text{ sf})$$

> **Note**
>
> Since $F(c) = 0.75$, c is the upper quartile of the distribution (Section 11.2).

ii $P(X > d) = 1 - F(d)$

$$= 1 - (1 - e^{-6d})$$

$$= e^{-6d} \qquad \leftarrow \text{You could quote Result 2 directly.}$$

So, $e^{-6d} = 0.005$

$$e^{6d} = \frac{1}{0.005} = 200$$

$$6d = \ln 200$$

$$d = \frac{\ln 200}{6} = 0.8830\cdots = 0.883 \ (3 \text{ sf})$$

b i $f(t) = \dfrac{d}{dt} F(t)$

$$= \frac{d}{dt}(1 - e^{-6t})$$

$$= 6e^{-6t}$$

So, $f(t) = \begin{cases} 6e^{-6t} & t \geq 0 \\ 0 & \text{otherwise} \end{cases}$

> **Note**
>
> This is an exponential distribution with $\lambda = 6$.

ii Standard deviation $= \dfrac{1}{\lambda} = \dfrac{1}{6}$

Exercise 1

1 The continuous random variable X has an exponential distribution with probability density function defined by

$$f(x) = \frac{1}{2}e^{-\frac{1}{2}x}, x \geq 0$$

Find

a $P(X > 2)$ **b** $P(X < 1.4)$ **c** $P(0.5 < X < 2.5)$

d $P(X = 6)$ **e** $E(X)$ **f** $\text{Var}(X)$

2 The continuous random variable Y has an exponential distribution, where

$$f(y) = 0.1e^{-0.1y}, y \geq 0$$

Find the probability that

a Y lies between 9 and 15

b Y is greater than 6

c Y is less than 10

d Y is not equal to the mean.

3 The continuous random variable X has a probability density function defined by

$$f(x) = 8e^{-8x}, x \geq 0$$

a State the value of the mean.

b Calculate the value of the median.

c Find the probability that X lies between the median and the mean.

4 The continuous random variable X has a probability density function defined by

$$f(x) = \lambda e^{-\lambda x}, x \geq 0$$

The standard deviation of X is $\frac{2}{3}$.

Find

a the value of λ **b** $P(X > 4)$

c $P(X < E(X))$ **d** the median of X

e the value of the constant c such that $P(X > c) = 0.05$.

5 A continuous random variable T has cumulative distribution function $F(t)$ defined by

$$F(t) = \begin{cases} 0 & t < 0 \\ 1 - e^{-4t} & t \geq 0 \end{cases}$$

Find

a $P(0.5 < T < 1)$

b the value of q such that $P(T < q) = 0.25$

c the probability density function $f(t)$

d $\text{Var}(T)$.

6 The continuous random variable X can be modelled by an exponential distribution with

$$P(X < x) = 1 - e^{-0.4x}, \ x \geq 0$$

a Determine

   i $P(X < 3)$,   **ii** $P(4 \leq X \leq 6)$

   iii $P(X > 2)$ **iv** the median of X

   v the quartiles and interquartile range of X.

b **i** Find f(x), the probability density function of X.

   ii Write down the mean of X.

12.2 Link between Poisson and exponential distributions

The lengths of intervals between successive events in a Poisson distribution can be shown to follow an exponential distribution.

For example, suppose that the discrete random variable X is the number of vehicles passing an electronic sensor used to monitor traffic flow and that the vehicles pass one at a time, independently and randomly, at a constant average rate of λ per minute. The conditions for a Poisson distribution are satisfied, so $X \sim \text{Po}(\lambda)$.   See Section 10.1

The Poisson variable for the number of vehicles passing per minute is *discrete*. However, since time is a *continuous* variable, a probability model for the time, in minutes, between the arrival of one vehicle and the next will need to be a continuous function.

The average number of vehicles passing the sensor in 1 minute is λ, so the average number of vehicles passing the sensor in t minutes is λt.

If Y is the number of vehicles passing the sensor in t minutes, then $Y \sim \text{Po}(\lambda t)$.

Now   P(no vehicles pass the sensor in t minutes) = $P(Y = 0) = e^{-\lambda t}$ See Section 10.1

so,   P($at\ least\ one$ vehicle passes the sensor in t minutes) = $1 - e^{-\lambda t}$

Let T be the length of time, in minutes, between successive vehicles.

So $P(T < t)$ = P(at least one vehicle passes the sensor in t minutes)

$$= 1 - e^{-\lambda t}$$

But $P(T < t) = F(t)$,

so $F(t) = 1 - e^{-\lambda t}$

> **Note**
>
> If the time is less than t minutes, you know that at least one vehicle passes in t minutes.

Differentiating with respect to t:

$$f(t) = F'(t)$$

$$= \frac{d}{dt}(1 - e^{-\lambda t})$$

$$= \lambda e^{-\lambda t}$$

This is the probability density function for an exponential distribution with parameter λ.

So, if X is the *number of vehicles* passing the sensor in *one minute*, where
   $X \sim \text{Po}(\lambda)$,
then T, *the length of time in minutes* between successive vehicles,
can be modelled by an exponential distribution with *same parameter* λ,
so $T \sim \text{Exp}(\lambda)$.

In general:

> The length of intervals between Poisson events with parameter λ has an exponential distribution also with parameter λ.

> If $X \sim \text{Po}(\lambda)$,
> where X is the number of occurrences of an event in a specified unit of time or space,

> then $T \sim \text{Exp}(\lambda)$,
> where T is the length of intervals between successive occurrences.

> **Note**
>
> The value of λ is the *same* in both distributions.

The unit for the lengths of intervals in the exponential distribution is the same as the unit in the corresponding Poisson distribution.

For example:

▶ If X is the number of emergency call-outs *per hour*,
   then T is the time in *hours* between successive call-outs.
▶ If X is the number of flaws in *a metre* of fabric,
   then T is the distance in *metres* between successive flaws.

Notice the relationship between the means:

If $X \sim \text{Po}(\lambda)$, mean $= \lambda$
If $T \sim \text{Exp}(\lambda)$, mean $= \dfrac{1}{\lambda}$

For example:

▶ If the mean number of call-outs per *hour* is 10,
   the mean time between call-outs is $\frac{1}{10}$ of an *hour* $= 6$ minutes.
▶ If the mean number of flaws in 1 *metre* of fabric is 5,
   the mean length between flaws is $\frac{1}{5}$ of a *metre* $= 20$ cm.

Example 7

The number of particle emissions per hour from a radioactive source has a Poisson distribution with mean 5.

a   Find the probability that there are at least 9 emissions in a 1-hour period.

b   State the distribution of the length of time, T hours, between successive emissions.

c   Find the mean time between emissions, giving your answer in minutes.

d   Find the probability that the time between emissions is

   i   exactly 12 minutes

   ii less than 20 minutes

   iii between 15 minutes and 40 minutes.

a   Let X be the number of emissions in 1 *hour*, where $X \sim \text{Po}(\lambda)$.

Since the mean is 5, $\lambda = 5$, so $X \sim \text{Po}(5)$

$P(X \geq 9) = 1 - P(X \leq 8)$ ← Use Table 2 with $\lambda = 5.0$ See Section 10.1

$\qquad\qquad = 1 - 0.9319$

$\qquad\qquad = 0.0681 \text{ (3 sf)}$

b   If T is the time *in hours* between emissions, then T has an exponential distribution with $\lambda = 5$, so $T \sim \text{Exp}(5)$

c $\text{mean} = \dfrac{1}{\lambda} = \dfrac{1}{5}$ so the mean time is $\dfrac{1}{5}$ h $= 12$ minutes

d i   Since T is continuous, $P\left(T = \dfrac{1}{5}\right) = 0$ See Section 11.1

ii $20 \min = \dfrac{1}{3}$ h

$\qquad P\left(T < \dfrac{1}{3}\right) = 1 - e^{-5 \times \frac{1}{3}}$ Result 1

$\qquad\qquad\qquad = 0.8111\cdots = 0.811 \text{ (3 sf)}$

iii $15 \min = \dfrac{1}{4}$ h, $40 \min = \dfrac{2}{3}$ h

$\qquad P\left(\dfrac{1}{4} < T < \dfrac{2}{3}\right) = e^{-5 \times \frac{1}{4}} - e^{-5 \times \frac{2}{3}}$ Result 3

$\qquad\qquad\qquad\qquad = 0.2508\cdots = 0.251 \text{ (3 sf)}$

Example 8

Weak spots occur in a cable independently and randomly, at an average rate of 0.025 per metre. The number, X, of weak spots in a metre length of cable has a Poisson distribution.

a   State the distribution of Y, the distance, in metres, between successive weak spots.

b   Find

   i $P(Y > 20)$ ii $P(60 < Y < 100)$

   iii   the mean distance, in metres, between weak spots.

a   If X is the number of weak spots in 1 *metre*, then
$X \sim \text{Po}(\lambda)$ with $\lambda = 0.025$, so $X \sim \text{Po}(0.025)$

If Y is the distance in *metres* between weak spots, then

$Y \sim \text{Exp}(\lambda)$ with $\lambda = 0.025$, so $Y \sim \text{Exp}(0.025)$

b i $P(Y > 20) = e^{-0.025 \times 20}$ Result 2

$\qquad\qquad\quad = 0.6065\cdots = 0.607 \text{ (3 sf)}$

ii $P(60 < Y < 100) = e^{-0.025 \times 60} - e^{-0.025 \times 100}$ Result 3

$\qquad\qquad\qquad\qquad = 0.1410\cdots = 0.141 \text{ (3 sf)}$

iii $\text{mean} = \dfrac{1}{\lambda} = \dfrac{1}{0.025} = 40$

The mean distance between weak spots is 40 m.

Example 9

The continuous random variable T minutes, the time between successive emails to Caleb's office computer, may be modelled by an exponential distribution with mean 10.

a Find the probability that the time between the arrival of an email and the next is between 6 and 12 minutes.

b State the distribution of Y, the number of emails received in an hour.

c Find the probability that Caleb receives more than 8 emails between 10.00 am and 11.00 am.

a T is the time, in *minutes*, between successive emails, where $T \sim \text{Exp}(\lambda)$.

Since mean $= \dfrac{1}{\lambda}$, $10 = \dfrac{1}{\lambda}$, so $\lambda = 0.1$ and $T \sim \text{Exp}(0.1)$

$P(6 < T < 12) = e^{-0.1 \times 6} - e^{-0.1 \times 12}$

$\qquad\qquad\qquad = e^{-0.6} - e^{-1.2}$

$\qquad\qquad\qquad = 0.2476\cdots = 0.248$ (3 sf)

b If X is the number of emails in 1 *minute*, then $X \sim \text{Po}(\lambda)$,

so $X \sim \text{Po}(0.1)$

If Y is the number of emails in 1 hour, then $Y \sim \text{Po}(0.1 \times 60)$,

so $Y \sim \text{Po}(6)$

c $P(Y > 8) = 1 - P(Y \leq 8)$ $\qquad \leftarrow$ Use Table 2 with $\lambda = 6.0$

$\qquad\qquad = 1 - 0.8472$

$\qquad\qquad = 0.1528 = 0.153$ (3 sf)

12.3 'No memory' property

As you have seen, the lengths of intervals between Poisson events have an exponential distribution. The events are random, and this leads to a remarkable property of the exponential distribution, known as the **'no memory' rule**.

> **If the continuous random variable X has an exponential distribution with parameter λ, then**
>
> $$P(X > a+b \mid X > a) = P(X > b)$$

This means that, even if you know that no events have occurred in the *first a units* of time or space, the probability that no events occur in the *next b units* is independent of everything that has gone before.

Proof

When $X \sim \text{Exp}(\lambda)$,

$P(X > a) = e^{-\lambda a}$, $P(X > b) = e^{-\lambda b}$ and $P(X > a + b) = e^{-\lambda(a+b)}$

Now $P(X > a+b \mid X > a) = \dfrac{P(X > a+b \text{ and } X > a)}{P(X > a)}$

$$= \frac{P(X > a+b)}{P(X > a)}$$

See AS Textbook section 13.3

$$= \frac{e^{-\lambda(a+b)}}{e^{-\lambda a}}$$

$$= e^{-\lambda b}$$

$$= P(X > b)$$

So, for example,

$P(X > 10 \mid X > 3) = P(X > 7)$

$P(X > 0.9 \mid X > 0.3) = P(X > 0.6)$

Similarly, it can be shown that

$$\mathbf{P(X < a+b \mid X > a) = P(X < b)}$$

For example

$P(X < 8 \mid X > 2) = P(X < 6)$

$P(X < 7.5 \mid X > 7) = P(X < 0.5)$

Example 10

Question

Selah is buying lengths of curtain material. She finds a roll of material that has been reduced in price because there are faults along the edge occurring randomly throughout the roll.

The length, X metres, between successive faults has an exponential distribution with parameter 0.4.

a   Selah notices that there are no faults in the first metre of the roll. What is the probability that there will not be any faults in the first 3 metres of the roll?

b   Selah buys 2.5 metres of material. Given that there are no faults in the first 2 metres of the roll, find the probability that there is at least one fault in the piece of material that she buys.

Answer

X is the length in metres between successive faults in the roll of material, where $X \sim \text{Exp}(\lambda)$ with $\lambda = 0.4$, that is $X \sim \text{Exp}(0.4)$.

a   P(no faults in first 3 metres, given that there are no faults in first metre)

   = P(there are no faults in next 2 metres)

   = P($X > 2$)   'No memory' rule

   = $e^{-0.4 \times 2}$

   = $e^{-0.8}$

   = $0.4493\cdots = 0.449$ (3 sf)   Result 2

(continued)

(continued)

b P(at least one fault in the 2.5 metre length of material, given there are no faults in the first 2 metres)

= P(there is at least one fault in the last 0.5 metres)

= P($X < 0.5$) 　　　　　'No memory' rule

= $1 - e^{-0.4 \times 0.5}$

= $1 - e^{-0.2}$

= $0.1812 \cdots = 0.181$ (3 sf) 　　Result 1

Example 11

Sadia arranges to meet Arlene at a coffee bar on Saturday evenings at 8.00 pm. Past experience suggests that Arlene will arrive at the coffee bar X minutes after 8.00 pm, where X may be modelled by an exponential distribution with parameter 0.05.

a Find the mean and standard deviation of the number of minutes after 8.00 pm that Arlene will arrive at the coffee bar.

b If Sadia arrives at 8.20 pm, find the probability that Arlene will already have arrived.

c Sadia arrives at 8.20 pm and finds that Arlene has not yet arrived. Find the probability that Arlene will arrive after 8.30 pm.

　　　　　　　　　　　　　　　　　　　　AQA SS05 June 2009

a X is the number of minutes past 8.00 pm that Arlene arrives.

$\lambda = 0.05$, so $X \sim \text{Exp}(0.05)$

mean $= \dfrac{1}{\lambda} = \dfrac{1}{0.05} = 20$,　standard deviation $= \dfrac{1}{0.05} = 20$

> **Note**
>
> mean = s.d. = $\dfrac{1}{\lambda}$

b P(Arlene will have arrived by 8.20 pm)

= P($X < 20$)

= $1 - e^{-0.05 \times 20}$

= $1 - e^{-1}$

= $0.6321 \cdots = 0.632$ (3 sf)

c P(Arlene will arrive after 8.30 pm given that she has not arrived by 8.20 pm)

= P($X > 30 \mid X > 20$)

= P($X > 10$)　　'No memory' rule

= $e^{-0.05 \times 10}$

= $e^{-0.5}$

= $0.6065 \cdots = 0.607$ (3 sf)

Exercise 2

1 The number of calls per hour to an emergency control centre has a Poisson distribution with mean 10.

　a Find the probability that there are exactly 6 calls between 2.30 pm and 3.30 pm.

　b State the distribution of T, the time in hours, between calls.

c A call is received at 11.15 am. Find the probability that

  i no more calls have been received by 11.35 am

  ii the next call is received between 11.25 am and 11.30 am.

2 Call-outs to a car breakdown service occur at an average rate of 5 per 8-hour shift. The number of call-outs may be modelled by a Poisson distribution.

  a **i** State the mean number of call-outs per hour.

   ii Find the probability that there will be exactly 2 call-outs in a randomly chosen hour.

  b **i** Find the mean time, in hours, between call-outs.

   ii State the distribution of T, the time in hours between call-outs.

  c Find the probability that the time between successive call-outs is

   i exactly 1 hour

   ii between 30 minutes and 2 hours.

3 The continuous random variable X has probability density function f(x) where

$$f(x) = \frac{1}{3}e^{-\frac{1}{3}x}, x \geq 0$$

Find

  a $P(X > 5)$ **b** $P(X < 2)$

  c $P(X > 9 \,|\, X > 4)$ **d** $P(X < 7 \,|\, X > 5)$

4 The lifetime, T hours, of a certain type of electrical component can be modelled by a continuous random variable with probability density function given by

  $f(t) = 0.001e^{-0.001t}, t \geq 0$

  a Find the mean lifetime, in hours, of a component of this type.

  b It is known that a particular component has lasted for at least 900 hours. Find the probability that the lifetime of the component is at least 1200 hours.

  c It is known that a particular component has lasted for at least 1500 hours. Find the probability that the lifetime of the component is less than 2500 hours.

5 A fluorescent light develops a fault which causes it to flash randomly.

  The continuous random variable, T minutes, is the time between flashes, where T has an exponential distribution with parameter $\frac{1}{4}$.

  a Elena notices that the light flashes at 10.30 am and that it has not flashed again by 10.32 am.

   i Find the probability that the light will not have flashed by 10.35 am.

   ii Find the probability that the light will have flashed by 10.36 am.

  b Find the probability that the light will flash exactly 3 times in a 5-minute interval.

6 The continuous random variable X has probability density function $f(x)$ where

$f(x) = \lambda e^{-\lambda x}, x \geq 0$

a Show that $P(X < a + b \mid X > a) = P(X < b)$.

b If $\lambda = 0.1$, find the probability that

    i X is less than 7, given that X is greater than 5

    ii X is greater than 7, given that X is greater than 5.

12.4 Further applications

Example 12

The continuous random variable X has an exponential distribution with probability density function $f(x)$, where

$$f(x) = \begin{cases} 20e^{-20x} & x \geq 0 \\ 0 & \text{otherwise} \end{cases}$$

a Derive the cumulative distribution function of X.

b Find the probability that

    i $X < 0.005$ **ii** $X \neq 0.5$

    iii X lies between 0.02 and 0.09.

c Find the *exact* value of the constant c such that $P(X < c) = 0.8$.

a When $x < 0$, $F(x) = 0$

When $x \geq 0$,

$F(x) = P(X \leq x)$

$\qquad = \displaystyle\int_0^x 20e^{-20t} \, dt \qquad \leftarrow$ Use $f(t) = 20e^{-20t}$

$\qquad = \left[-e^{-20t} \right]_0^x$

$\qquad = -e^{-20x} - (-e^0)$

$\qquad = 1 - e^{-20x}$

So, $\quad F(x) = \begin{cases} 0 & x < 0 \\ 1 - e^{-20x} & x \geq 0 \end{cases}$

b i $P(X < 0.005) = F(0.005) \leftarrow$ Substitute $x = 0.005$ directly into $F(x)$

$\qquad\qquad\qquad = 1 - e^{-20 \times 0.005}$

$\qquad\qquad\qquad = 1 - e^{-0.1}$

$\qquad\qquad\qquad = 0.09516\cdots = 0.0952 \text{ (3 sf)}$

ii Since it is certain that X is not an exact value, $P(X \neq 0.5) = 1$

iii $P(0.02 < X < 0.09)$

$\qquad\qquad = F(0.09) - F(0.02)$

$\qquad\qquad = 1 - e^{-20 \times 0.09} - (1 - e^{-20 \times 0.02})$

$\qquad\qquad = e^{-0.4} - e^{-1.8} \leftarrow$ You could quote this straight away using Result 3

$\qquad\qquad = 0.5050\cdots = 0.505 \text{ (3 sf)}$

(continued)

c $P(X < c) = 0.8$, so $F(c) = 0.8$

$1 - e^{-20c} = 0.8$

$e^{-20c} = 0.2$

$e^{20c} = 5$

$20c = \ln 5$

$c = \dfrac{1}{20}\ln 5$ ← This answer is *exact*.

Example 13

Rowena's sewing machine occasionally misses a stitch. When Rowena sews a seam, the distance, X metres, to the first missed stitch may be modelled by an exponential distribution with mean 0.8.

a   Show that the parameter, λ, of this distribution is 1.25.

b   Find the probability that the machine misses a stitch before Rowena has sewn 0.5 metres of a seam.

c   Rowena has to sew a seam of length 1.4 m.

  i   Find the probability that no stitches have been missed when she has sewn half of this seam.

  ii   Assuming that no stitches were missed in the first half of the seam, find the probability that there is at least one missed stitch in the completed seam.

AQA SS05 June 2010

a X is the distance, in m, to the first missed stitch and X may be modelled by an exponential distribution with mean 0.8.

If the parameter of this distribution is λ,

then   mean $= \dfrac{1}{\lambda} = 0.8$ so $\lambda = \dfrac{1}{0.8} = 1.25$

b $P(X < 0.5) = 1 - e^{-1.25 \times 0.5}$

$= 1 - e^{-0.625}$

$= 0.4647... = 0.465 \text{ (3 sf)}$

c **i**   P(no missed stitches in first half of seam)

$= P(X > 0.7)$

$= e^{-1.25 \times 0.7}$

$= e^{-0.875}$

$= 0.4168... = 0.417 \text{ (3 sf)}$

  ii   P(at least one missed stitch in completed seam | none in first half)

$= P(X < 1.4 \mid X > 0.7)$

$= P(X < 0.7)$ 'No memory' rule

$= 1 - 0.4168...$ From part **i**

$= 0.5831... = 0.583 \text{ (3 sf)}$

Example 14

The time, T minutes, between calls to an emergency control centre has an exponential distribution given by

$$f(t) = 0.5e^{-0.5t}, \; t \geq 0$$

a Find the mean time, in minutes, between calls.

The number of calls per minute to the control centre has a Poisson distribution.

b Find the mean number of calls per minute to the control centre.

c Find the probability that there are

  i at most 3 calls to the control centre in a 1-minute interval

  ii exactly 6 calls to the control centre in a 7-minute interval.

a T has an exponential distribution with $\lambda = 0.5$, so $T \sim \text{Exp}(0.5)$,

so mean $= \dfrac{1}{\lambda} = \dfrac{1}{0.5} = 2$

The mean time between calls is 2 minutes.

b Since the mean time between calls is 2 minutes, the mean number of calls in a 1-minute interval is 0.5.

c **i** If X is the number of calls in a 1-minute interval, then $X \sim \text{Po}(0.5)$

    $P(X \leq 3) = 0.9982 = 0.998$ (3 sf)   Use Table 2 with $\lambda = 0.5$

  ii If Y is the number of calls in a 7-minute interval, then

    $Y \sim \text{Po}(7 \times 0.5)$, so $Y \sim \text{Po}(3.5)$ See Section 10.1

    $P(Y = 6) = e^{-3.5} \dfrac{3.5^6}{6!}$ $\lambda = 3.5$ is not in Table 2.

              $= 0.07709\cdots = 0.0771$ (3 sf)

Example 15

Imran has recently retired and rarely wears a suit. He owns a dark suit which he wears for formal occasions such as weddings. The time, in days, before he next wears the suit may be modelled by an exponential distribution with parameter 0.0045.

a Find:

  i the mean of this exponential distribution;

  ii the probability that Imran will wear this suit during the next 100 days;

  iii the probability that Imran will not wear the suit for at least a year (365 days).

b The number of occasions per year on which Imran wears the suit will follow a Poisson distribution. Find the mean of this distribution.

c Imran also owns a light-coloured suit which he wears for social occasions. The number of occasions per year on which he wears the light-coloured suit may be modelled by a Poisson distribution with mean 1.72.

State the distribution of the number of occasions per year on which Imran wears either his dark suit or his light-coloured suit.

AQA SS05 June 12

a Let T be the time in days before Imran next wears his dark suit.

Then $T \sim \text{Exp}(\lambda)$ with $\lambda = 0.0045$.

i mean $= \dfrac{1}{\lambda} = \dfrac{1}{0.0045} = 222.2\cdots = 222$ days (3 sf)

ii $P(T < 100) = 1 - e^{-0.0045 \times 100} = 0.3623\cdots = 0.362$ (3 sf)

iii $P(T > 365) = e^{-0.0045 \times 365} = 0.1934\cdots = 0.193$ (3 sf)

b If X is the number of occasions in t days on which Imran wears his dark suit then $X \sim \text{Po}(\lambda t)$.

So in a year (365 days), $X \sim \text{Po}(0.0045 \times 365)$, so $X \sim \text{Po}(1.6425)$

Mean number of occasions per year $= 1.6425 = 1.64$ (2 dp)

c Let Y be the number of occasions per year on which Imran wears his light-coloured suit, where $Y \sim \text{Po}(1.72)$.

The number of occasions per year on which Imran wears either his dark suit or his light-coloured suit is given by $X + Y$.

Now $X + Y \sim \text{Po}(1.64 + 1.72)$

so $X + Y \sim \text{Po}(3.36)$

See Section 10.4

The number of occasions per year on which Imran wears either his dark suit or his light-coloured suit has a Poisson distribution with mean 3.36.

Summary

If $X \sim \text{Exp}(\lambda)$:

Probability density function f(x)

$$f(x) = \begin{cases} \lambda e^{-\lambda x} & x \geq 0 \\ 0 & \text{otherwise} \end{cases}$$

Cumulative distribution function F(x)

$$F(x) = \begin{cases} 0 & x < 0 \\ 1 - e^{-\lambda x} & x \geq 0 \end{cases}$$

Mean, variance and standard deviation

$$E(X) = \frac{1}{\lambda}, \text{Var}(X) = \frac{1}{\lambda^2}, \text{standard deviation} = \frac{1}{\lambda}$$

Probabilities

$P(X < a) = 1 - e^{-\lambda a}$ Result 1

$P(X > a) = e^{-\lambda a}$ Result 2

$P(a < X < b) = e^{-\lambda a} - e^{-\lambda b}$ Result 3

Link between Poisson and exponential distributions.

If $X \sim \text{Po}(\lambda)$,

where X is the number of occurrences of an event in a specified unit of time or space,

then $T \sim \text{Exp}(\lambda)$,

where T is the length of intervals between successive occurrences.

'No memory' rule

$P(X > a + b \mid X > a) = P(X > b)$

$P(X < a + b \mid X > a) = P(X < b)$

Review

1 The continuous random variable X has an exponential distribution with probability density function f given by

$$f(x) = \begin{cases} \dfrac{1}{8}e^{-\frac{1}{8}x} & x \geq 0 \\ 0 & \text{otherwise} \end{cases}$$

a Find

   i $P(X < 6)$

   ii $P(4 < X < 10)$

   iii $P(X > 10 \mid X > 3)$

b Find

   i $P(X > E(X))$

   ii $\text{Var}(X)$

2 The continuous random variable Y has an exponential distribution with probability density function $f(y)$, where

$$f(y) = \begin{cases} \dfrac{1}{4}e^{-\frac{1}{4}y} & y \geq 0 \\ 0 & \text{otherwise} \end{cases}$$

a State the value of the parameter λ.

b Find the mean of Y.

c Derive the cumulative distribution function of Y.

d **i** Find the value of the constant c if $P(Y > c) = 0.15$.

    ii Find the value of the constant d if $P(Y < d) = 0.15$.

3 A continuous random variable X has an exponential distribution with cumulative distribution function $F(x)$ defined by

$$F(x) = \begin{cases} 0 & x < 0 \\ 1 - e^{-x} & x \geq 0 \end{cases}$$

a Derive the probability density function $f(x)$ and state the value of the parameter λ.

b Find $P(2 < X < 3)$.

c Find the median of X.

d Find the lower quartile of X.

4 In a particular county, the distance between successive potholes on country roads may be modelled by an exponential distribution with mean 125 metres.

Determine the probability that the distance between successive potholes is:

a less than 100 metres;

b between 100 metres and 300 metres.

AQA MBS7 January 2004

5 The lifetime of a HiPower light is T thousand hours, where T has an exponential distribution with parameter $\lambda = 0.8$.

   a Find:

     i the mean value of T;

     ii the mean lifetime of a HiPower light;

     iii the probability that a HiPower light has a lifetime of less than 1000 hours.

   b HiPower lights are used throughout a large office building.

     i A light has been used for 500 hours. Find the probability that it lasts for at least another 500 hours.

     ii Three new lights are installed in one of the offices. Find the probability that at least one of them fails within 500 hours.

AQA SS05 June 2006

6 Between 8 am and 6 pm, the time, T minutes, between successive arrivals of vehicles for fuel at a village garage may be assumed to have an exponential distribution with a mean of 8.

   a Write down the numerical value for the standard deviation of T.

   b Calculate the probability that the time between successive arrivals of vehicles for fuel at the garage is between 5 minutes and 15 minutes.

   c Between 6 pm and 9 pm, the time, S minutes, between successive arrivals of vehicles for fuel at the garage may be assumed to be independent of T and to have an exponential distribution with a mean of 15.

   Calculate the probability that no vehicles arrive for fuel at the garage between 5.45 pm and 6.15 pm.

AQA MBS7 June 2004

7 The number of customers arriving per minute at a particular shop has a Poisson distribution with mean 0.4.

   a Find the probability that

     i exactly 3 customers arrive between 9.30 am and 9.40 am

     ii more than 12 customers arrive during a particular period of 30 minutes.

   b The time, in minutes, between the arrival of successive customers is denoted by T.

     i State the distribution of T.

     ii Find $P(T \leq 4)$.

     iii Find $P(6 < T < 10)$.

     iv A customer arrives at 10.00 am. Given that no more customers have arrived by 10.10 am, find the probability that no more customers arrive before 10.15 am.

8 A continuous random variable X has cumulative distribution function $F(x)$ defined by

$$F(x) = \begin{cases} 0 & x < 0 \\ 1 - e^{-\lambda x} & x \geq 0 \end{cases}$$

a Show that the median is $\dfrac{\ln 2}{\lambda}$.

b Find the quartiles and show that the interquartile range is $\dfrac{\ln 3}{\lambda}$.

c Show that the standard deviation is less than the interquartile range.

Assessment

1 Adrian is a skilful badminton player. When he serves low, the height, in centimetres, at which the shuttle crosses over the top of the net may be modelled by an exponential distribution with parameter $\lambda = 0.4$.

a For one of Adrian's low serves, find the probability that the height at which the shuttle crosses over the top of the net is:

   i less than 2 cm; [2]

   ii between 2 cm and 5 cm. [2]

b Verify that, for Adrian's low serves, the median height at which the shuttle crosses over the net is between 1.7 cm and 1.8 cm. [4]

AQA SS05 June 2007

2 The lifetime, X **thousand** hours of a component in a photocopier can be modelled by an exponential distribution with

$$P(X \leq x) = 1 - e^{-\frac{x}{6}}, x \geq 0$$

Determine the probability that the lifetime of a component is:

a less than 3000 hours; [2]

b between 5000 hours and 10000 hours. [3]

AQA MBS7 January 2002

3 The continuous random variable T has probability density function $f(t)$, where

$$f(t) = \begin{cases} 5e^{-5t} & t \geq 0 \\ 0 & \text{otherwise} \end{cases}$$

a Derive the cumulative distribution function of T. [4]

b Find the probability that $T > E(T)$. [1]

c Find the value of the constant c such that $P(T > c) = 0.05$. [2]

AQA MS04 June 2014

4 The time to failure, in hours, of a drill bit used in tunnelling machinery may be modelled by an exponential distribution with parameter $\lambda = 0.02$. Drill bits are used continuously until they fail.

a Find the mean time to failure of a drill bit. [2]

b Find the probability that a drill bit will fail during an eight-hour shift. [3]

c Find the probability of a drill bit **not** failing during 5 consecutive eight-hour shifts. [3]

d It is suggested that the risk of a drill bit failing during a shift could be reduced by always using a new one at the start of each shift. Comment on this suggestion. [2]

e The number of drill bits which fail during 5 consecutive eight-hour shifts may be modelled by a Poisson distribution.

    i Find the mean of this Poisson distribution. [2]

    ii Hence find the probability of no drill bits failing during 5 consecutive eight-hour shifts. [1]

AQA SS05 June 2008

5 The time, D days, between successive accidents at a factory can be modelled by an exponential distribution with mean 16.

a Write down the numerical value of the standard deviation of D. [1]

b Calculate the probability that the time between successive accidents at the factory is:

    i more than 20 days;

    ii between 20 and 30 days. [5]

c Given that there are no accidents during a 20-day period, determine the probability that there are no accidents during the next 20 days. Justify your answer. [3]

d Given that the factory is open five days each week, specify the distribution of the **weekly number** of accidents in the factory. [2]

AQA MBS7 January 2003

6 Trains stop at a particular station for a minimum of 1 minute.

The time, T minutes, **in excess of 1 minute**, that trains stop at the station may be modelled by an exponential distribution with mean 2.

a State the probability that the train stops at the station for at least 1 minute. [1]

b Calculate the probability that the train stops at the station for more than 5 minutes. [3]

c A train has been stopped at the station for 3 minutes.

Calculate the probability that it stops at the station for **a total of** less than 5 minutes. [4]

d A random sample of 5 trains each stop at the station for more than 5 minutes.

Indicate why this casts doubt on the above model. [2]

AQA MBS7 January 2005

The Normal Distribution

Introduction

One of the most important continuous random variables in statistics is the normal variable. Its distribution, known as the normal distribution, is a good model for many situations, especially in the natural sciences. It is also used extensively in hypothesis tests.

In a normal distribution there are relatively few very small or very large values, with the bulk of the distribution being concentrated around the middle value.

Here are some variables that might be modelled by a normal distribution:

▶ The height of adult males in a particular country.
▶ The time taken by students to complete a task.
▶ The volume of coffee dispensed into cups by a coffee machine.
▶ The birth weight of babies.

Objectives

By the end of this chapter you should know how to...

▶ Recognise the shape and symmetry of a normal distribution.
▶ Use a table of cumulative probabilities for the standard normal variable.
▶ Find probabilities for any normal variable by standardising.
▶ Find unknown values including mean and standard deviation.
▶ Solve problems involving the sum of independent normal random variables.

Recap

You will need to remember...

Continuous random variables   Chapter 11

▶ Probabilities are given by the area under the curve of $y = f(x)$.
▶ $P(X = x) = 0$, for all values of x so, for example,
$P(X < x)$ is the same as $P(X \le x)$
▶ $F(x) = P(X \le x) = \int_{-\infty}^{x} f(t)\, dt$
$P(X < a) = F(a)$
$P(X > a) = 1 - F(a)$
$P(a < X < b) = F(b) - F(a)$
▶ For independent variables X and Y
$E(X + Y) = E(X) + E(Y)$
$Var(X + Y) = Var(X) + Var(Y)$
▶ For n independent observations of X
$E(X_1 + X_2 + \cdots + X_n) = nE(X)$
$Var(X_1 + X_2 + \cdots + X_n) = nVar(X)$

13.1 The normal distribution

The **normal variable** is a special continuous random variable with parameters μ and σ, where μ is the mean and σ is the standard deviation.

> If the continuous random variable X follows a *normal distribution*,
> with parameters μ and σ, you can write
>
> $X \sim N(\mu, \sigma^2)$
>
>   mean   variance
>
> This is read as
>
> X has a normal distribution with mean μ and variance σ^2.

Notice that the variance, σ^2, is used in the description.

The probability density function of a normal variable is very complicated and is included here just for interest. It is given by

$$f(x) = \frac{1}{\sigma\sqrt{2\pi}}e^{-\frac{1}{2\sigma^2}(x-\mu)^2} \qquad -\infty < x < \infty$$

A very important feature of the normal variable is the graph of $y = f(x)$. It is known as a **normal distribution curve** and has these features:

▶ It is bell shaped.
▶ It is symmetrical about the mean μ.
▶ It extends from $-\infty$ to ∞.
▶ The total area under the curve is 1.

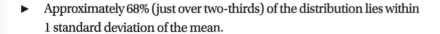

In theory, the possible values of x are from $-\infty$ to ∞. However, in practice:

▶ Approximately 68% (just over two-thirds) of the distribution lies within 1 standard deviation of the mean.

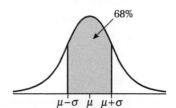

▶ Approximately 95% of the distribution lies within 2 standard deviations of the mean.

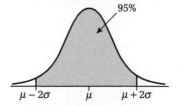

▶ Almost all (99.7%) of the distribution lies within 3 standard deviations of the mean.

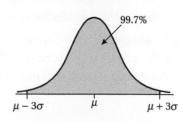

Here are some normal distribution curves, each drawn to the same scale.

i $X \sim N(0, 1^2)$

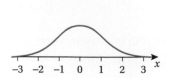

$\mu = 0, \sigma = 1$

ii $X \sim N\left(4, \left(\dfrac{1}{2}\right)^2\right)$

$\mu = 4, \sigma = \dfrac{1}{2}$

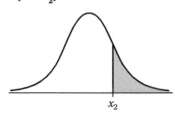

iii $X \sim N(50, 2^2)$

$\mu = 50, \sigma = 2$

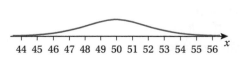
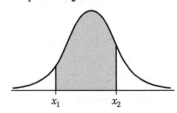
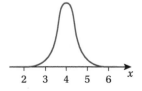
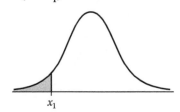

Notice that:

▶ Changes in the mean μ alter the position of the curve along the x-axis.
▶ Changes in the standard deviation σ alter the spread of the curve about the mean.

Finding probabilities

Probabilities are given by areas under the normal curve.

$P(X < x_1)$ $P(X > x_2)$ $P(x_1 < X < x_2)$

Since X is continuous, the probability that X takes an *exact* value is zero, so $P(X = x) = 0$ for *any* value of x. This means, for example, that $P(X < x_1)$ is the same as $P(X \le x_1)$.

See Section 11.1

You saw in Chapter 11 that, when X is continuous, probabilities can be found by one of these methods:

▶ Finding the area under the curve, usually by integrating the probability density function $f(x)$ between required limits.
▶ Substituting appropriate values into the cumulative distribution function, $F(x)$.

The second method is the one used for the normal variable.

In general, to find the cumulative distribution function $F(x)$ for a continuous variable X, you would find

$$F(x) = P(X \le x) = \int_{-\infty}^{x} f(t)\,dt$$

See Section 11.2

However, this integral is *very* difficult to evaluate so, instead, a cumulative probability table is available and you are able to read off values from it.

This table is written for a specific variable, called the **standard normal variable**, and it can then be used for any normal variable by a process called **standardising**, described later in Section 13.3.

13.2 The standard normal variable, Z

The **standard normal variable** is given the special symbol Z, rather than X.

It has a mean of 0 and a variance of 1,

so $Z \sim N(0, 1)$

The standard deviation of Z is also 1.

The normal distribution table

Cumulative probabilities for the standard normal variable are given in a **normal distribution table**. For various values of z, this gives $P(Z \le z)$.

However, instead of writing $F(z)$ for the cumulative probability, the special notation $\Phi(z)$ is used, where

$$\Phi(z) = P(Z \le z) = P(Z < z)$$

An extract from **Table 3: Normal Distribution Function** in the *Formulae and Statistical Tables* booklet is printed below.

The shaded area is labelled p where $p = \Phi(z)$.

Notice that the probabilities are given to 5 decimal places.

> **Note**
>
> Φ is the Greek upper case letter phi read as 'fie'.

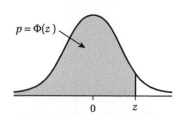

$p = \Phi(z)$

z	0.00	0.01	0.02	0.03	0.04	0.05	0.06	0.07	0.08	0.09	z
0.0	0.50000	0.50399	0.50798	0.51197	0.51595	0.51994	0.52392	0.52790	0.53188	0.53586	0.0
0.1	0.53983	0.54380	0.54776	0.55172	0.55567	0.55962	0.56356	0.56749	0.57142	0.57535	0.1
0.2	0.57926	0.58317	0.58706	0.59095	0.59483	0.59871	0.60257	0.60642	0.61026	0.61409	0.2
0.3	0.61791	0.62172	0.62552	0.62930	0.63307	0.63683	0.64058	0.64431	0.64803	0.65173	0.3
0.4	0.65542	0.65910	0.66276	0.66640	0.67003	0.67364	0.67724	0.68082	0.68439	0.68793	0.4
0.5	0.69146	0.69497	0.69847	0.70194	0.70540	0.70884	0.71226	0.71566	0.71904	0.72240	0.5
0.6	0.72575	0.72907	0.73237	0.73565	0.73891	0.74215	0.74537	0.74857	0.75175	0.75490	0.6
0.7	0.75804	0.76115	0.76424	0.76730	0.77035	0.77337	0.77637	0.77935	0.78230	0.78524	0.7
0.8	0.78814	0.79103	0.79389	0.79673	0.79955	0.80234	0.80511	0.80785	0.81057	0.81327	0.8
0.9	0.81594	0.81859	0.82121	0.82381	0.82639	0.82894	0.83147	0.83398	0.83646	0.83891	0.9
1.0	0.84134	0.84375	0.84614	0.84849	0.85083	0.85314	0.85543	0.85769	0.85993	0.86214	1.0
1.1	0.86433	0.86650	0.86864	0.87076	0.87286	0.87493	0.87698	0.87900	0.88100	0.88298	1.1
1.2	0.88493	0.88686	0.88877	0.89065	0.89251	0.89435	0.89617	0.89796	0.89973	0.90147	1.2
1.3	0.90320	0.90490	0.90658	0.90824	0.90988	0.91149	0.91309	0.91466	0.91621	0.91774	1.3
1.4	0.91924	0.92073	0.92220	0.92364	0.92507	0.92647	0.92785	0.92922	0.93056	0.93189	1.4
1.5	0.93319	0.93448	0.93574	0.93699	0.93822	0.93943	0.94062	0.94179	0.94295	0.94408	1.5
1.6	0.94520	0.94630	0.94738	0.94845	0.94950	0.95053	0.95154	0.95254	0.95352	0.95449	1.6

Finding a probability when $z \ge 0$

For $z \ge 0$, you can find $P(Z < z)$ by reading values directly from the table.

$P(Z < 0.16) = \Phi(0.16)$ ← row 0.1, column 0.06

         $= 0.56356$

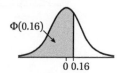

$\Phi(0.16)$

If the z-value has more than two decimal places, round it to 2 decimal places first.

$P(Z < 0.237) = \Phi(0.24)$ ← row 0.2, column 0.04

         $= 0.59483$

To find $P(Z > z)$, use the fact that the total area
under the curve is 1, so

$$P(Z > z) = 1 - \Phi(z)$$

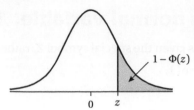

Example 1

Question

If $Z \sim N(0, 1)$, find

a $P(Z < 0.85)$

b $P(Z > 0.85)$

Answer

a

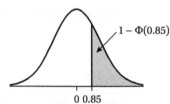

$P(Z < 0.85) = \Phi(0.85)$

$\quad\quad\quad\quad\quad = 0.80234$

b

$P(Z > 0.85) = 1 - \Phi(0.85)$

$\quad\quad\quad\quad\quad = 1 - 0.80234$

$\quad\quad\quad\quad\quad = 0.19766$

Finding a probability when *z* < 0

When z is negative, use the symmetrical property of the normal curve.

In these illustrations, a is positive.

P(Z<−a)

$P(Z < -a) = \Phi(-a)$

$\quad\quad\quad\quad = P(Z > a)$

$\quad\quad\quad\quad = 1 - \Phi(a)$

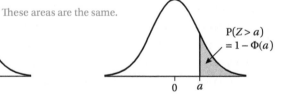

These areas are the same.

$P(Z < -a) = \Phi(-a)$

$P(Z > a) = 1 - \Phi(a)$

P(Z>−a)

$P(Z > -a) = P(Z < a)$

$\quad\quad\quad\quad = \Phi(a)$

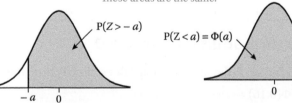

These areas are the same.

$P(Z > -a)$

$P(Z < a) = \Phi(a)$

Summarising:

$$P(Z < -a) = 1 - \Phi(a)$$
$$P(Z > -a) = \Phi(a)$$

Example 2

If $Z \sim N(0, 1)$ find

a $P(Z < 1.38)$ **b** $P(Z > -1.38)$

c $P(Z > 1.38)$ **d** $P(Z < -1.38)$

a

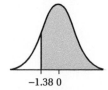

$$P(Z < 1.38) = \Phi(1.38)$$
$$= 0.91621 = 0.916 \text{ (3 sf)}$$

b

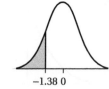

$$P(Z > -1.38) = \Phi(1.38)$$
$$= 0.91621 = 0.916 \text{ (3 sf)}$$

c

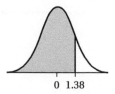

$$P(Z > 1.38) = 1 - \Phi(1.38)$$
$$= 1 - 0.91621$$
$$= 0.08379 = 0.0838 \text{ (3 sf)}$$

d

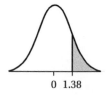

$$P(Z < -1.38) = 1 - \Phi(1.38)$$
$$= 1 - 0.91621$$
$$= 0.08379 = 0.0838 \text{ (3 sf)}$$

Important probability results

In these illustrations, a and b are positive and $a < b$.

Result 1 (upper-tail sandwich)

$$P(a < Z < b) = P(Z < b) - P(Z < a)$$
$$= \Phi(b) - \Phi(a)$$

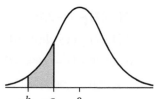

Result 2 (lower-tail sandwich)

$$P(-b < Z < -a) = P(Z < -a) - P(Z < -b)$$
$$= \Phi(-a) - \Phi(-b)$$
$$= 1 - \Phi(a) - (1 - \Phi(b))$$
$$= 1 - \Phi(a) - 1 + \Phi(b)$$
$$= \Phi(b) - \Phi(a)$$

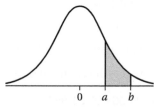

Result 3 (middle sandwich)

$$P(-a < Z < b) = P(Z < b) - P(Z < -a)$$
$$= \Phi(b) - \Phi(-a)$$
$$= \Phi(b) - (1 - \Phi(a))$$

Expanding the bracket gives

$$P(-a < Z < b) = \Phi(a) + \Phi(b) - 1$$

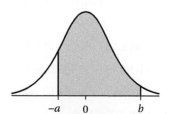

Result 4 (symmetrical middle sandwich)

$$P(|Z| < a) = P(-a < Z < a)$$
$$= \Phi(a) - (1 - \Phi(a)) \quad \leftarrow \text{Expand the bracket}$$
$$= \Phi(a) + \Phi(a) - 1$$
$$= 2\Phi(a) - 1$$

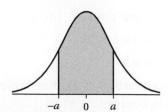

Summarising:

Result 1 $P(a < Z < b) = \Phi(b) - \Phi(a)$

Result 2 $P(-b < Z < -a) = \Phi(b) - \Phi(a)$

Result 3 $P(-a < Z < b) = \Phi(b) - (1 - \Phi(a))$

$\qquad\qquad\qquad\quad = \Phi(a) + \Phi(b) - 1$

Result 4 $P(-a < Z < a) = 2\Phi(a) - 1$

> **Note**
>
> It is important that you understand when and how to use these.

Example 3

Question

If $Z \sim N(0, 1)$, find

a $P(0.5 < Z < 2)$

b $P(-1.4 < Z < -0.6)$

Answer

a

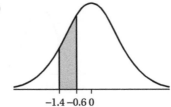

0 0.5   2

$P(0.5 < Z < 2) = \Phi(2) - \Phi(0.5)$ Result 1

$\qquad\qquad\qquad = 0.97725 - 0.69146$

$\qquad\qquad\qquad = 0.28579 = 0.286 \text{ (3 sf)}$

b

−1.4 −0.6 0

$P(-1.4 < Z < -0.6) = \Phi(1.4) - \Phi(0.6)$ Result 2

$\qquad\qquad\qquad\qquad = 0.91924 - 0.72575$

$\qquad\qquad\qquad\qquad = 0.19349 = 0.193 \text{ (3 sf)}$

Example 4

Question

If $Z \sim N(0, 1)$

a   find $P(-2.696 < Z < 1.865)$

b   show that the central 95% of the distribution lies between $z = -1.96$ and $z = 1.96$.

Answer

a $P(-2.696 < Z < 1.865) \leftarrow$ Round to 2 dp

$\qquad = \Phi(1.87) - \Phi(-2.70)$

$\qquad = \Phi(1.87) - (1 - \Phi(2.70))$ Result 3

$\qquad = 0.96926 - (1 - 0.99653)$

$\qquad = 0.96579 = 0.966 \text{ (3 sf)}$

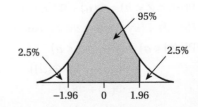

−2.70   0   1.87

Alternatively, using the expanded version

$P(-2.696 < Z < 1.865)$

$\qquad = \Phi(2.70) + \Phi(1.87) - 1$

$\qquad = 0.99653 + 0.96926 - 1$

$\qquad = 0.96579 = 0.966 \text{ (3 sf)}$

b $P(-1.96 < Z < 1.96)$

$\qquad = 2\Phi(1.96) - 1$ Result 4

$\qquad = 2 \times 0.97500 - 1$

$\qquad = 0.95$

95%

2.5%   2.5%

−1.96   0   1.96

The central 95% of the distribution lies between $z = -1.96$ and $z = 1.96$, that is, between $z = \pm1.96$.

Example 5

Given that $P(Z > z_1) = 0.65$ and $P(Z > z_2) = 0.25$, find $P(z_1 < Z < z_2)$.

It is very important that you are able to locate the positions of z_1 and z_2 on a sketch.

Think carefully about whether the z-value is in the upper tail (to the right of 0) or in the lower tail (to the left of 0).

If $P(Z > z_1) = 0.65$, more than half the area is to the right of z_1, so z_1 must be to the left of the middle, that is in the lower tail.

Note that $\Phi(z_1) = 0.35$.

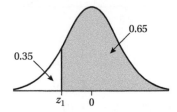

If $P(Z > z_2) = 0.25$, less than half the area is to the right of z_2, so z_2 must be to the right of the middle, that is in the upper tail.

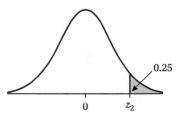

Now put the information on one diagram

$P(z_1 < Z < z_2)$

$\quad = 1 - (0.25 + 0.35)$

$\quad = 0.4$

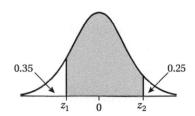

Exercise 1

In all these questions $Z \sim N(0, 1)$.

Always draw sketches to illustrate the probabilities and also check that your answer is sensible. For example, if your probability is negative or if it is greater than 1, then your answer is definitely wrong.

1 Find

  a $P(Z < 0.87)$ **b** $P(Z > -0.87)$

  c $P(Z > 0.87)$ **d** $P(Z < -0.87)$

2 Find

  a $P(Z > 1.80)$ **b** $P(Z < -0.65)$

  c $P(Z \geq -2.46)$ **d** $P(Z \leq 1.36)$

  e $P(Z > 2.58)$ **f** $P(Z > -2.37)$

  g $P(Z < 1.86)$ **h** $P(Z \leq -0.72)$

3 Find the probabilities represented by the shaded areas in the diagrams.

a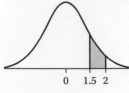

b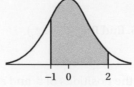

c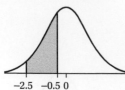

d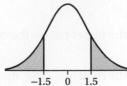

4 Find

   a $P(0.83 < Z < 1.83)$ **b** $P(-2.56 < Z < 0.13)$

   c $P(-1.762 \le Z \le -0.246)$ **d** $P(0 < Z < 1.73)$

   e $P(-2.05 < Z < 0)$ **f** $P(-2.08 < Z < 2.08)$

   g $P(1.764 < Z < 2.567)$ **h** $P(-1.65 \le Z < 1.722)$

   i $P(-0.98 < Z < -0.16)$ **j** $P(Z < -1.97 \text{ or } Z > 2.5)$

5 Find

   a $P(-1.78 < Z < 1.78)$ **b** $P(-1.65 < Z < 1.65)$

6 It is given that $P(Z < a) = 0.3$ and $P(a < Z < b) = 0.6$.

Find

   a $P(Z < b)$ **b** $P(Z > a)$

7 It is given that $P(Z < a) = 0.7$ and $P(Z > b) = 0.45$.

Find

   a $\Phi(b)$ **b** $P(b < Z < a)$

8 It is given that $P(-a < Z < a) = 0.8$.

Find

   a $P(Z < a)$ **b** $P(Z < -a)$

13.3 Standardising any normal variable

To use the tables for a normal variable X with mean μ and standard deviation σ, the variable

   $X \sim N(\mu, \sigma^2)$

is **standardised** (linearly scaled) to Z, the standard normal variable, with mean 0 and standard deviation 1 where

   $Z \sim N(0, 1)$

The process is illustrated:

Figure 1:

This shows $y = f(x)$ where

   $X \sim N(\mu, \sigma^2)$

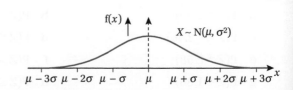

Figure 2:

Translate the curve μ units to the left, so that the mean is 0.

You now have

$$X - \mu \sim N(0, \sigma^2)$$

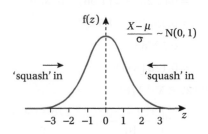

Figure 3:

Divide by the standard deviation σ. This reduces the horizontal scale by a factor of σ, so the standard deviation (and variance) is 1.

You now have

$$Z = \frac{X - \mu}{\sigma} \sim N(0, 1)$$

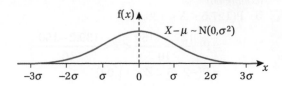

To standardise any normal variable $X \sim N(\mu, \sigma^2)$

▶   subtract the mean μ

▶   then divide by the standard deviation σ

so $Z = \dfrac{X - \mu}{\sigma}$ where $Z \sim N(0,1)$

In general, the standardised value z tells you how many standard deviations the x-value is *above* the mean when z is positive, or *below* the mean when z is negative.

For example, a standardised value of $z = 2$ tells you that the corresponding x-value is 2 standard deviations above the mean of X.

A standardised z-value of -1.5 tells you that the corresponding x-value is 1.5 standard deviations below the mean of X.

Example 6

A shop sells end-of-roll remnants of fabric. The length of remnants, X cm, may be modelled by a normal distribution with a mean of 150 and a standard deviation of 10.

Determine

a $P(X < 165)$

b $P(127.5 < X < 139.2)$

c $P(122 < X < 178)$

d   i $P(X = 153.5)$

   ii $P(X \neq 153.5)$

> **Note**
>
> Although X and Z have different means and spreads, it is convenient to show the corresponding values for them on *one* sketch.

X is the length, in cm, of a remnant, where

$X \sim N(150, 10^2)$ $\mu = 150, \sigma = 10$

a $P(X < 165)$

$$= P\left(Z < \frac{165 - 150}{10} \right)$$

$$= P(Z < 1.5)$$

$$= \Phi(1.5)$$

$$= 0.93319$$

$$= 0.933 \text{ (3 sf)}$$

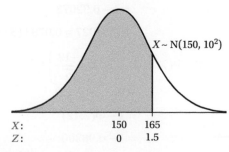

(continued)

b $P(127.5 < X < 139.2)$

$$= P\left(\frac{127.5-150}{10} < Z < \frac{139.2-150}{10}\right)$$

$$= P(-2.25 < Z < -1.08)$$

$$= \Phi(2.25) - \Phi(1.08) \quad \text{Result 2}$$

$$= 0.98778 - 0.85993$$

$$= 0.12785$$

$$= 0.128 \text{ (3 sf)}$$

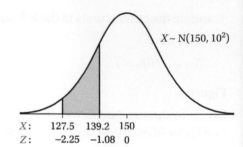

$X \sim N(150, 10^2)$

| X: | 127.5 | 139.2 | 150 |
| Z: | −2.25 | −1.08 | 0 |

c $P(122 < X < 178)$

$$= P\left(\frac{122-150}{10} < Z < \frac{178-150}{10}\right)$$

$$= P(-2.8 < Z < 2.8)$$

$$= 2\Phi(2.8) - 1 \quad \text{Result 4}$$

$$= 2 \times 0.99744 - 1$$

$$= 0.99488 = 0.995 \text{ (3 sf)}$$

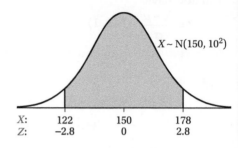

$X \sim N(150, 10^2)$

| X: | 122 | 150 | 178 |
| Z: | −2.8 | 0 | 2.8 |

d i $P(X = 153.5) = 0$ It is impossible for X to take an exact value.

   ii $P(X \neq 153.5) = 1$ It is certain that X does not take an exact value.

Example 7

Every day Bruno jogs around the park. The time he takes, in minutes, follows a normal distribution with mean 12 and variance 2.

a Find the probability that Bruno takes

  i longer than 14 minutes

  ii less than 9 minutes

  iii between 10 and 13 minutes.

b Five days are selected at random. Find the probability that Bruno takes between 10 and 13 minutes on all 5 days.

> **Note**
>
> Always check whether you are given the variance or the standard deviation.

X is the time, in minutes, that Bruno takes to jog around the park.

$X \sim N(12, 2)$ $\mu = 12$, $\sigma^2 = 2$ so $\sigma = \sqrt{2}$

> **Note**
>
> Define the variable in words and state its distribution.

a i $P(X > 14) = P\left(Z > \frac{14-12}{\sqrt{2}}\right)$

$$= P(Z > 1.41)$$

$$= 1 - \Phi(1.41)$$

$$= 1 - 0.92073$$

$$= 0.07927 = 0.0793 \text{ (3 sf)}$$

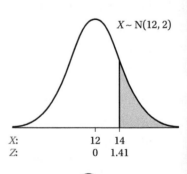

$X \sim N(12, 2)$

| X: | | 12 | 14 |
| Z: | | 0 | 1.41 |

 ii $P(X < 9) = P\left(Z < \frac{9-12}{\sqrt{2}}\right)$

$$= P(Z < -2.12)$$

$$= 1 - \Phi(2.12)$$

$$= 1 - 0.98300$$

$$= 0.01700 = 0.0170 \text{ (3 sf)}$$

(continued)

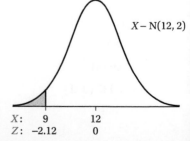

$X \sim N(12, 2)$

| X: | 9 | 12 |
| Z: | −2.12 | 0 |

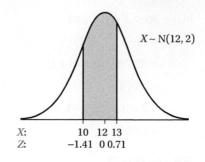

(continued)

iii $P(10 < X < 13) = P\left(\dfrac{10-12}{\sqrt{2}} < Z < \dfrac{13-12}{\sqrt{2}}\right)$

$= P(-1.41 < Z < 0.71)$

$= \Phi(1.41) + \Phi(0.71) - 1$ Result 3

$= 0.92073 + 0.76115 - 1$

$= 0.68188 = 0.682 \ (3\text{ sf})$

b $P(10 < X < 13) = 0.68188$

So P(time is between 10 and 13 minutes on all 5 days)

$= 0.68188^5 = 0.1474\cdots = 0.147 \ (3\text{ sf})$

Example 8

The normal variable X has mean μ and standard deviation σ.

1000 observations are taken from the distribution.

How many would you expect to be between $\mu - \sigma$ and $\mu + \sigma$?

$X \sim N(\mu, \sigma^2)$

$\mu - \sigma$ is 1 standard deviation below the mean, so $z = -1$.

$\mu + \sigma$ is 1 standard deviation above the mean, so $z = 1$.

$P(\mu - \sigma < X < \mu + \sigma)$

$= P(-1 < Z < 1)$

$= 2\Phi(1) - 1$ Result 4

$= 2 \times 0.84134 - 1$

$= 0.68268$

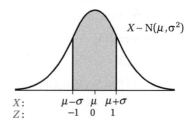

Now $1000 \times 0.68268 = 682.68 \approx 683$

So you would expect 683 observations to be between $\mu - \sigma$ and $\mu + \sigma$.

Important results

$P(\mu - \sigma < X < \mu + \sigma)$

$= P(-1 < Z < 1)$

$\approx 68\%$ (just over two-thirds)   See Example 8

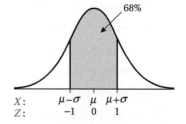

$P(\mu - 2\sigma < X < \mu + 2\sigma)$

$= P(-2 < Z < 2)$

$= 2\Phi(2) - 1$

$= 2 \times 0.97725 - 1$

$= 0.9945$

$\approx 95\%$

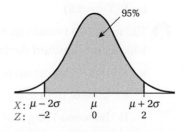

$P(\mu - 3\sigma < X < \mu + 3\sigma)$

$= P(-3 < Z < 3)$

$\approx 99.7\%$ (almost all)

$= 2\Phi(3) - 1$

$= 2 \times 0.99865 - 1$

$= 0.9973$

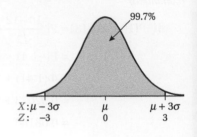

Exercise 2

Always draw sketches and check that your answer is sensible.

1 X has a normal distribution with mean 300 and standard deviation 5,
so $X \sim N(300, 5^2)$

Find the probabilities represented by the shaded areas in the diagrams.

a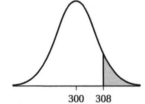
b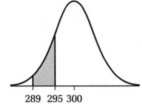
c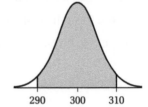

2 A machine packs sugar into bags. The mass, X g, of sugar in a bag is
normally distributed with a mean of 200 and a standard deviation of 2.

Determine

a $P(X < 197)$ b $P(X < 200.5)$

c $P(198.5 < X < 199.5)$ d $P(X = 197)$

e $P(202 < X < 206)$

3 The heights of boys at a particular age follow a normal distribution with
mean 150.3 cm and standard deviation 5 cm. Find the probability that the
height of a boy picked at random from this age group is

a less than 153.2 cm   b more than 158 cm

c exactly 152.6 cm   d between 150 cm and 158 cm

e more than 10 cm from the mean height.

4 The random variable X is distributed normally with a mean of 50 and
a variance of 20. Find

a $P(X > 60.3)$ b $P(X < 59.8)$ c $P(X \neq 50)$

5 $X \sim N(-8, 12)$. Find

a $P(X < -9.8)$ b $P(X > -8.2)$ c $P(-7 \leq X \leq 0.5)$

6 The mass of a certain type of cabbage is normally distributed with mean
1000 g and standard deviation 150 g.

a Find the proportion of cabbages with mass

i greater than 850 g

ii between 750 g and 1290 g.

b Estimate the number of cabbages in a batch of 800 cabbages with mass
less than 900 g or more than 1375 g.

7 The number of hours of life of a certain type of torch battery is normally distributed with mean 150 and standard deviation 12.

In a quality control test two batteries are chosen at random from a batch and if both batteries have a life less than 120 hours, the batch is rejected. Find the probability that the batch is rejected.

8 Cartons of milk from a particular supermarket are advertised as containing 1 litre of milk, but in fact the volume of the milk in a carton is normally distributed with mean 1012 ml and standard deviation 5 ml.

   a Find the probability that a randomly chosen carton contains more than 1010 ml.

   b Find the probability that exactly 3 cartons in a sample of 10 cartons contain more than 1012 ml.

   c Estimate how many cartons in a batch of 1000 cartons contain less than the advertised volume of milk.

9 The lifetime, in hours, of a "Glow" light bulb is normally distributed with mean 2000 and standard deviation 120. Find the probability that the lifetime of a "Glow" bulb will be

   a greater than 2150 hours

   b greater than 1910 hours

   c between 1850 hours and 2090 hours.

10 An intelligence test used in a certain country has scores which are normally distributed with mean 100 and standard deviation 15. Find the probability that a randomly selected person from the country has a score

   a higher than 140   b below 120

   c between 100 and 110   d between 85 and 90.

11 The mass of vegetable marrows supplied to retailers by a wholesaler has a normal distribution with mean 1.5 kg and standard deviation 0.02 kg.

The wholesaler supplies three sizes of marrow:

Size 1   under 1.48 kg

Size 2   from 1.48 kg to 1.53 kg

Size 3   over 1.53 kg

Find, to three decimal places, the proportions of marrows in the three sizes.

12 A normal variable X has mean μ and standard deviation σ.

Find

   a $P(\mu - 1.5\sigma < X < \mu + 1.5\sigma)$

   b $P(\mu - 2.5\sigma < X < \mu + 2.5\sigma)$

   c $P(X < \mu - 3\sigma \text{ or } X > \mu + 3\sigma)$

13.4 Finding the *z*-value that gives a known probability

If you are given a probability and have to relate it to a *z*-value, you need to use
Table 4 of the *Formulae and Statistical Tables* booklet.

The table gives the values of *z* satisfying $P(Z \leq z) = p$, where $Z \sim N(0, 1)$.

This notation is useful:

If $\quad \Phi(z) = p$

then $\quad z = \Phi^{-1}(p)$

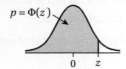

$p = \Phi(z)$

$\Phi^{-1}(p)$ is read as 'phi to the minus 1 of *p*'.

Table 4: Percentage points of the normal distribution

p	0.00	0.01	0.02	0.03	0.04	0.05	0.06	0.07	0.08	0.09	*p*
0.5	0.0000	0.0251	0.0502	0.0753	0.1004	0.1257	0.1510	0.1764	0.2019	0.2275	0.5
0.6	0.2533	0.2793	0.3055	0.3319	0.3585	0.3853	0.4125	0.4399	0.4677	0.4958	0.6
0.7	0.5244	0.5534	0.5828	0.6128	0.6433	0.6745	0.7063	0.7388	0.7722	0.8064	0.7
0.8	0.8416	0.8779	0.9154	0.9542	0.9945	1.0364	1.0803	1.1264	1.1750	1.2265	0.8
0.9	1.2816	1.3408	1.4051	1.4758	1.5548	1.6449	1.7507	1.8808	2.0537	2.3263	0.9
p	0.000	0.001	0.002	0.003	0.004	0.005	0.006	0.007	0.008	0.009	*p*
0.95	1.6449	1.6546	1.6646	1.6747	1.6849	1.6954	1.7060	1.7169	1.7279	1.7392	0.95
0.96	1.7507	1.7624	1.7744	1.7866	1.7991	1.8119	1.8250	1.8384	1.8522	1.8663	0.96
0.97	1.8808	1.8957	1.9110	1.9268	1.9431	1.9600	1.9774	1.9954	2.0141	2.0335	0.97
0.98	2.0537	2.0749	2.0969	2.1201	2.1444	2.1701	2.1973	2.2262	2.2571	2.2904	0.98
0.99	2.3263	2.3656	2.4089	2.4573	2.5121	2.5758	2.6521	2.7478	2.8782	3.0902	0.99

▶ The table gives values for $p \geq 0$ where $p = \Phi(z)$.

▶ For $p < 0.5$, use the fact that the normal curve is symmetrical.

▶ For $0.5 \leq p < 0.95$, you can input *p*-values only to 2 decimal places, so you may need to round first.

▶ For $p \geq 0.95$, you can input *p*-values to 3 decimal places so, again, you may need to round first.

▶ *z*-values in the main body of the table are given to 4 decimal places.

Using the table:

a $\Phi(z) = 0.94$

$\quad z = \Phi^{-1}(0.94) \leftarrow$ row $p = 0.9$, column 0.04

$\quad\quad = 1.5548$

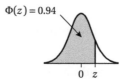

$\Phi(z) = 0.94$

b $\Phi(z) = 0.7289$ You need to round 0.7289 to 2 dp

$\quad z = \Phi^{-1}(0.73) \leftarrow$ row $p = 0.7$, column 0.03

$\quad\quad = 0.6128$

c $\Phi(z) = 0.975$

$\quad z = \Phi^{-1}(0.975) \leftarrow$ row $p = 0.97$, column 0.005

$\quad\quad = 1.9600$

d $\Phi(z) = 0.98235$ Round to 3 dp

$\qquad z = \Phi^{-1}(0.982) \leftarrow$ row $p = 0.98$, column 0.002

$\qquad = 2.0969$

When finding z-values:

▶ Always draw a sketch.
▶ Think carefully about where to position z, considering whether z is in the upper tail (z positive), lower tail (z negative) or centre ($z = 0$).

Example 9

Given $Z = N(0, 1)$, find a where

a $P(Z < a) = 0.9693$ **b** $P(Z > a) = 0.38$

c $P(Z > a) = 0.7367$ **d** $P(Z < a) = 0.08$

a $P(Z < a) = 0.9693$

$\qquad \Phi(a) = 0.9693 = 0.969 \text{ (3 dp)}$

$\qquad\qquad a = \Phi^{-1}(0.969) \quad \leftarrow$ row 0.96, column 0.09

$\qquad\qquad = 1.8663$

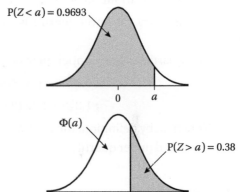

b $P(Z > a) = 0.38$

$\qquad \Phi(a) = 1 - 0.38$

$\qquad\qquad = 0.62$

$\qquad\qquad a = \Phi^{-1}(0.62) \quad \leftarrow$ row 0.6, column 0.02

$\qquad\qquad = 0.3055$

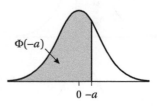

c $P(Z > a) = 0.7367$

Since the probability is greater than 0.5, a must be negative and therefore $-a$ is positive.

Using symmetry

$\qquad \Phi(-a) = 0.7367 = 0.74 \text{ (2 dp)}$

$\qquad\qquad -a = \Phi^{-1}(0.74)$

$\qquad\qquad = 0.6433$

$\qquad\qquad a = -0.6433$

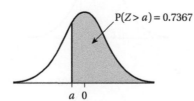

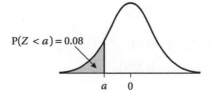

d $P(Z < a) = 0.08$

Since the probability is less than 0.5, a must be negative.

Using symmetry

$\qquad \Phi(-a) = 1 - 0.08$

$\qquad\qquad = 0.92$

$\qquad\qquad -a = \Phi^{-1}(0.92)$

$\qquad\qquad = 1.4051$

$\qquad\qquad a = -1.4051$

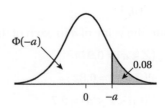

Example 10

Given $Z = N(0, 1)$, find

a the upper quartile of the distribution

b the interquartile range.

a If q is the upper quartile, then 75% of the
distribution lies to the left of q,

so $P(Z < q) = 0.75$

$\Phi(q) = 0.75$

$q = 0.6745$

The upper quartile is 0.6745.

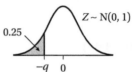

b 25% of the distribution lies to the left of the lower quartile.

By symmetry, the lower quartile is -0.6745.

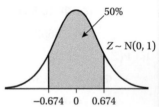

Interquartile range = upper quartile – lower quartile

$= 0.6745 - (-0.6745)$

$= 1.349 = 1.35$ (3 sf)

Note that, by definition, the central 50% of the distribution lies between the
lower and upper quartiles.

Exercise 3

1 $Z \sim N(0, 1)$.

Find the value of z in each of these.

a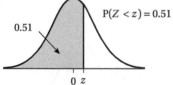
$P(Z < z) = 0.51$

b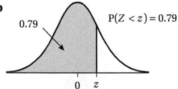
$P(Z < z) = 0.79$

c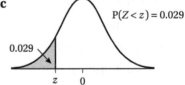
$P(Z < z) = 0.029$

d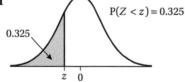
$P(Z < z) = 0.325$

e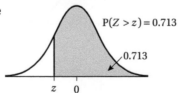
$P(Z > z) = 0.713$

f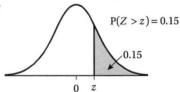
$P(Z > z) = 0.15$

2 $Z \sim N(0, 1)$.

Find the value of a in each of these.

a $(Z < a) = 0.974$

b $P(Z \le a) = 0.24$

c $P(Z > a) = 0.82$

d $P(Z > a) = 0.2351$

e $P(-a < Z < a) = 0.7$

13.5 Finding x for any normal variable

If $X \sim N(\mu, \sigma^2)$, to find an x-value when you know a probability:
▶ Find the corresponding z-value from Table 4.
▶ Substitute this z-value into the standardising formula $z = \dfrac{x - \mu}{\sigma}$.
▶ Re-arrange to make x the subject.

Example 11

The time, X minutes, taken by pupils in a particular class to do their mathematics homework may be modelled by a normal distribution with a mean of 25 and a standard deviation of 7.

The teacher says that 85% of pupils take less than x minutes.

Find the value of x, correct to 2 significant figures.

X is the time taken, in minutes.

$X \sim N(25, 7^2)$ $\mu = 25$ and $\sigma = 7$

You are given

$\qquad P(X < x) = 0.85$

so $P(Z < z) = 0.85$ where $z = \dfrac{x - 25}{7}$

$\qquad\qquad z = \Phi^{-1}(0.85)$

$\qquad\qquad\quad = 1.0364$

So $1.0364 = \dfrac{x - 25}{7}$

$\quad 1.0364 \times 7 = x - 25$

$\qquad\qquad x = 25 + 1.0364 \times 7$

$\qquad\qquad\quad = 32.2... = 32 \ (2 \ sf)$

> **Note**
>
> Show the information on a sketch.

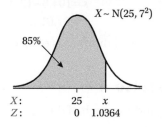

$X \sim N(25, 7^2)$

85%

| X: | 25 | x |
| Z: | 0 | 1.0364 |

Example 12

Ada sells apples from her orchard at a roadside stall. The mass, in grams, of apples from her orchard is a random variable with distribution N(104.8, 70).

a   Ada describes the heaviest 5% of the apples as 'extra-large'. Find the minimum mass of one of Ada's 'extra-large' apples.

b   Ada describes the lightest 2% of apples as 'extra-small'. Find the maximum mass of one of Ada's 'extra small' apples.

> **Note**
>
> Take care: In this example, variance = 70, so $\sigma = \sqrt{70}$.

a   Let X be the mass, in grams, of apples from the orchard.

$\qquad X \sim N(104.8, 70)$ $\mu = 104.8, \ \sigma = \sqrt{70}$

If m is the minimum mass of an extra-large apple,

then $P(X > m) = 0.05$

$\qquad\qquad P(X < m) = 0.95$

So $P(Z < z) = 0.95$ where $z = \dfrac{m - 104.8}{\sqrt{70}}$

Now $z = \Phi^{-1}(0.95)$

$\qquad\qquad\quad = 1.6449$

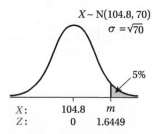

$X \sim N(104.8, 70)$
$\sigma = \sqrt{70}$

5%

| X: | 104.8 | m |
| Z: | 0 | 1.6449 |

(continued)

(continued)

So $\quad 1.6449 = \dfrac{m-104.8}{\sqrt{70}}$

$\qquad m = 104.8 + 1.6449 \times \sqrt{70} \quad \leftarrow$ Input $\sqrt{70}$ directly on your calculator

$\qquad = 118.56...$

The minimum mass of an extra-large apple is 119 g (3 sf).

b If s is the maximum mass of an extra-small apple

then $\quad$ P$(X < s) = 0.02$

so $\quad$ P$(Z < z) = 0.02 \quad$ where $\quad z = \dfrac{s-104.8}{\sqrt{70}}$

that is $\quad \Phi(z) = 0.02$

$\qquad \Phi(-z) = 1 - 0.02$

$\qquad\qquad = 0.98$

$\qquad -z = \Phi^{-1}(0.98)$

$\qquad\qquad = 2.0537$

$\qquad z = -2.0537$

So $\quad -2.0537 = \dfrac{s-104.8}{\sqrt{70}}$

$\qquad s = 104.8 - (2.0537) \times \sqrt{70}$

$\qquad = 87.61...$

The maximum mass of an extra-small apple is 87.6 g (3 sf).

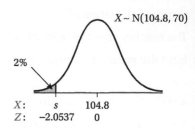

$X \sim \mathrm{N}(104.8, 70)$

$$
\begin{array}{lcc}
 & 2\% & \\
X: & s & 104.8 \\
Z: & -2.0537 & 0
\end{array}
$$

Exercise 4

1 Find x in each of these.

a $X \sim \mathrm{N}(60, 5^2)$

P$(X < x) = 0.997$

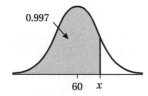

b $X \sim \mathrm{N}\left(5, \dfrac{4}{9}\right)$

P$(X < x) = 0.3$

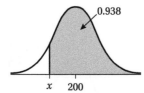

c $X \sim \mathrm{N}(200, 6^2)$

P$(X > x) = 0.938$

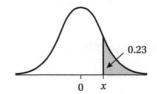

d $X \sim \mathrm{N}(0, 4)$

P$(X > x) = 0.23$

2. The heights of female students at a particular college are normally distributed with mean 169 cm and standard deviation 9 cm.

   a Given that 80% of these female students have a height less than h cm, find the value of h.

   b Given that 60% of these female students have a height greater than s cm, find the value of s.

3. The masses of lettuces sold at a market stall are normally distributed with mean 600 g and standard deviation 20 g.

   a Find the mass exceeded by 10% of the lettuces.

   b 5% of the lettuces have a mass less than a g. Find a.

4. $X \sim N(400, 64)$.

   a Find the limits within which the central 95% of the distribution lies.

   b Find the interquartile range of the distribution.

5. The lengths of metal rods produced by a machine are normally distributed with mean 120 cm and standard deviation 10 cm.

   a Find the probability that a rod selected at random has a length within 5 cm of the mean.

   b Rods shorter than l cm are rejected.

      Estimate the value of l if 1% of all rods are rejected.

6. The lifetime, in hours, of a certain make of batteries may be modelled by a normal distribution with mean 160 and standard deviation 30.

   Calculate the interval, symmetrical about the mean, within which 70% of the battery lifetimes lie.

13.6 Finding an unknown mean or standard deviation or both

Example 13

The time, X minutes, Wijono spends at the gym each day follows a normal distribution with mean μ and standard deviation 5. The probability that he spends more than 52 minutes at the gym is 0.25.

a Find the value of μ, giving your answer to 4 significant figures.

b Find $P(40 < X < 46)$.

$X \sim N(\mu, 5^2)$ μ is unknown, $\sigma = 5$

a $P(X > x) = 0.25$

    so $P(Z > z) = 0.25$ where $z = \dfrac{52 - \mu}{5}$

           $\Phi(z) = 1 - 0.25$

               $= 0.75$

           $z = \Phi^{-1}(0.75)$ Use Table 4

               $= 0.6745$

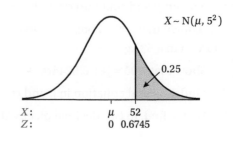

(continued)

(continued)

Answer

So $\dfrac{52-\mu}{5} = 0.6745$

$52 - \mu = 0.6745 \times 5$

$\mu = 52 - 0.6745 \times 5$

$= 48.6275 = 48.63\ (4\ \text{sf})$

b $X \sim N(48.63, 5^2)$ $\mu = 48.63,\ \sigma = 5$

$P(40 < X < 46) = P\left(\dfrac{40-48.63}{5} < Z < \dfrac{46-48.63}{5}\right)$

$= P(-1.73 < Z < -0.53)$

$= \Phi(1.73) - \Phi(0.53)$ Use Table 3

$= 0.95818 - 0.70194$

$= 0.25624 = 0.256\ (3\ \text{sf})$

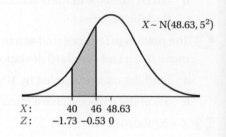

$X \sim N(48.63, 5^2)$

X: 40 46 48.63
Z: −1.73 −0.53 0

Example 14

Question

The random variable X is normally distributed with a mean of 100. It is known that $P(X < 106) = 0.9$.

Find the standard deviation of X.

Answer

$X \sim N(100, \sigma^2)$ $\mu = 100,\ \sigma$ is unknown

$P(X < 106) = 0.9$

so $P(Z < z) = 0.9$ where $z = \dfrac{x-100}{\sigma}$

$\Phi(z) = 0.9$

$z = \Phi^{-1}(0.9) = 1.2816$

So $1.2816 = \dfrac{106-100}{\sigma}$

$= \dfrac{6}{\sigma}$

$1.2816\sigma = 6$

$\sigma = \dfrac{6}{1.2816} = 4.681... = 4.68\ (3\ \text{sf})$

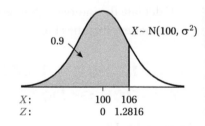

$X \sim N(100, \sigma^2)$

0.9

X: 100 106
Z: 0   1.2816

Example 15

Question

The mass, X kg, of boxes of oranges may be modelled by a normal distribution with mean μ and standard deviation σ.

It is known that 30% of boxes weigh more than 4.00 kg and 20% of boxes weigh more than 4.53 kg.

a Show that $4.00 = \mu + 0.5244\sigma$.

b Find a second equation in μ and σ.

c Hence find, to two decimal places, the mean and standard deviation of X.

X is the mass in kg of a box of oranges.

$X \sim N(\mu, \sigma^2)$ μ is unknown, σ is unknown

a $P(X > 4.00) = 0.3$

so $P(Z > z_1) = 0.3$ where $z_1 = \dfrac{4.00 - \mu}{\sigma}$

$\Phi(z_1) = 0.7$

$z_1 = \Phi^{-1}(0.7)$

$= 0.5244$

So $\dfrac{4.00 - \mu}{\sigma} = 0.5244$

$4.00 - \mu = 0.5244\sigma$

$4.00 = \mu + 0.5244\sigma$ \qquad (1)

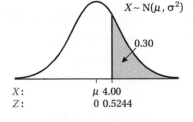

$X \sim N(\mu, \sigma^2)$

0.30

X: μ 4.00
Z: 0 0.5244

b $P(X > 4.53) = 0.2$

$P(Z > z_2) = 0.2$ where $z_1 = \dfrac{4.53 - \mu}{\sigma}$

$\Phi(z_2) = 0.8$

$z_2 = \Phi^{-1}(0.8)$

$= 0.8416$

So $\dfrac{4.53 - \mu}{\sigma} = 0.8416$

$4.53 - \mu = 0.8416\sigma$

$4.53 = \mu + 0.8416\sigma$ \qquad (2)

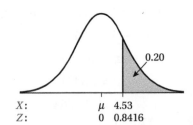

0.20

X: μ 4.53
Z: 0 0.8416

c Solve the simultaneous equations to find μ and σ.

$4.53 = \mu + 0.8416\sigma$ \qquad (2)

$4.00 = \mu + 0.5244\sigma$ \qquad (1)

$(2) - (1)$

$0.53 = 0.3172\sigma$

$\sigma = \dfrac{0.53}{0.3172} = 1.670$ $\leftarrow$ Store in calculator memory

Substitute for σ in (1)

$4.00 = \mu + 0.5244 \times 1.670...$

$\mu = 4.00 - 0.5244 \times 1.670...$

$= 3.123...$

So $\mu = 3.12$ (2 dp) and $\sigma = 1.67$ (2 dp) $\leftarrow$ You can check your solution by substituting into (2)

Example 16

The random variable X is normally distributed with mean μ and variance σ^2.

It is given that $P(X > 200) = 0.017$ and $P(X > 170) = 0.85$.

Find the values of μ and σ.

$X \sim N(\mu, \sigma^2)$ μ is unknown, σ is unknown

$P(X > 200) = 0.017$

So $P(Z > z_1) = 0.017$ where $z_1 = \dfrac{200 - \mu}{\sigma}$

$\quad \Phi(z_1) = 1 - 0.017 = 0.983$

$\qquad z_1 = \Phi^{-1}(0.983)$

$\qquad\quad = 2.1201$

So $\dfrac{200 - \mu}{\sigma} = 2.1201$

$\quad 200 - \mu = 2.1201\sigma$ $\hspace{3cm}$ (1)

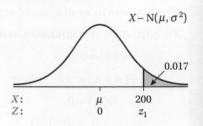

$P(X > 170) = 0.85$

So $P(Z > z_2) = 0.85$ where $z_2 = \dfrac{170 - \mu}{\sigma}$

$\quad \Phi(-z_2) = 0.85$

$\qquad -z_2 = \Phi^{-1}(0.85)$

$\qquad\quad = 1.0364$

$\qquad z_2 = -1.0364$

So $\dfrac{170 - \mu}{\sigma} = -1.0364$

$\quad 170 - \mu = -1.0364\sigma$ $\hspace{2.5cm}$ (2)

Now solve the simultaneous equations to find μ and σ.

$\quad 200 - \mu = 2.1201\sigma$ $\hspace{2.5cm}$ (1)

$\quad 170 - \mu = -1.0364\sigma$ $\hspace{2.2cm}$ (2)

$(1) - (2)\ 30 = 3.1565\sigma$

$\qquad \sigma = \dfrac{30}{3.1565} = 9.504$ $\quad \leftarrow$ Store in calculator memory

Substitute in (1)

$200 - \mu = 2.1201 \times 9.504...$

$\qquad \mu = 200 - 2.102 \times 9.504... = 179.85...$

So $\mu = 180$ (3 sf) and $\sigma = 9.50$ (3 sf)

Exercise 5

You are advised to draw sketches and check that your answer is reasonable.

1 The random variable X is normally distributed with mean μ and standard deviation 25.

   a Find μ if $P(X < 67.5) = 0.70$.

   b Find μ if $P(X < 108) = 0.20$.

2 The random variable X is distributed as $N(\mu, 12)$. It is given that $P(X > 32) = 0.84$.

   a Find the value of μ.

   b Find the probability that a random observation of X lies between 34.5 and 35.3.

Answer

3 Tea is sold in boxes marked 750 g. The mass of tea in a box is normally distributed with mean 755 g. The probability that the mass of tea in a box is more than 752 g is 0.975.

   a Find the standard deviation of the mass of the tea in a box.

   b Find the probability that the mass of the tea is less than the amount marked on the box.

4 The heights, in metres, of 500 people are normally distributed with standard deviation 0.080. Given that the heights of 130 of these people are greater than the mean height but less than 1.806 m, estimate the mean height.

5 The random variable X is normally distributed with mean 45 and standard deviation σ.

   Given that $P(X < 50) = 0.954$, find the value of σ.

6 The volume of orange juice in cartons sold by a particular company is normally distributed with mean 333 ml. It is known that 20% of the cans contain more than 340 ml.

   a Find the standard deviation of the volume of orange juice in a can.

   b Find the percentage of cans that contain less than 330 ml.

7 For a certain make of car, the distance travelled, X km, before a tyre is replaced has a normal distribution with mean 25 000.

   The manufacturer guarantees to pay compensation to anyone whose tyre does not last for 20 000 km. It is thought that 5% of all tyres sold will qualify for compensation.

   Find the standard deviation of X.

8 A machine is used to fill cans of soup with a nominal volume of 0.5 litres. The quantity of soup actually delivered is normally distributed with mean μ litres and standard deviation σ litres. It is required that no more than 1% of cans should contain less than the nominal volume.

   a Find the least value of μ which will comply with the requirement when $\sigma = 0.003$.

   b Find the greatest value of σ which will comply with the requirement when $\mu = 0.506$.

9 In a normal distribution with mean μ and standard deviation σ,

   $P(X < 16.2) = 0.5$ and $P(X > 18.3) = 0.10$.

   a Write down the value of μ and calculate the value of σ.

   b Find $P(14.5 < X < 15.9)$.

10 The mass of a box of apples from a certain supplier is normally distributed such that 20% of the boxes are heavier than 5.51 kg and 15% of the boxes are heavier than 5.62 kg.

   Find the mean and standard deviation of the mass of a box of apples.

11. Metal rods produced by a machine have lengths that are normally distributed. It is known that 2% of the rods are rejected as being too short and 5% are rejected as being too long.

   a Given that the least and greatest acceptable lengths of the rods are 6.32 cm and 7.52 cm, calculate the mean and variance of the lengths of the rods.

   b Ten rods are chosen at random from a batch produced by the machine. Find the probability that exactly three of them are rejected for being too long.   *Use the binomial distribution.*

12. The random variable X is distributed as $N(\mu, \sigma^2)$.
   $P(X < 35) = 0.2$ and $P(35 < X < 45) = 0.65$.
   Find μ and σ.

13. A farmer cuts hazel twigs to make into bean poles to sell at the market. He advertises them as being 240 cm long but in fact the lengths of the sticks are normally distributed such that 55% of the sticks are longer than 240 cm and 10% are longer than 250 cm.

   Find the probability that a randomly selected stick is shorter than 235 cm.

14. In a large consignment of bags of sugar, it is found that 5% weigh less than 515 g and 2% weigh less than 510 g. The mass of a bag of sugar is normally distributed.

   Find the mean and the standard deviation of the mass of a bag of sugar.

15. The random variable X has distribution $N(\mu, \sigma^2)$, where $P(X > 53) = 0.96$ and $P(X < 65) = 0.97$.

   Find the interquartile range of the distribution.

13.7 Sum of independent normal random variables

If X and Y are independent normal variables, where $X \sim N(\mu_1, \sigma_1^2)$ and $Y \sim N(\mu_2, \sigma_2^2)$, then $X + Y$ is also normally distributed.

Now $E(X + Y) = E(X) + E(Y) = \mu_1 + \mu_2$

$Var(X + Y) = Var(X) + Var(Y) = \sigma_1^2 + \sigma_2^2$ See Chapter 11

So, if $X \sim N(\mu_1, \sigma_1^2)$ and $Y \sim N(\mu_2, \sigma_2^2)$,

then $X + Y \sim N(\mu_1 + \mu_2, \sigma_1^2 + \sigma_2^2)$

The result can be extended to the sum of n independent normal variables.

Example 17

A coffee machine is installed in a students' common room. It dispenses white coffee by first releasing a quantity of black coffee and then adding a quantity of milk. The amount of black coffee dispensed is normally distributed with

(continued)

(continued)

mean 122.5 ml and standard deviation 7.5 ml. Independently, the amount of milk dispensed is normally distributed with mean 30 ml and standard deviation 5 ml.

Each cup is marked to a level of 137.5 ml. If this level is not attained, the customer receives the drink free of charge.

Show that approximately 5% of cups of white coffee will be given free of charge.

Let B be the amount, in ml, of black coffee dispensed, where $B \sim N(122.5, 7.5^2)$.

Let M be the amount, in ml, of milk dispensed, where $M \sim N(30, 5^2)$.

Let W be the amount, in ml, of white coffee dispensed, made by combining the black coffee and the milk, where $W = B + M$.

$E(W) = E(B) + E(M) = 122.5 + 30 = 152.5$

$Var(W) = Var(B) + Var(M) = 7.5^2 + 5^2 = 81.25$

Since B and M are independent normal variables, W is also a normal variable.

So $W \sim N(152.5, 81.25)$ $\quad \mu = 152.5$, $\sigma = \sqrt{81.25}$

The drink is free of charge if $W < 137.5$.

$P(\text{free of charge}) = P(W < 137.5)$

$$= P\left(Z < \frac{137.5 - 152.5}{\sqrt{81.25}} \right)$$

$$= P(Z < -1.66)$$

$$= 1 - \Phi(1.66)$$

$$= 1 - 0.95154$$

$$= 0.04846$$

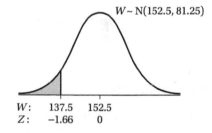

$W \sim N(152.5, 81.25)$

| W: | 137.5 | 152.5 |
| Z: | −1.66 | 0 |

Approximately 5% of the cups of white coffee will be given free of charge.

Example 18

Four runners, Åsmund, Bob, Chris and Dai, train to take part in a relay race in which Åsmund is to run the first 100 m, Bob 200 m, Chris 500 m and Dai 800 m.

During training their individual times, recorded in seconds, follow normal distributions as follows:

	Åsmund	Bob	Chris	Dai
Mean	10.8	23.7	62.8	121.2
Standard deviation	0.2	0.3	0.9	2.1

Find the probability that they will run the relay race in less than 222 seconds.

With obvious notation,

$A \sim N(10.8, 0.2^2)$, $B \sim N(23.7, 0.3^2)$, $C \sim N(62.8, 0.9^2)$ and $D \sim N(121.2, 2.1^2)$

Let T be the total time, in seconds, for the relay race, where $T = A + B + C + D$

$E(T) = E(A) + E(B) + E(C) + E(D)$

$= 10.8 + 23.7 + 62.8 + 121.2$

$= 218.5$

(continued)

(continued)

$\text{Var}(T) = \text{Var}(A) + \text{Var}(B) + \text{Var}(C) + \text{Var}(D)$

$\qquad = 0.2^2 + 0.3^2 + 0.9^2 + 2.1^2$

$\qquad = 5.35$

So $T \sim N(218.5, 5.35)$, with $\mu = 218.5$ and $\sigma = \sqrt{5.35}$

$P(T < 222) = P\left(Z < \dfrac{222 - 218.5}{\sqrt{5.35}}\right)$

$\qquad = P(Z < 1.51)$

$\qquad = \Phi(1.51)$

$\qquad = 0.93448 = 0.934 \text{ (3 sf)}$

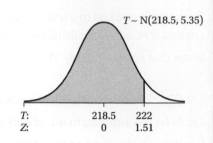

T: 218.5   222
Z: 0   1.51

Independent observations of the same normal variable X

Now consider the special case when $X_1, X_2, ..., X_n$ are n independent observations from the *same* normal distribution X, where $X \sim N(\mu, \sigma^2)$.

The sum, $X_1 + X_2 + \cdots + X_n$, is also normally distributed.

$E(X_1 + X_2 + \cdots + X_n) = \mu + \mu + \cdots + \mu = n\mu$

$\text{Var}(X_1 + X_2 + \cdots + X_n) = \sigma^2 + \sigma^2 + \cdots + \sigma^2 = n\sigma^2$

> If $X \sim N(\mu, \sigma^2)$, then the sum of n independent observations of X is also normally distributed, where
>
> $\qquad X_1 + X_2 + \cdots + X_n \sim N(n\mu, n\sigma^2)$

Example 19

A baker makes digestive biscuits and packs them into boxes containing 12 biscuits. The mass of a biscuit is normally distributed with mean 20 g and standard deviation 2 g. Find the probability that the total mass of the 12 biscuits in a box is greater than 230 g.

Let B be the mass, in grams, of a biscuit, where $B \sim N(20, 2^2)$.

Let T be the total mass, in grams, of 12 biscuits.

Then $T = B_1 + B_2 + \cdots + B_{12}$

$\qquad E(T) = 12E(B) = 12 \times 20 = 240$

$\qquad \text{Var}(T) = 12\text{Var}(B) = 12 \times 2^2 = 48$

So $T \sim N(240, 48)$ $\mu = 240$, $\sigma = \sqrt{48}$

$P(T > 230) = P\left(Z > \dfrac{230 - 240}{\sqrt{48}}\right)$

$\qquad = P(Z > -1.44)$

$\qquad = \Phi(1.44)$

$\qquad = 0.92507 = 0.925 \text{ (3 sf)}$

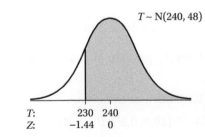

T: 230   240
Z: −1.44   0

Example 20

The maximum load a lift can carry is 450 kg.

In a certain country, the mass of a man is normally distributed with mean 60 kg and standard deviation 10 kg. The mass of a woman is normally distributed with mean 55 kg and standard deviation 5 kg.

A total of 5 men and 2 women enter a lift. Assuming that their masses are independent, find the probability that the lift will be overloaded.

Let M be the mass, in kilograms, of a man, where $M \sim N(60, 10^2)$.

Let W be the mass, in kilograms, of a woman, where $W \sim N(55, 5^2)$.

The lift is overloaded if $T > 450$,

where $T = M_1 + M_2 + M_3 + M_4 + M_5 + W_1 + W_2$

$$E(T) = 5E(M) + 2E(W)$$
$$= 5 \times 60 + 2 \times 55$$
$$= 410$$

$$Var(T) = 5Var(M) + 2Var(W)$$
$$= 5 \times 10^2 + 2 \times 5^2$$
$$= 550$$

Since M and W are normally distributed, T is also normally distributed,

so $T \sim N(410, 550)$ $\qquad \mu = 410, \ \sigma = \sqrt{550}$

$$P(T > 450) = P\left(Z > \frac{450 - 410}{\sqrt{550}}\right)$$
$$= P(Z > 1.71)$$
$$= 1 - \Phi(1.71)$$
$$= 1 - 0.95637$$
$$= 0.04363 = 0.0436 \text{ (3 sf)}$$

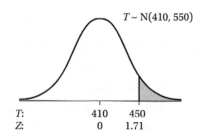

$T \sim N(410, 550)$

T:	410	450
Z:	0	1.71

Exercise 6

① X and Y are independent random variables, where $X \sim N(20, 3^2)$ and $Y \sim N(30, 3.5^2)$.

   a Find the mean and standard deviation of the distribution of $X + Y$.

   b State the distribution of $X + Y$.

   c Find   **i** $P(X + Y < 60)$ **ii** $P(38 < X + Y < 45)$

② A manufacturer produces jars of honey. In the production process, the mass of honey dispensed into a jar is normally distributed with mean 345 g and standard deviation 3 g. The mass of an empty jar is normally distributed with mean 180 g and standard deviation 2 g and the mass of a jar lid is normally distributed with mean 5 g and standard deviation 0.5 g.

   a Find the distribution of the total mass of a jar of honey (honey, jar and lid).

   b A jar of honey is selected from the jars ready for despatch. Find the probability that it weighs

      i more than 525 g

      ii between 532 g and 538 g.

3 The random variable X has the distribution $N(24.3, 3.5^2)$. The sum of 10 independent observations of X is denoted by Y.

   a Find $E(Y)$ and $Var(Y)$.

   b Find $P(Y > 260.1)$.

   c Given that $P(Y < c) = 0.95$, find c.

4 The mass of a certain grade of apple is normally distributed with mean 120 g and standard deviation 10 g.

   a An apple of this grade is selected at random.

     Find the probability that its mass lies between 100.5 g and 124 g.

   b Four apples of this grade are selected at random.

     Find the probability that their total mass exceeds 505 g.

5 The mass of a Chocolate Delight cake is normally distributed with mean 20 g and standard deviation 2 g. The cakes are sold in boxes containing 6 cakes. The mass of the box is normally distributed with mean 30 g and standard deviation 4 g.

   a Find the probability that the total mass of 6 Chocolate Delight cakes is less than 110 g.

   b Find the probability that the total mass of a box containing 6 cakes is

     i more than 162 g

     ii less than 137 g

     iii between 140 g and 153 g.

6 In a physics laboratory there are two types of resistors, A and B.

The resistance of Type A may be modelled by a normal variable with mean 4 ohms and standard deviation 0.12 ohms.

The resistance of Type B can be modelled by a normal variable with mean 2 ohms and standard deviation 0.05 ohms.

   a Matt selects a resistor of each type at random and connects them to make a 6-ohm resistor. Find the probability that it has a resistance greater than 6.25 ohms.

   b Molly selects 3 Type B resistors at random and connects them to make a 6-ohm resistor. Find the probability that it has a resistance greater than 6.25 ohms.

7 Each weekday Mr Georgiou, a retired statistician, walks to his local library to read the newspapers and then walks back home going by a different route. He proposes this model.

The time he takes to walk to the library has a normal distribution with mean 15 minutes and standard deviation 2 minutes.

The time he takes to walk back home from the library has a normal distribution with mean 20 minutes and standard deviation 3 minutes.

The time he spends in the library has a normal distribution with mean 25 minutes and standard deviation 4 minutes.

Assume that the model is true.

Mr Georgiou sets out to walk to the library at 11.20 am. His wife prepares lunch for 12.30 pm. Find the probability that he is late for lunch.

8 Else, Carly, Jorid and Vicky are in the 4×100 m freestyle relay team, each swimming 100 m.

Their individual times to swim 100 m, recorded in seconds, may be modelled by normal variables with the parameters:

	Else	Carly	Jorid	Vicky
Mean (s)	52.5	52.0	53.5	51.5
Standard deviation (s)	0.3	0.6	1.2	0.6

Calculate the probability that, in a particular race,

a Else will swim her leg in less than 52.5 seconds

b the relay team will take less than 3 minutes 31.3 seconds to swim the race.

9 The time taken to carry out a standard service on a car of Type A may be modelled by a normal variable with mean 1 hour and standard deviation 10 minutes.

Assume that only one car is serviced at a time.

a Find the probability that the total time taken to service 6 Type A cars is more than 6.5 hours.

The time taken to carry out a standard service on a car of Type B is a normal variable with mean 1.5 hours and standard deviation 15 minutes.

b Find the probability that the total time taken to service 3 Type A cars and 4 Type B cars is less than 8 hours.

13.8 Further applications

Example 21

The weight, X grams, of talcum powder in a tin may be modelled by a normal distribution with mean 253 and standard deviation σ.

a Given that $\sigma = 5$, determine:

i $P(X < 250)$;

ii $P(245 < X < 250)$;

iii $P(X = 245)$.

b Assuming that the value of the mean remains unchanged, determine the value of σ necessary to ensure that 98% of tins contain more than 245 grams of talcum powder.

AQA MS1B June 2009

X is the weight, in g, of talcum powder in a tin.

a $X \sim N(253, 5^2)$ $\mu = 253, \sigma = 5$

i $P(X < 250) = P\left(Z < \dfrac{250 - 253}{5}\right)$

$= P(Z < -0.6)$

$= 1 - \Phi(0.6)$

$= 1 - 0.72575$

$= 0.27425 = 0.274 \text{ (3 sf)}$

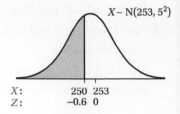

$X \sim N(253, 5^2)$

X: 250 253
Z: −0.6 0

ii $P(245 < X < 250) = P\left(\dfrac{245 - 253}{5} < Z < \dfrac{250 - 253}{5}\right)$

$= P(-1.6 < Z < -0.6)$

$= \Phi(1.6) - \Phi(0.6)$

$= 0.94520 - 0.72575$

$= 0.21945 = 0.219 \text{ (3 sf)}$

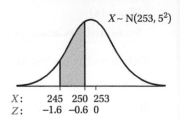

$X \sim N(253, 5^2)$

X: 245 250 253
Z: −1.6 −0.6 0

iii $P(X = 245) = 0$

b $X \sim N(253, \sigma^2)$ $\mu = 253, \sigma$ is unknown

$P(X > 245) = 0.98$

that is $P(Z > z) = 0.98$ where $z = \dfrac{245 - 253}{\sigma} = -\dfrac{8}{\sigma}$

so $\Phi(-z) = 0.98$

$-z = \Phi^{-1}(0.98)$

$-z = 2.0537$

$z = -2.0537$

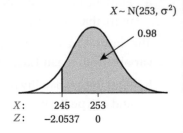

$X \sim N(253, \sigma^2)$

0.98

X: 245   253
Z: −2.0537   0

So, $-2.0537 = -\dfrac{8}{\sigma}$

$2.0537\sigma = 8$

$\sigma = \dfrac{8}{2.0537}$

$= 3.895... = 3.90 \text{ (3 sf)}$

Example 22

The volume, V litres, of *Cleanall* washing-up liquid in a 5-litre container may be modelled by a normal distribution with a mean, μ, of 5.028 and a standard deviation of 0.015.

a Determine the probability that the volume of *Cleanall* in a randomly selected 5-litre container is:

i less than 5.04 litres;

ii more than 5 litres.

b Determine the value of v such that $P(\mu - v < V < \mu + v) = 0.95$.

AQA MS1A January 2013

Answer

Question

V is the volume, in litres, of washing-up liquid in a bottle.

$V \sim N(5.028, 0.015^2)$ $\mu = 5.028, \sigma = 0.015$

a i $P(V < 5.04) = P\left(Z < \dfrac{5.04 - 5.028}{0.015} \right)$

$= P(Z < 0.8)$

$= \Phi(0.8) = 0.78814 = 0.788 \, (3 \text{ sf})$

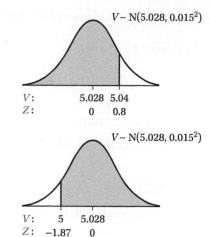

$V \sim N(5.028, 0.015^2)$

V: 5.028 5.04
Z: 0   0.8

ii $P(V > 5) = P\left(Z > \dfrac{5 - 5.028}{0.015} \right)$

$= P(Z > -1.87)$

$= \Phi(1.87)$

$= 0.96926 = 0.969 \, (3 \text{ sf})$

b $P(\mu - v < V < \mu + v) = 0.95$

that is $P(-z < Z < z) = 0.95$

where $z = \dfrac{(\mu + v) - \mu}{0.015} = \dfrac{v}{0.015}$

$\Phi(z) = 0.975$

$z = \Phi^{-1}(0.975)$

$= 1.96$

So, $1.96 = \dfrac{v}{0.015}$

$v = 1.96 \times 0.015 = 0.0294 \, (3 \text{ sf})$

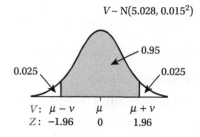

$V \sim N(5.028, 0.015^2)$

V: 5   5.028
Z: −1.87   0

$V \sim N(5.028, 0.015^2)$

0.95

0.025   0.025

V: $\mu - v$ μ $\mu + v$
Z: −1.96   0   1.96

Example 23

Huynh should start work each week day at 9.00 am. To get to work, he walks to the bus stop and then catches a bus to his office.

The duration of Huynh's walk to the bus stop is normally distributed with mean 16 minutes and standard deviation 3.6 minutes. The time he spends waiting at the bus stop is normally distributed with mean 6 minutes and standard deviation 2.4 minutes. The duration of his bus journey is normally distributed with mean 32 minutes and standard deviation 6.8 minutes.

On a particular day, Huynh leaves home at 8.00 am. Determine the probability that he is late for work.

Let W minutes be the duration of his walk to the bus stop, where $W \sim N(16, 3.6^2)$.

Let S minutes be the time he waits for a bus, where $S \sim N(6, 2.4^2)$.

Let B minutes be the duration of his bus journey, where $B \sim N(32, 6.8^2)$.

If T minutes is the total time for his journey, then $T = W + S + B$

$E(T) = E(W) + E(S) + E(B)$

$= 16 + 6 + 32$

$= 54$

$Var(T) = Var(W) + Var(S) + Var(B)$

$= 3.6^2 + 2.4^2 + 6.8^2$

$= 64.96$

(continued)

(continued)

So $T \sim N(54, 64.96)$

Huynh will be late for work if $T > 60$.

$$P(T > 60) = P\left(Z > \frac{60 - 54}{\sqrt{64.96}}\right)$$

$$= P(Z > 0.74)$$

$$= 1 - \Phi(0.74)$$

$$= 1 - 0.77035$$

$$= 0.22965 = 0.230 \text{ (3 sf)}$$

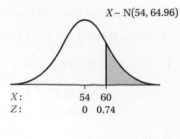

$X \sim N(54, 64.96)$

$X:$ 54   60
$Z:$ 0   0.74

Summary

Standard normal variable Z

$Z \sim N(0, 1)$ mean $= 0$, variance $= 1$, standard deviation $= 1$

Normal variable X

$X \sim N(\mu, \sigma^2)$ mean $= \mu$, variance $= \sigma^2$, standard deviation $= \sigma$

To standardise X, use $Z = \dfrac{X - \mu}{\sigma}$

Finding probabilities

To find **probabilities** use the normal distribution function, **Table 3**.

In these illustrations $a > 0$, $b > 0$, $a < b$

$P(Z < a) = \Phi(a)$

$\Phi(a)$

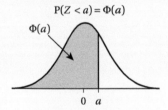

$P(Z > a) = 1 - \Phi(a)$

$1 - \Phi(a)$

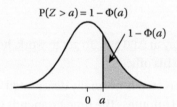

$P(Z > -a) = \Phi(a)$

$\Phi(a)$

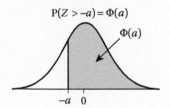

$P(Z < -a) = 1 - \Phi(a)$

$1 - \Phi(a)$

$P(a < Z < b) = \Phi(b) - \Phi(a)$

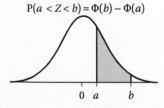

$P(-b < Z < -a) = \Phi(b) - \Phi(a)$

$$P(-a < Z < b) = \Phi(b) - (1 - \Phi(a))$$
$$= \Phi(a) + \Phi(b) - 1$$

$P(-a < Z < a) = 2\Phi(a) - 1$

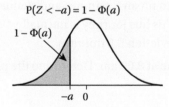

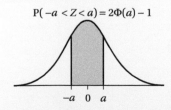

Finding z-values

To find **z-values** use percentage points of the normal distribution, **Table 4**.

If $\quad P(Z < z) = p$,

that is $\quad \Phi(z) = p$,

then $\quad z = \Phi^{-1}(p)$

Sum of independent normal variables

For independent normal variables $X \sim N(\mu_1, \sigma_1^2)$ and $Y \sim N(\mu_2, \sigma_2^2)$,

$$X + Y \sim N(\mu_1 + \mu_2, \sigma_1^2 + \sigma_2^2)$$

The result can be extended to the sum of n independent normal variables.

In particular, when $X_1, X_2, ..., X_n$ are n independent observations of X, where $X \sim N(\mu, \sigma^2)$,

$$X_1 + X_2 + \cdots + X_n \sim N(n\mu, n\sigma^2)$$

Review

1 The continuous random variable X has a normal distribution with mean 200 and standard deviation 30.

Determine

  a $\quad P(X < 260)$ **b** $\quad P(X > 180)$ **c** $\quad P(X < 140)$

  d $\quad P(X > 300)$ **e** $\quad P(X = E(X))$

2 The normal variable Y is such that $Y \sim N(25, 4^2)$.

Determine

  a $\quad P(28.5 \leq Y < 30.2)$ **b** $\quad P(19.3 < Y < 24.2)$ **c** $\quad P(23.7 < Y < 27.8)$

3 Draught excluder for door and windows is sold in rolls of nominal length 10 metres.

The actual length, X metres, of draught excluder on a roll may be modelled by a normal distribution with mean 10.2 and standard deviation 0.15.

  a Determine:

    i $\quad P(X < 10.5)$;   **ii** $\quad P(10.0 < X < 10.5)$.

  b A customer randomly selects six 10-metre rolls of draught excluder.

    Calculate the probability that all six rolls selected contain more than 10 metres of draught excluder.

AQA MS1B January 2010

4 The time spent by customers in a particular supermarket is normally distributed with mean 16.3 minutes and standard deviation 4.2 minutes.

  a Find the probability that the time a customer spends in the supermarket

    i is less than 5 minutes

    ii is not exactly 6 minutes.

  b A random sample of 6 customers is selected. Find the probability that at least one spends more than 25 minutes in the supermarket.

5 The random variable X is distributed $N(100, 16)$.

   a Find the value of x such that

     i $P(X < x) = 0.96$

     ii $P(X > x) = 0.005$

     iii $P(X \leq x) = 0.15$

   b Find the interquartile range of X.

6 The lengths of certain items follow a normal distribution with mean μ mm and standard deviation 6.0 mm. It is known that 2.5% of the items are longer than 82.0 mm.

   a Find the value of μ.

   b Find the probability that the length of a randomly selected item is between 63.7 mm and 71.3 mm.

7 The diameters of bolts produced by a machine follow a normal distribution with mean 1.34 cm and standard deviation 0.04 cm.

A bolt is rejected if its diameter is less than 1.24 cm or more than 1.40 cm.

   a Find the percentage of bolts which are accepted.

The setting of the machine is altered so that the mean diameter changes but the standard deviation remains the same. With the new setting, 3% of the bolts are rejected because they are too large in diameter.

   b Find the new mean diameter of bolts produced by the machine.

   c Find the percentage of bolts that are now rejected because they are too small in diameter.

8 A machine dispenses raisins into bags so that the mass of the raisins in a bag is normally distributed.

   a The standard deviation is 1.25 g. It is found that 2.5% of bags contain less than 826 g of raisins. Find the mean mass of raisins in a bag.

   b After the machine is serviced, the mean mass of the raisins in a bag is 828.1 g and 0.1% of bags contain more than 830 g. Find the standard deviation of the mass of raisins in a bag after the service.

9 Chopped lettuce is sold in bags nominally containing 100 grams.

The weight, X grams, of chopped lettuce, delivered by the machine filling the bags, may be assumed to be normally distributed with mean μ and standard deviation 4.

   a Assuming that $\mu = 106$, determine the probability that a randomly selected bag of chopped lettuce:

     i weighs less than 110 grams;

     ii is underweight.

   b Determine the minimum value of μ so that at most 2 per cent of bags of chopped lettuce are underweight. Give your answer to one decimal place.

AQA MS1A January 2005

10 The speeds of cars passing a checkpoint on a certain motorway follow a normal distribution. Observations show that 95% of the cars passing the checkpoint are travelling at less than 136 km/h and 10% are travelling at less than 88 km/h.

   a Find the mean speed of cars passing the checkpoint.

   b Find the proportion of cars travelling at more than 112 km/h.

11 The volume of *Fizzer* lemonade in a can, X millilitres, may be modelled by a normal distribution with a mean of 340 and a standard deviation of 3.2.

   a Determine:

      i $P(X > 345)$;

      ii $P(X < 333)$;

      iii $P(333 < X < 345)$;

      iv $P(X = 340)$.

   b The volume of *Fizzer* cola in a can, Y millilitres, may be modelled by a normal distribution with mean μ and standard deviation σ.

      It is required that $P(Y > 330) = 0.975$ and $P(Y < 345) = 0.975$.

      i   State why $\mu = 337.5$.

      ii Determine the value of σ.

AQA MS1A June 14

12 The continuous random variable X is distributed $N(\mu, \sigma^2)$ where $P(X < 95.8) = 0.2$ and $P(X > 93.6) = 0.9$. Find the values of μ and σ.

13 The random variable X has a normal distribution with mean μ and standard deviation σ. If $2\mu = 1.5\sigma$, find $P(X < 3\mu)$.

14 At a health centre, consultations with a nurse take X minutes, where X may be modelled by a normal distribution with mean 5.4 and standard deviation 1.3. Consultations with a doctor take Y minutes, where Y may be modelled by a normal distribution with mean 9.2 and standard deviation 2.4.

Hassan has a consultation with a nurse, followed immediately by a consultation with a doctor.

Find the probability that the total duration of Hassan's consultations is:

   a more than 20 minutes

   b less than 15 minutes

   c between 11.5 minutes and 14 minutes

   d exactly 17 minutes.

15 Plastic rods are produced in two lengths, 'short' and 'long'.

S cm is the length of a short rod, where $S \sim N(5, 0.25)$.

L cm is the length of a long rod, where $L \sim N(10, 1)$.

Rods of either sort are joined to give longer lengths. Find the probability that

   a the length of 2 short rods and 4 long rods is greater than 52 cm

   b the length of 2 long rods and 3 short rods is between 33 cm and 36 cm.

Assessment

1 The volume of *Everwhite* toothpaste in a pump-action dispenser may be modelled by a normal distribution with a mean of 106 ml and a standard deviation of 2.5 ml.

Determine the probability that the volume of *Everwhite* in a randomly selected dispenser is:

a less than 110 ml; [3]

b more than 100 ml; [2]

c between 104 ml and 108 ml; [3]

d **not** exactly 106 ml. [1]

AQA MS1B January 2013

2 a The baggage loading time, X minutes, of a chartered aircraft at its UK airport may be modelled by a normal random variable with mean 55 and standard deviation 8.

Determine:

i $P(X < 60)$; [3]

ii $P(55 < X < 60)$. [2]

b The baggage loading time, Y minutes, of a chartered aircraft at its overseas airport may be modelled by a normal random variable with mean μ and standard deviation 16.

Given that $P(Y < 90) = 0.95$, find the value of μ. [4]

AQA MS1A January 2006

3 To supplement her pension, Cordelia bakes cakes which are then sold in Amir's nearby shop.

a The weight, X grams, of her square fruit cake may be modelled by a normal random variable with a mean of 2200 and a standard deviation of 160.

Determine:

i $P(X < 2500)$; **ii** $P(X > 2000)$; **iii** $P(2000 < X < 2500)$. [7]

b The weight, Y grams, of Cordelia's round fruit cake may be modelled by a normal random variable with a mean of 1125 and a standard deviation of σ.

If Amir requests that at most 10 per cent of these cakes should weigh less than 1000 grams, find the maximum value for σ. [4]

AQA MS1A January 2012

4 A machine produces ice hockey pucks whose weights may be modelled by a normal distribution with a mean of 165 grams amd a standard deviation of σ grams.

a Given that $\sigma = 2.5$, determine the probability that the weight of a randomly selected puck is:

i less than 167 grams; [3]

ii more than 162 grams. [2]

b An ice hockey club purchases a box of 12 pucks produced by the machine.

Assuming that the pucks in any box represent a random sample, calculate the probability that all 12 pucks weigh less than 167 grams. [2]

c An ice hockey confederation requires that at most 1 per cent of pucks have weights outside the range 160 grams to 170 grams.

Assuming that the value of the mean remains unchanged at 165 grams, calculate, to two decimal places, the maximum value of σ which meets this requirement. [4]

AQA MS1A June 2011

5 In a large-scale tree-felling operation, a machine cuts down trees, strips off the branches and then cuts the trunks into logs of length X metres for transporting to a sawmill.

It may be assumed that values of X are normally distributed with mean μ and standard deviation 0.16, where μ can be set to a specific value.

a Given that μ is set to 3.3, determine:

  i $P(X < 3.5)$;   **ii** $P(X > 3.0)$;   **iii** $P(3.0 < X < 3.5)$.

b The sawmill now requires a batch of logs such that there is a probability of 0.025 that any given log will have a length less than 3.1 metres.

Determine, to two decimal places, the new value of μ.

AQA MS1B January 2008

6 a The weight, X grams, of soup in a carton may be modelled by a normal random variable with mean 406 and standard deviation 4.2.

Find the probability that the weight of soup in a carton:

  i is less than 400 grams; [3]

  ii is between 402.5 grams and 407.5 grams. [4]

b The weight, Y grams, of chopped tomatoes in a tin is a normal random variable with mean μ and standard deviation σ.

  i Given that $P(Y < 310) = 0.975$, explain why:

$$310 - \mu = 1.96\sigma \quad [3]$$

  ii Given that $P(Y < 307.5) = 0.86$, find, to two decimal places, values for μ and σ. [4]

AQA MS1B January 2006

7 The quarterly consumption, $X\,\text{m}^3$, of water by two-person households may be modelled by a normal distribution with mean μ and variance σ^2.

From an analysis of records, it is found that $P(X < 45) = 0.98$.

a Show that

$$45 - \mu = 2.0537\sigma \quad [3]$$

b Given also that $P(X > 30) = 0.95$, find, to one decimal place, values for μ and σ. [4]

AQA MS1A January 2009

8 The thickness, X cm, of a hardback book may be modelled by a normal distribution with mean 4.9 and variance 1.920. The thickness, Y cm, of a paperback book may be modelled by a normal distribution with mean 2.0 and variance 0.730.

A randomly selected hardback book and a randomly selected paperback book are placed together on a shelf. Find the probability that the thickness of the two books is between 4.47 cm and 5.92 cm. [4]

9 A certain school holds cake sales to raise money for charity. It is found that the amount of money raised by a cake sale at the school may be modelled by a normal distribution with mean $23.50 and standard deviation $3.40.

The students in Class 4T want to raise $40 for some sports equipment and decide to hold two cake sales. Find the probability that they will raise at least $40. [4]

10 The process of painting the bodywork of a mass-produced truck consists of giving it one coat of Paint A, three coats of Paint B and two coats of Paint C.

A record of the volume of paint used for each truck has been kept for many years. It is found that, for each type of paint, the volume, in litres, of paint in one coat is normally distributed with mean and standard deviation as shown in the table.

	Mean (litres)	Standard deviation (litres)
One coat of Paint A	3.7	0.42
One coat of Paint B	1.3	0.15
One coat of Paint C	1.0	0.12

The volume required of each type of paint is independent of the volume required of any other paint.

a Find the mean and standard deviation of the total volume of paint used on a truck. [3]

b Find the percentage of trucks receiving

    i   less than 8.5 litres of paint. [2]

    ii   more than 10.0 litres of paint. [2]

14 Estimation

Introduction

Suppose you are studying a large colony of birds and want to know the average mass of one-week old chicks. You would not be able to weigh every chick, so it would seem sensible to make an estimate by selecting a sample of chicks and finding their average mass.

Estimation is the process of using information from a sample to make a statement about the population as a whole and this chapter considers ways of doing this, based on statistical theory.

Recap

You will need to remember...

Expectation algebra

▶ $E(aX) = aE(X)$
▶ $E(\sum X_i) = \sum E(X_i)$
▶ $\text{Var}(aX) = a^2\text{Var}(X)$
▶ $\text{Var}(\sum X_i) = \sum \text{Var}(X_i)$

Using normal distribution tables

▶ To find probabilities.
▶ To find an x-value, given a probability.

Finding probabilities and the mean and variance for

▶ A binomial distribution.
▶ A Poisson distribution.
▶ The distribution of a continuous random variable.
▶ An exponential distribution.

Objectives

By the end of this chapter you should be able to able to...

▶ Distinguish between a population parameter and a sample statistic.
▶ Understand the concept of a simple random sample.
▶ Find unbiased estimates of a population mean and variance.
▶ Determine probabilities using the sampling distribution of the mean of a random sample from a normal distribution.
▶ Understand when to apply the Central Limit Theorem.
▶ Determine probabilities using the normal distribution as an approximation to the sampling distribution of the mean of a large sample from any distribution.

14.1 Sampling

In a statistical enquiry you often need information about a particular group. This group is known as the **population** and it could be any size: small, large or even infinite. Note, however, that the statistical term 'population' does not necessarily mean 'people'.

Here are some examples of populations:

▶ Students in a class.
▶ Pebbles on a beach.
▶ People aged over 75 in a particular city.
▶ Cans of soft drink produced in a factory.

▶ Fish in Lake Ontario.
▶ Rational numbers between 0 and 10.

To collect information about a population you could either carry out a census or take a sample from the population.

In a **census**, every member of the population is surveyed. When the population is small, this could be straightforward. For example, it would be easy to find out how each student in a class travelled to school on a particular morning.

However, when the population is large, surveying every member of the population would be very time-consuming and difficult to do with accuracy. On some occasions it would not be sensible to survey every member. For example, if you were to carry out a census to establish the lifetime of a particular brand of light bulb, you would have to test every bulb until it failed, thus destroying the population.

So, as it may be time-consuming, expensive or even impossible to investigate an entire population, it is usual to collect information by selecting some of the members of the population, thereby forming a **sample**.

Provided that the sample is representative of the population and unbiased (fair), it can give an indication of the population characteristic being studied.

Note

Large samples give more reliable information than small samples.

Random samples

To eliminate bias, you would take a **random sample**.

A random sample of size n is a sample chosen such that:
▶ Every member of the population has an equal chance of being selected.
▶ All possible samples of size n have an equal chance of being selected.

If the item being selected is replaced into the population before the next item is selected, then it can appear more than once in the sample. This is known as **sampling with replacement**.

If the item selected is not replaced into the population before the next item is selected, this is known as **sampling without replacement**.

Estimating population parameters

To define
▶ a binomial distribution, you need to know n and p
▶ a Poisson distribution, you need to know λ
▶ an exponential distribution, you need to know λ
▶ a normal distribution, you need to know μ and σ^2.
These are known as the **population parameters** of the distributions.

If a particular population parameter of the distribution of the random variable X is not known, you could take a random sample of size n from the population and use it to estimate the population parameter.

If you take several random samples of size n, then the **sample statistics** obtained are themselves random variables and are called **estimators**. An estimator is said to be unbiased if the expected value of the estimator is equal to the true value of the parameter.

Unbiased estimators of the population mean and variance

This chapter considers ways of estimating the mean μ and variance σ^2 of a random variable X.

n independent observations, $X_1, X_2, ..., X_n$ are taken from the distribution of X and **unbiased estimators** of μ and σ^2 are formed as follows.

An unbiased estimator of the population mean μ is $\bar{X}$, where

$$\bar{X} = \frac{X_1 + X_2 + \cdots + X_n}{n} = \frac{1}{n}\sum X_i, \text{ for } i = 1, 2, \quad , n$$

An unbiased estimator of the population variance σ^2 is S^2, where

$$S^2 = \frac{\sum(X_i - \bar{X})^2}{n-1} = \frac{1}{n-1}\left(\sum X_i^2 - \frac{\left(\sum X_i\right)^2}{n}\right)$$

It can be shown that both the estimators $\bar{X}$ and S^2 are unbiased,

that is $E(\bar{X}) = \mu$ and $E(S^2) = \sigma^2$ See Section 14.2

The value that an unbiased estimator gives for a population parameter is called an **unbiased estimate**.

If the sample values are $x_1, x_2, ..., x_n$,
▶ an unbiased estimate of μ is $\bar{x}$, where

$$\bar{x} = \frac{x_1 + x_2 + \cdots + x_n}{n} = \frac{1}{n}\sum x_i$$

▶ an unbiased estimate of σ^2 is s^2, where

$$s^2 = \frac{\sum(x_i - \bar{x})^2}{n-1} = \frac{1}{n-1}\left(\sum x_i^2 - \frac{\left(\sum x_i\right)^2}{n}\right)$$

Alternative notation:

$\hat{\mu}$ is used for the estimate of μ, where $\hat{\mu} = \bar{x}$

$\hat{\sigma}^2$ is used for the estimate of σ^2, where $\hat{\sigma}^2 = s^2$

Note

Upper case letters are used for $X_1, X_2, ...,$ because $X_1, X_2, ...$ are random variables.

Note

These are given in the examination formulae booklet.

Note

Lower case letters are used for $x_1, x_2, ...,$ because $x_1, x_2, ...$ denote particular values of a random variable.

Example 1

The random variable X grams, the mass of a bag of a particular brand of flour, has mean μ and standard deviation σ.

A random sample of 150 bags are weighed and the masses, x grams, are summarised by

$$\Sigma x = 112\,200, \quad \Sigma(x-\bar{x})^2 = 1944.45 \text{ where } \bar{x} \text{ is the sample mean.}$$

Calculate unbiased estimates of the population mean μ and the population variance σ^2.

$$\hat{\mu} = \bar{x} = \frac{\Sigma x}{n} = \frac{112200}{150} = 748$$

$$s^2 = \frac{\Sigma(x-\bar{x})^2}{n-1} = \frac{1944.45}{149} = 13.05$$

Example 2

The random variable Y is the height, in cm, of men undergoing a particular medical examination. A random sample of 250 men is taken and their heights measured, with the results:

$$\Sigma y = 43\,205, \quad \Sigma y^2 = 7\,469\,107$$

Calculate values for $\bar{y}$ and s, where s^2 denotes the unbiased estimate of σ^2.

$$\hat{\mu} = \bar{y} = \frac{\Sigma y}{n} = \frac{43205}{250} = 172.82$$

$$s^2 = \frac{1}{n-1}\left(\Sigma y^2 - \frac{(\Sigma y)^2}{n}\right) = \frac{1}{249}\left(7469107 - \frac{43205^2}{250}\right) = 9.7144\ldots$$

$$s = \sqrt{9.7144\ldots} = 3.116\ldots = 3.12 \text{ (3 sf)}$$

Calculator note

If you know the actual data, rather than just a summary, you can use your calculator in statistical mode to get the values of $\bar{x}$ and s directly. Input the data as described in the AS Textbook Section 14.1.

The key for s is usually marked $\boxed{\sigma_{n-1}}$ or similar. You then need to square this value to obtain s^2.

Example 3

The concentrations, in milligrams per litre, of a trace element in 7 randomly chosen samples of water from a spring were as follows:

240.8	237.3	236.7	236.6	234.2	233.9	232.5

Find unbiased estimates of the mean and variance of the concentration of the trace element per litre of water from the spring.

Using a calculator in statistical mode

$$\hat{\mu} = \bar{x} = 236$$

$$s = 2.753..., \text{ so } s^2 = (2.753...)^2 = 7.58$$

Note that if you are given *sample data*, either in raw form or summarised, and are asked to calculate the *sample variance*, you should find s^2, with divisor $(n-1)$ as described above.

The formula with divisor n, introduced in the AS Textbook in Section 14.1, is used to calculate the variance or standard deviation when you have the complete data for a *whole population*.

Example 4

A proofreader suggests that X, the number of errors on a page of manuscript, has a Poisson distribution.

50 pages were selected at random from the manuscript and the number of errors on each page was noted. The results are summarised as follows:

$$\sum x = 130, \quad \sum(x - \bar{x})^2 = 85 \text{ where } \bar{x} \text{ is the sample mean.}$$

a   Find the sample mean and variance.

b   Use these to decide whether the data support the proofreader's suggestion.

a   From the sample,

$$\bar{x} = \frac{\sum x}{n} = \frac{130}{50} = 2.6$$

$$s^2 = \frac{\sum(x - \bar{x})^2}{n-1} = \frac{85}{49} = 1.734... = 1.73 \,(3\text{ sf})$$

b   If $X \sim \text{Po}(\lambda)$, then mean $= \lambda$ and variance $= \lambda$, so in a Poisson distribution, the mean and the variance have the same value.

Now, since $\hat{\mu} = 2.6$ and $s^2 = 1.73$, mean $\neq$ variance.

So the sample data do not support the proofreader's suggestion that the number of errors on a page has a Poisson distribution.

Exercise 1

If you are given the values of the raw data, use a calculator in statistical mode.

1 Calculate unbiased estimates of the mean μ and variance σ^2 of the population from which the sample is taken.

   a $\sum x = 100, \sum(x - \bar{x})^2 = 308, n = 10$

   b $\sum x = 120, \sum x^2 = 2102, n = 8$

2 Calculate unbiased estimates of the mean μ and variance σ^2 of the population from which each of these samples is taken.

   a   46, 48, 51, 50, 45, 53, 50, 48

   b   1.684, 1.691, 1.687, 1.688, 1.689, 1.688, 1.690, 1.693, 1.685

   c

x	20	21	22	23	24	25
f	4	14	17	26	20	9

3 A measuring rule was used to measure the length of a rod of stated length 1 m. On 8 successive occasions these results, in mm, were obtained:

| 1000 | 999 | 999 | 1002 | 1001 | 1000 | 1002 | 1001 |

Calculate unbiased estimates of the population mean and variance of the *errors*, in mm, occurring when the rule is used for measuring a 1 m length.

4 80 adults in Ruritania were asked to measure their pulse rates, x beats per minute, when they woke up in the morning. The results were summarised as follows.

$$\Sigma x = 5520, \quad \Sigma x^2 = 382\,160$$

Stating an assumption about the 80 adults, calculate unbiased estimates of the mean and variance of the pulse rates upon waking of all adults in Ruritania.

5 A random sample of 100 plants of a certain type are chosen and their heights, x cm, measured. The results are summarised by

$$\Sigma x = 7600, \quad \Sigma x^2 = 592\,000$$

Calculate values for $\bar{x}$ and s, where s^2 denotes the unbiased estimate of σ^2.

6 The labelling on a packet of a certain brand of breakfast cereal reads 'Contents 1 kg'. A random sample of 100 packets of the breakfast cereal is taken and the mass, in grams, of the contents of each bag is recorded, with these results:

$$\Sigma x = 112\,200, \quad \Sigma(x - \bar{x})^2 = 1944.45 \quad n = 100$$

Calculate unbiased estimates of the mean and variance of the mass of the breakfast cereal in packets of this brand.

7 The random variable X is the height, in cm, of an adult male living in a certain city.
The height of each man in a random sample of 200 men living in the city was measured and these results were obtained.

$$\Sigma x = 35\,050, \quad \Sigma x^2 = 6\,163\,109$$

Calculate values for $\bar{x}$ and s, where s^2 denotes the unbiased estimate of σ^2.

8 The discrete random variable X is thought to follow a Poisson distribution with parameter 2.5.

A sample of size 100 is taken from the population of X, with these results:

$$\Sigma x = 254.3, \quad \Sigma x^2 = 901.87$$

a Calculate the sample mean and variance.

b State, with a reason, whether the sample data support the belief that X has a Poisson distribution with parameter 2.5.

14.2 The sample mean $\bar{X}$

When you take random samples from a population, it is useful to know how these sample values are distributed.

Imagine carrying out this procedure:

▶ Take a random sample of n independent observations from a population.
▶ Calculate the mean of the n sample values, that is the **sample mean**.
▶ Repeat the procedure until you have taken *all possible samples* of size n, calculating the mean of each sample.
▶ Form a distribution of all the sample means.

The distribution that would be formed is called the **distribution of the sample mean**.

Mean and variance of $\bar{X}$

Consider the random variable X with mean μ and variance σ^2.

When n independent observations, $X_1, X_2, ..., X_n$ are taken from X, the sample mean is $\bar{X}$, where

$$\bar{X} = \frac{1}{n}\Sigma X_i \quad \text{for } i = 1, 2, ..., n$$

$$E(\bar{X}) = E\left(\frac{1}{n}\Sigma X_i\right)$$

Section 11.3: $E(aX) = aE(X)$

$$= \frac{1}{n}E\left(\Sigma X_i\right)$$

$$= \frac{1}{n}\times n\mu \qquad E(X_1) = \mu,\ E(X_2) = \mu,\ ...,\ E(X_n) = \mu$$

$$= \mu$$

$$Var(\bar{X}) = Var\left(\frac{1}{n}\Sigma X_i\right)$$

Section 11.4: $Var(aX) = a^2 Var(X)$

$$= \frac{1}{n^2}Var\left(\Sigma X_i\right)$$

$$= \frac{1}{n^2}\times n\sigma^2 \qquad Var(X_1) = \sigma^2,\ Var(X_2) = \sigma^2,\ ...,\ Var(X_n) = \sigma^2$$

$$= \frac{\sigma^2}{n}$$

This gives the following very important results:

$$E(\bar{X}) = E(X) = \mu$$

$$Var(\bar{X}) = \frac{Var(X)}{n} = \frac{\sigma^2}{n}$$

The standard deviation of $\bar{X}$ is known as the **standard error of the mean**.

So, standard error of the mean $= \sqrt{\dfrac{\sigma^2}{n}} = \dfrac{\sigma}{\sqrt{n}}$

> **Note**
>
> If σ is unknown, the estimator for the standard error of the mean is $\dfrac{S}{\sqrt{n}}$.

Note that, since σ has been divided by $\sqrt{n}$, the standard deviation of $\bar{X}$ is smaller than the standard deviation of X, implying that the sample means are much more clustered around the mean than the population values are.

14.3 Distribution of the sample mean

There are two cases to be considered, depending on whether X is normally distributed or not.

Case 1: Distribution of $\bar{X}$ when X is normally distributed

If X is normally distributed, then $\bar{X}$ is also normally distributed.

If $X \sim N(\mu, \sigma^2)$,

then $\bar{X} \sim N\left(\mu, \dfrac{\sigma^2}{n}\right)$, where $\bar{X} = \dfrac{1}{n}\sum X_i$

> **Note**
>
> This very important result holds for **any** sample size n.

These diagrams illustrate the shape of the distribution of $\bar{X}$ when the sample size is 5 and 25, where samples are taken from X, a normal distribution with a mean of 100 and a standard deviation of 8.

You can see that the larger the sample size, the more clustered the sample means are around μ.

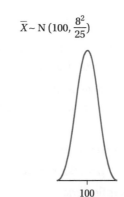

$$\bar{X} \sim N\left(100, \frac{8^2}{25}\right)$$

$$\bar{X} \sim N\left(100, \frac{8^2}{5}\right)$$

$$X \sim N(100, 8^2)$$

100	100	100
Distribution of X	**Distribution of $\bar{X}$** when $n = 5$	**Distribution of $\bar{X}$** when $n = 25$

Standardising $\bar{X}$

When finding probabilities, you will need to standardise $\bar{X}$ in order to use the normal distribution tables.

If $\bar{X} \sim N\left(\mu, \dfrac{\sigma^2}{n}\right)$, then $Z = \dfrac{\bar{X} - \mu}{\dfrac{\sigma}{\sqrt{n}}}$, where $Z \sim N(0, 1)$

See Section 13.3

Calculation note:

To avoid rounding errors, it is often better not to calculate the value of $\dfrac{\sigma}{\sqrt{n}}$ until the final stage. This is particularly the case when the value is not exact.

Example 5

Question

The time, X minutes, taken to complete a jumbo Sudoku puzzle can be modelled by a normal distribution with a mean of 52.7 minutes and a standard deviation of 13.5 minutes.

The times taken by a random sample of 10 people are noted and the mean time calculated. Find the probability that the sample mean is less than 54.

Answer

X is the time, in minutes, taken to solve the puzzle, where

$\quad X \sim N(52.7, 13.5^2)$ $\qquad\qquad \mu = 52.7, \sigma = 13.5$

$\bar{X}$ is the *mean* time, in minutes, taken by 10 people to solve the puzzle.

Since the distribution of X is normal, the distribution of $\bar{X}$ is also normal,

so $\quad \bar{X} \sim N\left(52.7, \dfrac{13.5^2}{10}\right)$ $\qquad$ s.d. $= \dfrac{13.5}{\sqrt{10}}$ $\quad \leftarrow$ Leave in uncalculated form

$P(\bar{X} < 54) = P\left(Z < \dfrac{54 - 52.7}{\dfrac{13.5}{\sqrt{10}}}\right)$

$\qquad\qquad = P(Z < 0.3045\ldots) \quad \leftarrow$ Round to 2 dp

$\qquad\qquad = \Phi(0.30) \quad \leftarrow$ Use Table 3 (Section 13.2)

$\qquad\qquad = 0.61791 = 0.618 \ (3\ \text{sf})$

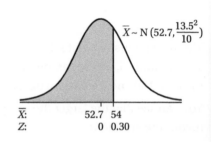

$\bar{X} \sim N\left(52.7, \dfrac{13.5^2}{10}\right)$

$\bar{X}$: $\qquad$ 52.7 54
Z: $\qquad$ 0 0.30

Example 6

Question

The heights of a new variety of sunflower can be modelled by a normal distribution with mean 200 cm and standard deviation 40 cm.

A random sample of 50 sunflowers is taken and the mean height calculated. Find the probability that the sample mean lies between 195 cm and 205 cm.

Answer

Let X be the height, in cm, of a sunflower, where

$\quad X \sim N(200, 40^2)$ $\qquad\qquad \mu = 200, \sigma = 40$

$\bar{X}$ is the *mean* height, in cm, of 50 sunflowers.

So $\quad \bar{X} \sim N\left(200, \dfrac{40^2}{50}\right)$ $\qquad$ s.d. $= \dfrac{40}{\sqrt{50}}$

$P(195 < \bar{X} < 205)$

$\quad = P\left(\dfrac{195 - 200}{\dfrac{40}{\sqrt{50}}} < Z < \dfrac{205 - 200}{\dfrac{40}{\sqrt{50}}}\right)$

$\quad = P(-0.88 < Z < 0.88)$

$\quad = 2\Phi(0.88) - 1 \quad \leftarrow$ Result 4 (Page 249)

$\quad = 2 \times 0.81057 - 1$

$\quad = 0.62114 = 0.621 \ (3\ \text{sf})$

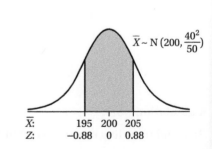

$\bar{X} \sim N\left(200, \dfrac{40^2}{50}\right)$

$\bar{X}$: $\qquad$ 195 200 205
Z: $\qquad$ −0.88 0 0.88

Example 7

At a college, the mass of a male student may be modelled by a normal distribution with mean 70 kg and standard deviation 5 kg.

a Find the probability that the mean mass of a random sample of 4 male students is less than 65 kg.

b A random sample of n male students is taken. The probability that their mean mass is less than 72.45 kg is 0.975. Find the value of n.

X is the mass, in kilograms, of a male student at the college, where

$$X \sim N(70, 5^2) \qquad \mu = 70, \sigma = 5$$

a $\bar{X}$ is the mean mass, in kilograms, of 4 male students at the college.

$$\bar{X} \sim N\left(70, \frac{5^2}{4}\right) \qquad \text{s.d.} = \frac{5}{\sqrt{4}}$$

$$P(\bar{X} < 65) = P\left(Z < \frac{65 - 70}{\frac{5}{\sqrt{4}}}\right)$$

$$= P(Z < -2)$$

$$= 1 - \Phi(2)$$

$$= 1 - 0.97725$$

$$= 0.02275 = 0.0228 \text{ (3 sf)}$$

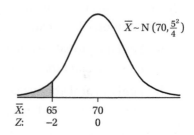

$\bar{X} \sim N(70, \frac{5^2}{4})$

$\bar{X}$:	65	70
Z:	-2	0

b $\bar{X}$ is the mean mass, in kilograms, of n male students at the college, where

$$\bar{X} \sim N\left(70, \frac{5^2}{n}\right) \qquad \text{s.d.} = \frac{5}{\sqrt{n}}$$

$$P(\bar{X} < 72.45) = P\left(Z < \frac{72.45 - 70}{\frac{5}{\sqrt{n}}}\right)$$

$$= P\left(Z < \frac{2.45\sqrt{n}}{5}\right)$$

$$= P\left(Z < 0.49\sqrt{n}\right)$$

So, $\quad P\left(Z < 0.49\sqrt{n}\right) = 0.975$

Now, if $\Phi(z) = 0.975$

$$z = \Phi^{-1}(0.975) \quad \leftarrow \text{Table 4, row 0.97, column 0.005 (Section 13.4)}$$

$$= 1.96$$

So $\quad 0.49\sqrt{n} = 1.96$

$$\sqrt{n} = \frac{1.96}{0.49} = 4$$

$$n = 16$$

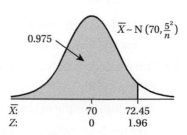

$\bar{X} \sim N(70, \frac{5^2}{n})$

0.975

$\bar{X}$:		70	72.45
Z:		0	1.96

Exercise 2

1. The volume of lemonade in bottles is normally distributed with a mean 758 ml and a standard deviation 12 ml. A random sample of 10 bottles is taken and the mean volume found. Find the probability that the sample mean is

   a   less than 750 ml

   b   between 752 ml and 760 ml.

2. A wholesaler sells carrots whose mass is normally distributed with mean 145 g and standard deviation 9 g.

   A sample of 16 carrots is taken. Stating a necessary assumption, find the probability that the mean mass of carrots in the sample is

   a   less than 148 g

   b   greater than 141 g

   c   between 143 g and 150 g.

3. In a certain city, men have heights which can be modelled by a normal distribution with mean 170 cm and standard deviation 10 cm.

   a   Find the probability that the mean height of three randomly selected men is greater than 178 cm.

   b   Find the probability that the mean height of a random sample of three men differs from the population mean height by more than 10 cm.

4. The length of a particular type of insect may be modelled by a normal distribution with mean 15.3 mm and standard deviation 2.6 mm.

   For a biology project, each student in a class of 30 students collects a random sample of 10 insects of this type and finds the mean length.

   a   Deena is a student in the class. Find the probability that the mean length of the insects in Deena's sample is between 13.5 mm and 15.1 mm.

   b   How many of the 30 samples collected by the students in the class would you expect to have a mean length greater than 16.0 mm?

5. The random variable X has a distribution N(30, 5).

   Random samples of size n are taken from the distribution.

   a   Find the probability that the sample mean exceeds 30.5

       i   when $n = 10$ **ii**   when $n = 40$ **iii**   when $n = 100$.

   b   Find the least value of n such that the probability that the sample mean exceeds 30.5 is less than 0.01.

6. The random variable Y follows a normal distribution with mean 25 and variance 340.

   A random sample of size n is taken from the distribution.

   Find the least value of n such that $P(\bar{Y} > 28) < 0.005$.

7. In a computer game, competitors have to find their way out of a maze. The time taken, in seconds, may be modelled by a normal distribution with mean 80 and standard deviation σ.

The probability that the mean time of a random sample of 100 competitors is greater than 82 seconds is 0.05. Find the value of σ.

8 The random variable X is distributed normally with mean 82.6 and standard deviation σ. The mean of 20 independent observations of X is $\bar{X}$. Given that $P(\bar{X} > 80) = 0.75$, find the value of σ.

9 The diameter, D millimetres, of an America pool ball may be modelled by a normal random variable with mean 57.15 and standard deviation 0.04.

  a Determine:

   i $P(D < 57.2)$;   [3]

   ii $P(57.1 < D < 57.2)$.   [2]

  b A box contains 16 of these pool balls. Given that the balls may be regarded as a random sample, determine the probability that:

   i   all 16 balls have diameter less than 57.2 mm;   [2]

   ii the mean diameter of the 16 balls is greater than 57.16 mm. [4]

                                               AQA MS1B June 2011

10 When a particular make of tennis ball is dropped from a vertical distance of 250 cm onto concrete, the height, X centimetres, to which it first bounces may be assumed to be normally distributed with a mean of 140 and a standard deviation of 2.5.

  a Determine:

   i $P(X < 145)$;   [3]

   ii $P(138 < X < 142)$.   [4]

  b Determine, to one decimal place, the maximum height exceeded by 85% of first bounces.   [4]

  c Determine the probability that, for a random sample of 4 first bounces, the mean height is greater than 139 cm.   [4]

                                               AQA MS1B June 2008

Case 2: Distribution of $\bar{X}$ when X is not normally distributed

When X is not normally distributed, provided that the sample size is *large*, the Central Limit Theorem can be applied.

The **Central Limit Theorem** (CLT) states that, for samples of size n taken from a population X with mean μ and variance σ^2 whose distribution is *not normal*, the distribution of $\bar{X}$ tends to a *normal* distribution as n tends to infinity.

This means that, when n is large ($n \geq 30$ say), the distribution of $\bar{X}$ is approximately normal.

For a non-normal population X with mean μ and variance σ^2, when n is large, ($n \geq 30$ say), by the Central Limit Theorem,

$$\bar{X} \sim N\left(\mu, \frac{\sigma^2}{n}\right) \text{ approximately.}$$

Note

The larger the sample size n, the better the approximation.

Example 8

Question

A manufacturer produces small chocolate cakes. The mass of the mixture in a cake has mean 52.3 grams and standard deviation 4.1 grams.

a Find the probability that the mean mass of the mixture in a random sample of 120 cakes is between 51.5 grams and 52.0 grams.

b Give a reason why it is not necessary to assume that the mass of the mixture in a cake is normally distributed in order to carry out the calculations in part **a**.

Answer

a Let X be the mass, in grams, of the mixture in a small chocolate cake, where $\mu = 52.3$, $\sigma = 4.1$ and the distribution of X is unknown.

$\bar{X}$ is the mean mass, in grams, of a random sample of 120 cakes.

$$E(\bar{X}) = 52.3, \quad Var(\bar{X}) = \frac{4.1^2}{120}$$

Since n is large, by the Central Limit Theorem,

$$\bar{X} \sim N\left(52.3, \frac{4.1^2}{120}\right) \text{ approximately,} \qquad s.d. = \frac{4.1}{\sqrt{120}}$$

$P(51.5 < \bar{X} < 52.0)$

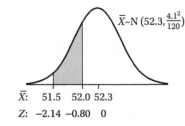

$$= P\left(\frac{51.5 - 52.3}{\frac{4.1}{\sqrt{120}}} < Z < \frac{52.0 - 52.3}{\frac{4.1}{\sqrt{120}}}\right)$$

$= P(-2.14 < Z < -0.80)$ ← Result 2 (page 249)

$= \Phi(2.14) - \Phi(0.80)$

$= 0.98382 - 0.78814$

$= 0.19568 = 0.196 \text{ (3 sf)}$

$\bar{X}$:   51.5   52.0 52.3

Z: −2.14 −0.80   0

b It is not necessary to assume that the mass of the mixture is normally distributed because, when the sample size is large ($n = 120$), by the Central Limit Theorem, the distribution of $\bar{X}$ is approximately normal.

Example 9

Question

Kasto arrives at random times at a bus stop. The time, X minutes, that he has to wait for the next bus is given by

$$f(x) = \begin{cases} \dfrac{1}{9} & 0 \le x \le 9 \\ 0 & \text{otherwise} \end{cases}$$

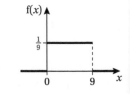

a Find the mean and variance of X.

b Find the probability that in a random sample of 75 occasions his mean waiting time is less than 5 minutes.

Answer

a By symmetry, $E(X) = 4.5$ See Section 11.3

$$E(X^2) = \frac{1}{9}\int_0^9 x^2 \, dx$$

$$= \frac{1}{9}\left[\frac{x^3}{3}\right]_0^9$$

(continued)

(continued)

$$= \frac{1}{27}(9^3 - 0)$$

$$= 27$$

$$\mathrm{Var}(X) = \mathrm{E}(X^2) - (\mathrm{E}(X))^2$$

$$= 27 - 4.5^2$$

$$= 6.75$$

b $\bar{X}$ is the mean of a random sample of 75 independent observations, so $n = 75$.

$$\mathrm{E}(\bar{X}) = 4.5, \quad \mathrm{Var}(\bar{X}) = \frac{6.75}{75}$$

Since n is large, by the Central Limit Theorem,

$$\bar{X} \sim \mathrm{N}\left(4.5, \frac{6.75}{75}\right) \text{ approximately,} \qquad \text{s.d.} = \sqrt{\frac{6.75}{75}} = 0.3$$

$$P(\bar{X} < 5) = P\left(Z < \frac{5 - 4.5}{0.3}\right)$$

$$= P(Z < 1.666..) \quad \leftarrow \text{Round to 2 dp.}$$

$$= \Phi(1.67)$$

$$= 0.95254 = 0.953 \text{ (3 sf)}$$

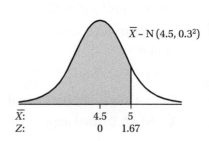

$\bar{X} \sim \mathrm{N}(4.5, 0.3^2)$

| $\bar{X}$: | 4.5 | 5 |
| Z: | 0 | 1.67 |

Example 10

The life, T hours, of an electrical component has an exponential distribution with probability density function given by

$$f(t) = \begin{cases} 0.005e^{-0.005t} & t \geq 0 \\ 0 & \text{otherwise} \end{cases}$$

a Determine the mean and variance of T.

b A random sample of 100 components is taken. Find the probability that the mean life of the sample is less than 180 hours.

a $\lambda = 0.005$ *Exponential distribution, see Section 12.1*

$$\mathrm{E}(T) = \frac{1}{\lambda} = \frac{1}{0.005} = 200$$

$$\mathrm{Var}(T) = \frac{1}{\lambda^2} = \frac{1}{0.005^2} = 40000$$

b $\mathrm{E}(\bar{T}) = 200, \quad \mathrm{Var}(\bar{T}) = \frac{40000}{100} = 400$

Since n is large, by the Central Limit Theorem,

$$\bar{T} \sim \mathrm{N}(200, 400) \text{ approximately,} \qquad \text{s.d.} = \sqrt{400} = 20$$

$$P(\bar{T} < 180) = P\left(Z < \frac{180 - 200}{20}\right)$$

$$= P(Z < -1)$$

$$= 1 - \Phi(1)$$

$$= 1 - 0.84134$$

$$= 0.15866 = 0.159 \text{ (3 sf)}$$

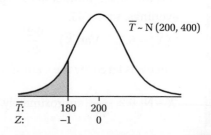

$\bar{T} \sim \mathrm{N}(200, 400)$

| $\bar{T}$: | 180 | 200 |
| Z: | -1 | 0 |

Examples 11 and 12 illustrate a surprising application of the Central Limit Theorem: that it holds even when the distribution of X is *discrete*.

Example 11

The random variable X has the distribution B(12, 0.4).

a Find E(X) and Var(X).

A random sample of 100 independent observations is taken.

b Determine the probability that the mean of the sample is greater than 5.

a $X \sim \text{B}(12, 0.4)$ $n = 12$, $p = 0.4$, where n is the number of trials.

$\text{E}(X) = np = 12 \times 0.4 = 4.8$ AS Textbook Section 15.1

$\text{Var}(X) = np(1 - p) = 12 \times 0.4 \times 0.6 = 2.88$

b $\bar{X}$ is the mean of a sample of 100 observations, so $n = 100$, where n is the sample size.

$\text{E}(\bar{X}) = 4.8$, $\text{Var}(\bar{X}) = \dfrac{2.88}{100} = 0.0288$

Since n is large, by the Central Limit Theorem,

$\bar{X} \sim \text{N}(4.8, 0.0288)$ approximately,   s.d. $= \sqrt{0.0288}$

$P(\bar{X} > 5) = P\left(Z > \dfrac{5 - 4.8}{\sqrt{0.0288}} \right)$

$= P(Z > 1.18)$

$= 1 - \Phi(1.18)$

$= 1 - 0.88100$

$= 0.11900 = 0.119 \ (3 \text{ sf})$

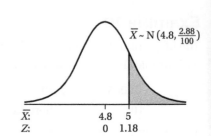

$\bar{X} \sim \text{N}\left(4.8, \frac{2.88}{100}\right)$

| $\bar{X}$: | 4.8 | 5 |
| Z: | 0 | 1.18 |

Example 12

The number of telephone calls, X, made to a counselling service in an evening may be modelled by a Poisson distribution with parameter 6.2.

a Find the mean and variance of X.

A random sample of 45 evenings is taken and the number of calls per evening noted.

b Find the probability that the mean number of calls per evening in the sample is less than 5.1.

a $X \sim \text{Po}(\lambda)$, with $\lambda = 6.2$

$\text{E}(X) = 6.2$, $\text{Var}(X) = 6.2$ See Section 10.1

b $\bar{X}$ is the mean number of calls per evening in a random sample of 45 evenings, so $n = 45$.

$\text{E}(\bar{X}) = 6.2$, $\text{Var}(\bar{X}) = \dfrac{6.2}{45}$

Since n is large, by the Central Limit Theorem,

$\bar{X} \sim \text{N}\left(6.2, \dfrac{6.2}{45} \right)$ approximately,   s.d. $= \sqrt{\dfrac{6.2}{45}} = \sqrt{0.1377\ldots}$

(continued)

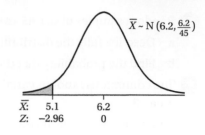

(continued)

$$P(\bar{X} < 5.1) = P\left(Z < \frac{5.1 - 6.2}{\sqrt{0.1377}}\right)$$

$$= P(Z < -2.96)$$

$$= 1 - \Phi(2.96)$$

$$= 1 - 0.99846$$

$$= 0.00154 \,(3\ \text{sf})$$

$\bar{X} \sim N\left(6.2, \frac{6.2}{45}\right)$

$\bar{X}$: 5.1   6.2
Z: −2.96   0

Exercise 3

1 In an examination taken by students in a particular country, the mean mark was 64.5 and the variance was 64.

A total of 100 examination scripts are selected at random. Find the probability that the mean mark of the scripts in the sample is

a higher than 65.5

b between 63.8 and 64.5.

2 Rubber balls are produced for sale in toy shops. The diameters X, in millimetres, of the balls have mean 62.5 and standard deviation 1.2.

The balls are packed in boxes of 50.

Given that the balls in a box may be regarded as a random sample, find the probability that the mean diameter of the balls in a box is

a less than 62.6 mm

b between 62.0 mm and 62.3 mm.

3 The random variable X has the distribution B(20, 0.3).

a Find the mean and variance of X.

A random sample of 50 observations of X is taken.

b Describe fully the distribution of the sample mean.

c Find the probability that the sample mean is

   i less than 6.7   **ii** more than 5.4.

4 The number of text messages Joe sends to his friend each day may be modelled by a Poisson distribution with mean 12.5.

Find the probability that, in a random sample of 40 days, the mean number of text messages Joe sends is greater than 14.

5 The continuous random variable X has a probability density function given by

$$f(x) = \begin{cases} \dfrac{3}{8}(1 + x^2) & -1 \le x \le 1 \\ 0 & \text{otherwise} \end{cases}$$

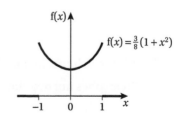

$f(x) = \frac{3}{8}(1 + x^2)$

a i *State* the value of E(X).

  ii Find Var(X).

b The mean of 50 independent observations of X is denoted by $\bar{X}$.

  i Find the mean and variance of $\bar{X}$.

  ii State the approximate distribution of $\bar{X}$.

  iii Find $P(0 < \bar{X} < 0.2)$.

6. Random samples of size 65 are taken from the distribution Po(5.8).

    a Describe fully the distribution of the sample mean.

    b Find the probability that the sample mean lies between 5.1 and 6.3.

7. The continuous random variable X has an exponential distribution with mean 3.

    A random sample of 200 observations is taken from the distribution. Find the probability that the sample mean is between 2.47 and 3.36.

8. The random variable Y has mean 20 and standard deviation 3. The random variable $\bar{Y}$ is the mean of a random sample of 80 independent observations from Y.

    a State the approximate distribution of $\bar{Y}$.

    b Determine

      i $P(20.16 < \bar{Y} < 20.83)$

      ii $P(\bar{Y} = 20.6)$.

14.4 Further applications

Example 13

A machine fills small cans with soda water. The volume of soda water delivered by the machine may be modelled by a normal random variable with a mean of 153 ml and a standard deviation of 1.6 ml.

Each can is able to hold a maximum of 155 ml of soda water.

Printed on each can is 'Contents 150 ml'.

a Determine the probability that the volume of soda water delivered by the machine:

    i does not cause a can to overflow;

    ii is less than that printed on the can.

b Following adjustment to the machine, the volume of soda water in a can may be modelled by a normal random variable with a mean of 152 ml and a standard deviation of 0.8 ml.

Given that packs of 12 cans may be assumed to be a random sample of cans filled by the machine, determine the probability that, in a pack, the mean volume of soda water per can is more than 152.5 ml.

AQA MS1A June 2010

a **i** X is the volume, in ml, of soda water in a can, where $X \sim N(153, 1.6^2)$

    P(can does not overflow) $= P(X < 155)$

$$= \left(P\left(Z < \frac{155 - 153}{1.6}\right) \right)$$

$$= P(Z < 1.25)$$

$$= \Phi(1.25)$$

$$= 0.89435 = 0.894 \text{ (3 sf)}$$

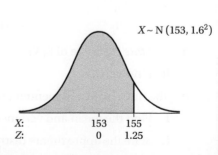

$X \sim N(153, 1.6^2)$

X: 153 155
Z: 0   1.25

(continued)

(continued)

ii $P(X < 150) = P\left(Z < \dfrac{150 - 153}{1.6}\right)$

$= P(Z < -1.875)$ ← Round to 2 dp

$= 1 - \Phi(1.88)$

$= 1 - 0.96995$

$= 0.03005 = 0.0301 \text{ (3 sf)}$

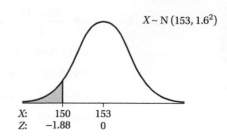

$X \sim N(153, 1.6^2)$

X:	150	153
Z:	-1.88	0

b Y is the volume, in ml, of soda water in a can, after adjustment, where

$\qquad Y \sim N(152, 0.8^2)$

$\bar{Y}$ is the mean volume, in ml, of 12 cans.

$\qquad \bar{Y} \sim N\left(152, \dfrac{0.8^2}{12}\right) \qquad$ s.d. $= \dfrac{0.8}{\sqrt{12}}$

$P(\bar{Y} > 152.5) = P\left(Z > \dfrac{152.5 - 152}{\dfrac{0.8}{\sqrt{12}}}\right)$

$= P(Z > 2.17)$

$= 1 - \Phi(2.17)$

$= 1 - 0.98500$

$= 0.01500 = 0.0150 \text{ (3 sf)}$

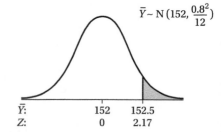

$\bar{Y} \sim N(152, \dfrac{0.8^2}{12})$

$\bar{Y}$:	152	152.5
Z:	0	2.17

Example 14

When Monica walks to work from home, she uses either route A or route B.

a Her journey time, X minutes, by route A may be assumed to be normally distributed with a mean of 37 and a standard deviation of 8.

Determine:

i $P(X < 45)$; **ii** $P(30 < X < 45)$.

b Her journey time, Y minutes, by Route B may be assumed to be normally distributed with a mean of 40 and a standard deviation of σ.

Given that $P(Y > 45) = 0.12$, calculate the value of σ.

c If Monica leaves home at 8.15 am to walk to work hoping to arrive by 9 am, state, with a reason, which route she should take.

d When Monica travels to work from home by car, her journey time, W minutes, has a mean of 18 and a standard deviation of 12.

Estimate the probability that, for a random sample of 36 journeys to work from home by car, Monica's mean time is more than 20 minutes.

e Indicate where, if anywhere, in this question you need to make use of the Central Limit Theorem.

AQA MS1B January 2007

a X minutes is the journey time by Route A, where $X \sim N(37, 8^2)$

i $P(X < 45) = P\left(Z < \dfrac{45 - 37}{8}\right)$

$= P(Z < 1)$

$= \Phi(1)$

$= 0.84134 = 0.841 \text{ (3 sf)}$

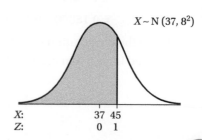

$X \sim N(37, 8^2)$

X:	37 45
Z:	0 1

(continued)

(continued)

ii $P(30 < X < 45) = P\left(\dfrac{30 - 37}{8} < Z < \dfrac{45 - 37}{8}\right)$

$\qquad\qquad\qquad = P(-0.88 < Z < 1)$

$\qquad\qquad\qquad = \Phi(0.88) + \Phi(1) - 1$

$\qquad\qquad\qquad = 0.81057 + 0.84134 - 1$

$\qquad\qquad\qquad = 0.65191 = 0.652 \text{ (3 sf)}$

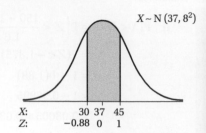

$X \sim N(37, 8^2)$

X: 30 37 45
Z: −0.88 0   1

b Y minutes is the journey time by Route B, where $Y \sim N(40, \sigma^2)$

If $\qquad\qquad P(Y > 45) = 0.12$

then $\qquad\qquad P(Y < 45) = 0.88$

So, $\quad P\left(Z < \dfrac{45 - 40}{\sigma}\right) = 0.88$

$\qquad\qquad \Phi\left(\dfrac{5}{\sigma}\right) = 0.88$

$\qquad\qquad \dfrac{5}{\sigma} = \Phi^{-1}(0.88) = 1.175 \quad \leftarrow \text{Use Table 4}$

$\qquad\qquad \sigma = \dfrac{5}{1.175}$

$\qquad\qquad\quad = 4.255... = 4.26 \text{ (3 sf)}$

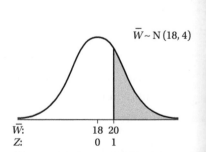

$Y \sim N(40, \sigma^2)$

Y: 40 45
Z: 0 1.175

c Monica wants to do the journey in less than 45 minutes.

$\quad$ P(time < 45 minutes by Route A) = $P(X < 45) = 0.841$

$\quad$ P(time < 45 minutes by Route B) = $P(Y < 45) = 0.88$

Since there is a greater probability that Monica will do the journey in less than 45 minutes if she goes by Route B rather than Route A, she should choose route B.

d W minutes is the journey time by car.

$\quad \mu = 18$, $\sigma = 12$ and the distribution of W is unknown.

$\overline{W}$ is the mean time, in minutes, of 36 journeys.

By the Central Limit Theorem, since n is large (> 30),

$\overline{W} \sim N\left(18, \dfrac{12^2}{36}\right)$, so $\overline{W} \sim N(18, 4)$ approximately, $\quad$ s.d. $= \sqrt{4} = 2$

$P(\overline{W} > 20) = P\left(Z > \dfrac{20 - 18}{2}\right)$

$\qquad\qquad\quad = P(Z > 1)$

$\qquad\qquad\quad = 1 - \Phi(1)$

$\qquad\qquad\quad = 1 - 0.84134$

$\qquad\qquad\quad = 0.15866 = 0.159 \text{ (3 sf)}$

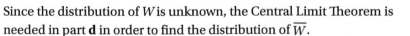

$\overline{W} \sim N(18, 4)$

$\overline{W}$:   18 20
Z: 0   1

e Since the distribution of W is unknown, the Central Limit Theorem is needed in part **d** in order to find the distribution of $\overline{W}$.

Example 15

John works from home. The number of business letters, X, that he receives on a weekday may be modelled by a Poisson distribution with mean 5.0.

The number of private letters, Y, he receives on a weekday may be modelled by a Poisson distribution with mean 1.5.

a Find, for a weekday:

  i $P(X < 4)$;   ii $P(Y = 4)$.

b i   Assuming that X and Y are independent random variables, determine the probability that, on a given weekday, John receives a **total** of more than 5 business letters and private letters.

  ii   Hence calculate the probability that John receives a **total** of more than 5 business and private letters on at least 7 out of 8 given weekdays.

c   The number of letters received by John's neighbour, Brenda, on 10 consecutive weekdays are

15	8	14	7	6	8	2	8	9	3

  i   Calculate the mean and variance of these data.

  ii   State, giving a reason based on your answers to part **c i**, whether or not a Poisson distribution might provide a suitable model for the number of letters received by Brenda on a weekday.

AQA MS2B June 2009

a i X is the number of business letters received on a weekday where $X \sim \text{Po}(5.0)$

       $P(X < 4) = P(X \leq 3)$ $\leftarrow$ Use Table 2 with $\lambda = 5.0$

              $= 0.2650 = 0.265$ (3 sf)

  ii Y is the number of private letters received on a weekday.

       $Y \sim \text{Po}(1.5)$

       $P(Y = 4) = e^{-1.5}\dfrac{1.5^4}{4!} = 0.04706\ldots = 0.0471$ (3 sf)

> **Note**
>
> Since $\lambda = 1.5$ is not in Table 2, use the formula.

b i $X + Y \sim \text{Po}(5.0 + 1.5)$ that is $X + Y \sim \text{Po}(6.5)$

       $P(X + Y > 5) = 1 - P(X + Y \leq 5)$

                 $= 1 - 0.3690$

                 $= 0.6310 = 0.631$ (3 sf)

  ii   Let W be the number of days in 8 when John received a total of more than 5 business and private letters, where $W \sim \text{B}(8, 0.631)$.

       $P(W \geq 7) = P(W = 7) + P(W = 8)$

              $= \dbinom{8}{7} \times 0.631^7 \times 0.369^1 + 0.631^8$ AS Textbook Chapter 15

              $= 0.1432\ldots = 0.143$ (3 sf)

> **Note**
>
> $P(X = x) = \dbinom{n}{x} p^x (1 - p)^{n-x}$

c i   Since the data are from a sample, you need to find s^2.

       Using a calculator in statistical mode:

       $\bar{x} = 8$, $s^2 = 16.88\ldots = 16.9$ (3 sf)

  ii   mean $\neq$ variance, so the Poisson distribution would not be a good model.

Summary

Random samples

A random sample of size n is a sample chosen such that:

▶ Every member of the population has an equal chance of being selected.

▶ All possible samples of size n have an equal chance of being selected.

Unbiased estimators

An estimator of a population parameter is unbiased if the expectation of the estimator is equal to the true value of the parameter.

Unbiased estimator for μ

$$\bar{X} = \frac{X_1 + X_2 + \cdots + X_n}{n} = \frac{1}{n}\sum X_i \text{ for } i = 1, 2, \dots, n$$

Unbiased estimator for σ^2

$$S^2 = \frac{\sum(X_i - \bar{X})^2}{n-1} = \frac{1}{n-1}\left(\sum X_i^2 - \frac{\left(\sum X_i\right)^2}{n}\right)$$

Unbiased estimates

These are *calculated* from the sample values, x_i

Unbiased estimate for μ

$$\bar{x} = \frac{1}{n}\sum x_i$$

Unbiased estimate for σ^2

$$S^2 = \frac{\sum(x_i - \bar{x})^2}{n-1} = \frac{1}{n-1}\left(\sum x_i^2 - \frac{\left(\sum x_i\right)^2}{n}\right)$$

The sample mean

$$E(\bar{X}) = E(X) = \mu$$

$$Var(\bar{X}) = \frac{Var(X)}{n} = \frac{\sigma^2}{n}$$

Standard deviation of $\bar{X} = \sqrt{\frac{\sigma^2}{n}} = \frac{\sigma}{\sqrt{n}}$

Distribution of the sample mean

Case 1: When X is normally distributed:

If $X \sim N(\mu, \sigma^2)$, then $\bar{X} \sim N\left(\mu, \frac{\sigma^2}{n}\right)$, for all values of n.

Case 2: When X is not normally distributed:

By the Central Limit Theorem, provided n is large, $(n \geq 30)$,

$$\bar{X} \sim N\left(\mu, \frac{\sigma^2}{n}\right), \text{ approximately.}$$

The larger the sample size n, the better the approximation.

Review

1. Each of a random sample of 50 UK one-pound coins was weighed. Their masses, x grams, are summarised as follows:

    $$\sum x = 474.51, \quad \sum x^2 = 4503.8276$$

    Calculate unbiased estimates of the mean and variance of the masses of UK one-pound coins.

2. Packets of nuts are filled by a machine.

    In a random sample of 10 packets, the masses, in grams, of the packets were as follows:

374.6	377.4	376.1	379.2	371.2
375.0	372.4	378.6	377.1	371.5

    Find unbiased estimates of the population mean and variance.

3. The mass of a bar of choclolate is distributed normally with a mean of 105.0 g and a standard deviation of 1.3 g. A random sample of 20 packets is taken.

    Find the probability that the mean mass of the packets in the sample is

    a   more than 105.6 g

    b   between 105.0 g and 105.3 g.

4. The mass, in kilograms, of articles produced by a machine has mean μ and standard deviation 4.55.

    A random sample of 100 articles from the machine is taken and the mean mass calculated.

    a   Find the probability that the sample mean will differ from μ by less than 0.8 kg.

    b   State, with a reason, whether it was necessary to use the Central Limit Theorem in part **a**.

5. The random variable W has a mean of 240 and a variance of 625.

    The random variable $\overline{W}$ is the mean of a random sample of 150 independent observations from W.

    a   State the approximate distribution of $\overline{W}$.

    b   Determine

    i $P(\overline{W} \le 245)$ ii $P(\overline{W} < 239)$

    iii $P(238.5 < \overline{W} < 243.5)$ iv $P(238 < \overline{W} < 242)$

6. The random variable X is distributed $\text{Po}(4.1)$.

    a   Find an approximate value of the probability that the mean of a random sample of 50 observations of X is less than 4.5.

    b   Explain whether it was necessary to use the Central Limit Theorem in your answer to part **a**.

7. The random variable X is distributed $B(30, 0.7)$. Find an approximate value of the probability that the mean of a random sample of 100 observations of X is between 20.5 and 21.5.

8 The continuous random variable X has probability density function defined by

$$f(x) = \begin{cases} kx(4 - x) & 0 \leq x \leq 4 \\ 0 & \text{otherwise} \end{cases}$$

a Find the value of the constant k.

b Find $E(X)$ and $Var(X)$.   Draw a sketch.

c Find the probability that the mean of a random sample of 50 observations of X is

   i exactly 2.1   **ii** greater than 1.82.

9 The heights of sunflowers may be assumed to be normally distributed with a mean of 185 cm and a standard deviation of 10 cm.

a Determine the probability that the height of a randomly selected sunflower:

   i is less than 200 cm;

   ii is more than 175 cm;

   iii is between 175 cm and 200 cm.

b Determine the probability that the mean height of a random sample of 4 sunflowers is more than 190 cm.

AQA MS1B June 2006

10 On a map, the symbol **P** indicates a car park. A geography student divided a map into 66 squares, each representing an area of 9 km².

a If the number of **P** symbols in a square could be modelled by a Poisson distribution with mean 0.6, find the probability of a square containing:

   i no **P** symbols;

   ii 3 or more **P** symbols.

b The student counted the number of **P** symbols in each of 66 squares. The results are shown in the table.

Number of *P* symbols	Number of squares
0	46
1	8
2	4
3	8

Calculate the mean and variance of the number of **P** symbols in a square.

c Give a reason why the Poisson distribution does **not** provide a good model for the number of **P** symbols in a square based on:

   i your calculations in part **a**;

   ii your calculations in part **b**;

   iii the likely distribution of car parks.

AQA SS02 June 2008

11 Chopped lettuce is sold in bags nominally containing 100 grams.

The weight, X grams, of chopped lettuce delivered by the machine filling the bags, may be assumed to be normally distributed with mean μ and standard deviation 4.

a Assuming that $\mu = 106$, determine the probability that a randomly selected bag of chopped lettuce:

i weighs less than 110 grams;

ii is underweight.

b Determine the minimum value of μ so that at most 2 per cent of bags of chopped lettuce are underweight.

c Boxes each contain 10 bags of chopped lettuces. The mean weight of a bag of chopped lettuce in a box is denoted by $\bar{X}$.

Given that $\mu = 108.5$:

i write down the values for the mean and variance of $\bar{X}$;

ii determine the probability that $\bar{X}$ exceeds 110.

AQA MS1B January 2005

12 The volume, X millilitres, of energy drink in a bottle can be modelled by a normal random variable with mean 507.5 and standard deviation 4.0.

a Find:

i $P(X < 515)$;

ii $P(500 < X < 515)$;

iii $P(X \neq 507.5)$.

b Determine the value of x such that $P(X < x) = 0.96$.

c The energy drink is sold in packs of 6 bottles. The bottles in each pack may be regarded as a random sample.

i Calculate the probability that the volume of energy drink in at least 5 of the 6 bottles in a pack is between 500 ml and 515 ml.

ii Determine the probability that the **mean** volume of energy drink in bottles in a pack is more than 505 ml.

AQA MS1A June 2013

Assessment

1 Cartons of orange juice are filled by a machine.

A sample of 10 cartons selected at random from the production contained the following volumes, in ml, of orange juice:

    201.2 205.0 209.1 202.3 204.6 206.4 210.1 201.9 203.7 207.3

Calculate unbiased estimates of the population mean and variance of the volume, in ml, of orange juice dispensed into the carton by the machine. [3]

2 The fuel consumption of a new model of car is being tested. In one trial, 50 cars chosen at random were driven under identical conditions and the distances x, in kilometres, covered on a litre of fuel were recorded. The results gave these totals:

$$\sum x = 525, \quad \sum(x - \bar{x})^2 = 112.5$$

Calculate values for $\bar{x}$ and s, where s^2 denotes the unbiased estimate of σ^2. [3]

3 The length, L centimetres, of *Slimline* bin liners may be modelled by a normal distribution with a mean of 69.5 and a standard deviation of 0.55.

a Determine:

   i $P(L < 70)$; [3]

   ii $P(69 < L < 70)$; [3]

   iii $P(L = 70)$. [1]

b Determine the maximum length exceeded by 90% of bin liners. [4]

c The bin liners are sold in packets of 20, and those in each packet may be considered to be a random sample.

Determine the probability that:

   i all the bin liners in a packet have lengths less than 70 cm; [2]

   ii the mean length of the bin liners in a packet is greater than 69.25 cm. [4]

AQA MS1A June 2008

4 Each day, Margo completes the crossword in her local morning newspaper. Her completion times, X minutes, can be modelled by a normal random variable with a mean of 65 and a standard deviation of 20.

a Determine:

   i $P(X < 90)$;

   ii $P(X > 60)$. [5]

b Given that Margo's completion times are independent form day to day, determine the probability that, during a particular period of 6 days:

   i she completes one of the six crosswords in exactly 60 minutes; [1]

   ii she completes each crossword in less than 60 minutes; [3]

   iii her mean completion time is less than 60 minutes. [4]

AQA MS1B June 2010

5 a Electra is employed by E & G Ltd to install electricity meters in new houses on an estate. Her time, X minutes, to install a meter may be assumed to be normally distributed with a mean of 48 and a standard deviation of 20.

Determine:

   i $P(X < 60)$; [2]

   ii $P(30 < X < 60)$; [3]

   iii the time, k minutes, such that $P(X < k) = 0.9$. [4]

b Gazali is employed by E & G Ltd to install gas meters in the same new houses. His time, Y minutes, to install a meter has a mean of 37 and a standard deviation of 25.

  i Explain why Y is unlikely to be normally distributed. [2]

  ii State why $\bar{Y}$, the mean of a random sample of 35 gas meter installations, is likely to be approximately normally distributed. [1]

  iii Determine $P(\bar{Y} > 40)$. [4]

AQA MS1B June 2007

6 A general store sells lawn fertiliser in 2.5 kg bags, 5 kg bags and 10 kg bags.

a The actual weight, W kilograms, of fertiliser in a 2.5 kg bag may be modelled by a normal random variable with mean 2.75 and standard deviation 0.15.

Determine the probability that the weight of fertiliser in a 2.5 kg bag is:

  i less than 2.8 kg;

  ii more than 2.5 kg. [5]

b The actual weight, X kilograms, of fertiliser in a 5 kg bag may be modelled by a normal random variable with mean 5.25 and standard deviation 0.20.

  i Show that $P(5.1 < X < 5.3) = 0.372$, correct to 3 decimal places. [2]

  ii A random sample of **four** 5 kg bags is selected. Calculate the probability that none of the four bags contains between 5.1 kg and 5.3 kg of fertiliser. [2]

c The actual weight, Y kilograms, of fertiliser in a 10 kg bag may be modelled by a normal random variable with mean 10.75 and standard deviation 0.50.

A random sample of **six** 10 kg bags is selected. Calculate the probability that the mean weight of fertiliser in the six bags is less than 10.5 kg. [4]

AQA MS1A June 2012

7 The number of computers, A, bought during one day from the Amplebuy computer store can be modelled by a Poisson distribution with a mean of 3.5.

The number of computers, B, bought during one day from the Bestbuy computer store can be modelled by a Poisson distribution with a mean of 5.0.

a **i** Calculate $P(A = 4)$. [2]

  ii Determine $P(B \leq 6)$. [1]

  iii Find the probability that a total of fewer than 10 computers is bought from these two stores on one particular day. [3]

b Calculate the probability that a total of fewer than 10 computers is bought from these two stores on at least 4 out of 5 consecutive days. [3]

Estimation 309

c The numbers of computers bought from the Choicebuy computer store over a 10-day period are recorded as

  8   12   6   6   9   15   10   8   6   12

  i Calculate the mean and variance of these data. [2]

  ii State, giving a reason based on your results in part **c i**, whether or not a Poisson distribution provides a suitable model for these data. [2]

AQA MS2B January 2007

8 The time, in minutes, that Michaela waits at a bus stop to catch a bus to work may be modelled by an exponential distribution with parameter $\lambda = 0.22$.

a Find the mean and the standard deviation of the time that Michaela waits for a bus. [2]

b Find the probability that Michaela will have to wait more than 5 minutes for a bus. [3]

c On Monday morning, Michaela, who has already been waiting at the bus stop for 3 minutes, is joined by a friend, Narinder.

  Find the probability that Narinder will have to wait less than 3 minutes for a bus. [3]

d Michaela records the time that she has to wait on each of 40 consecutive weekday mornings. Using the Central Limit Theorem, find, approximately, the probability that the mean of the times that she records exceeds 3 minutes. [4]

AQA SS05 June 2011

9 The continuous random variable, X, has probability density function defined by

$$f(x) = \begin{cases} k(9 - 2x) & 0 \le x \le 3 \\ 0 & \text{otherwise} \end{cases}$$

a Find the value of the constant k. [2]

b Find the mean and standard deviation of X. [4]

c Find an approximate value for the probability that the mean of a random sample of size 64 from this distribution exceeds 1.45. [4]

Introduction

A company has developed a new vaccine for malaria and claims that it is 80% effective. A professor of tropical medicine doubted that the vaccine would be this successful. She decided to test the vaccine on 20 people travelling to high-risk areas. On their return, 13 out of 20 people were symptom-free. The professor now has to decide whether this provides enough evidence to say that the company is overstating the effectiveness of the vaccine. This chapter explores how this decision can be made, backed by statistical theory.

One of the most important applications of statistics is in using a sample to test a claim about a population. This is known as hypothesis testing, and it has wide-ranging implications in many disciplines including medicine, science, economics and engineering.

Objectives

By the end of this chapter, you should know how to...

▶ Use the language of hypothesis testing:
  – Null and alternative hypotheses.
  – One-tailed and two-tailed tests.
  – Significance level.
  – Critical (rejection) region.
  – Test statistic.
  – Rejection rule.
  – Type I and Type II errors.

▶ Formulate hypotheses and carry out tests:
  – For a population proportion p using exact binomial probabilities.
  – For the mean λ of a Poisson distribution.
  – For the mean of a normal distribution with known variance (z-test).
  – For the mean of a distribution using a normal approximation (z-test).
  – For the mean of a normal distribution with unknown variance (t-test).

Recap

You will need to remember...

Binomial distribution: $X \sim B(n, p)$

▶ $P(X = x) = \binom{n}{x} p^x q^{n-x}$ for $x = 0, 1, 2, ..., n$ where $q = 1 - p$

▶ How to use cumulative binomial probabilities. (Table 1)

Poisson distribution: $X \sim Po(\lambda)$

▶ $P(X = x) = e^{-\lambda} \dfrac{\lambda^x}{x!}$ for $x = 0, 1, 2, ...,$

▶ How to use cumulative Poisson probabilities. (Table 2)

Normal distribution: $X \sim N(\mu, \sigma^2)$

▶ How to use standard normal tables. (Table 3)

▶ How to use percentage points of the normal distribution. (Table 4)

Estimation

▶ Unbiased estimator of population variance:
$$S^2 = \frac{\sum(X - \bar{X})^2}{n-1} = \frac{1}{n-1}\left(\sum X^2 - \frac{(\sum X)^2}{n}\right)$$

▶ Distribution of the sample mean.

15.1 Testing a population proportion, *p*

Anwar says that he can read people's thoughts.

To test this claim, a volunteer from the audience sits on the stage while Anwar sits in a separate room off stage. The volunteer chooses a card at random from a well-shuffled pack and concentrates on the card for five seconds. At the same time Anwar writes down the suit of the card: clubs, diamonds, hearts or spades. The card is replaced in the pack, the pack is shuffled and another card is drawn. The procedure is repeated until 20 cards have been drawn.

There are four suits, so if Anwar guesses the answer he has a one in four chance of writing down the correct suit. If he isn't guessing, you would expect him to get more than one in four correct. So in 20 trials he should get get more than 5 correct.

If he gets at most 5 out of 20, you would definitely say that he is just guessing.

If he gets as many as 19 or 20 you would have no hesitation in saying that he can read people's thoughts.

But what about other values? Would 12 correct answers be very unusual? What would you say if he got 10? What about 7?

Somehow you have to decide on a cut-off point, *c*. This is the smallest value you could find such that the probability of getting *c* or more correct answers is very small and it would be considered a rare event.

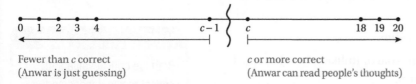

Fewer than *c* correct
(Anwar is just guessing)

c or more correct
(Anwar can read people's thoughts)

You could just choose a value for *c* that seems reasonable. However, if you carry out a **hypothesis** or **significance test** you will be able to back up your conclusion with statistical theory.

Is Anwar guessing?

Suppose that the discrete random variable *X* is the number of correct answers Anwar gives in the 20 trials.

If Anwar is guessing, the probability that he writes down the correct suit is 0.25. Since the card is put back and the pack is well shuffled each time, the trials are independent.

The conditions for a **binomial distribution** are satisfied,

so $X \sim B(n, p)$ with $n = 20$ and $p = 0.25$

See AS Textbook Chapter 15

The probability of Anwar getting all 20 correct if he is guessing is

$$P(X = 20) = 0.25^{20} = 0.000\,000\,000\,000\,909$$

This probability is *extremely small*.

Using binomial probabilities, you will find that the probability of 16 or more correct answers is still *very small indeed* (0.000 000 386).

Investigating further:

Using **Table 1:** Cumulative binomial probabilities

$$P(X \geq 13) = 1 - P(X \leq 12)$$

$$= 1 - 0.9998$$

$$= 0.0002 = 0.02\%$$

See AS Textbook Section 15.2

It would be *almost impossible* to get 13 or more correct answers just by guesswork.

Similarly, you will find that

$P(X \geq 12) \approx 0.09\%$	Getting 12 or more correct answers is a *very rare* event.
$P(X \geq 11) \approx 0.4\%$	Getting 11 or more correct answers is a *rare* event.
$P(X \geq 10) \approx 1\%$	It would be *very unlikely* for Anwar to get 10 or more correct answers, but it could happen on about 1 in every 100 occasions.
$P(X \geq 9) \approx 4\%$	It would still be *unlikely* for Anwar to get 9 or more correct answers.
$P(X \geq 8) \approx 10\%$	This probability is not that small. If Anwar is guessing, then on 10% of occasions he could give 8 or more correct answers.

You have to make a decision about the value of the probability that is considered to imply an unlikely or rare event. This probability is called the **significance level** of the test. As a guide, events that have a probability of less than 5% are generally regarded as *unlikely* and those having a probability of less than 1% are generally regarded as *very unlikely*. Often, a hypothesis test is carried out at the 5% significance level.

The cut-off point c is known as the **critical value**.

The set of observations considered to be unusual or unlikely (rare) events is called the **critical region** or **rejection region**.

Values which are not in the critical region are said to be in the **acceptance region**.

Suppose you choose a significance level of 5% to test Anwar's claim.

From the working above, $P(X \geq 8) \approx 10\%$. This is greater than 5%, so 8 is not in the critical region. Getting 8 correct answers by guessing would not be considered an unlikely or rare event.

But $P(X \geq 9) \approx 4\%$. Since this is less than 5%, any value of 9 or above is in the critical region and it would be considered an unlikely or rare event to get 9 correct answers by guessing.

To test Anwar's claim at a **5% significance level**:

▶ The critical value, c, is 9 correct answers.

▶ The critical region is $X \geq 9$, that is 9, 10, 11, 12, ..., 19 or 20 correct answers.

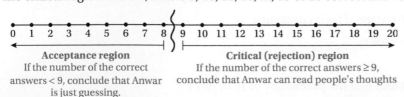

> **Note**
>
> Both the critical value and critical region depend on the significance level chosen.

The critical region can be shown on a vertical line graph. The probability distribution for $X \sim B(20, 0.25)$ has a long tail to the right (a positive skew), with some probabilities being too small to show on the diagram.

Since $P(X \geq 8) \approx 10\%$ and $P(X \geq 9) \approx 4\%$, the 5% boundary comes between 8 and 9.

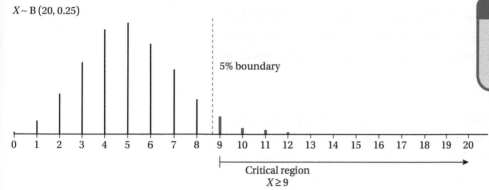

$X \sim B(20, 0.25)$

5% boundary

Critical region
$X \geq 9$

Carrying out the test

You would start by stating two **hypotheses**.

The assumption that Anwar is guessing, with a one in four chance of guessing correctly, is called the **null hypothesis** and it is denoted by H_0.

You would write

$H_0: p = 0.25$

If Anwar can read people's thoughts, he should get more than one in four correct and the probability he gives the correct suit will be greater than 0.25. This is called the **alternative hypothesis** and is denoted by H_1.

You would write

$H_1: p > 0.25$

The variable X, the number of correct answers, is the **test statistic**. Assuming the null hypothesis is true, $X \sim B(20, 0.25)$.

When the **significance level** has been decided, the **critical region** can be defined. In this example, since you are interested in whether the probability is *greater* than 0.25, the critical region is in the *upper tail* (right-hand end) of the distribution and the test is known as a **one-tailed (upper tail) test**.

The number of correct answers Anwar gives in the experiment is the **test value**.

If the test value is in the critical region, it is considered to be an unusual event when $p = 0.25$, so you would reject H_0 in favour of H_1. This means that you would reject that Anwar is guessing in favour of the alternative hypothesis that he can read people's thoughts, thus supporting Anwar's claim.

If the test value is in the acceptance region, you would not reject H_0. You do not have significant evidence that $p > 0.25$, so you would accept that he is guessing.

These are summarised in the **rejection rule**:

▶ If the test value is in the critical region, reject H_0 in favour of H_1.
▶ If the test value is not in the critical region, do not reject H_0.

Suppose Anwar gives 7 correct answers

You now have to find whether $x = 7$ is in the critical region. This can be approached in one of two ways.

Critical region method

Having set the critical region as described, check whether $x = 7$ is in it or not.

From the diagram, you can see that, at the 5% level of significance, 7 is not in the critical region. Therefore, you would not reject H_0 and would conclude that Anwar is guessing.

Probability method

An advantage of finding the critical region is that it gives a lot of information. However, in practice it could be time-consuming to set up. Instead, a **probability method** is often used which enables you to check whether the test value lies in the critical region without having to define the region first.

To test, at the 5% level, whether Anwar can read people's thoughts, you would find the probability that X is greater than or equal to 7 (the test value). If this probability is less than 5%, then the test value **is** in the upper tail 5% of the distribution (the critical region).

In this example, the rejection rule can be written:

This is the significance level of the test

▶ Reject H_0 if $P(X \geq x) < 0.05$, where x is the test value.

Calculate $P(X \geq 7)$, using Table 1 with $n = 20$, $p = 0.25$:

$$P(X \geq 7) = 1 - P(X \leq 6)$$

$$= 1 - 0.7858$$

$$= 0.2142 \qquad \leftarrow \text{This is known as the } \textbf{\textit{p}-value} \text{ in this test.}$$

Compare this with 5%:

Since $P(X \geq 7) > 5\%$, $x = 7$ is not in the critical region. As before, you would not reject H_0. You do not have enough evidence, at the 5% level, to say that Anwar can read people's thoughts.

When you are testing $x = 7$, it may seem strange that you have to work out $P(X \geq 7)$ rather than just $P(X = 7)$. However, this is necessary, since you are essentially considering the critical region to see whether the test value lies in it or not.

Types of test

In the above illustration, the alternative hypothesis, $H_1: p > 0.25$, is looking for an increase in p. There are, however, other types of alternative hypotheses.

In general, consider a test with null hypothesis H_0 such that

$\qquad H_0: p = p_0 \qquad \leftarrow p_0$ is a specific value such as 0.25.

In each of these illustrations, the significance level is 5%.

One-tailed test

A **one-tailed test** is carried out when the alternative hypothesis looks for an *increase* or a *decrease* in p.

In a **one-tailed upper tail test**, H_1 is looking for an *increase* in p

$H_1: p > p_0$

The critical region is in the upper tail and consists of values *greater than or equal to c* such that $P(X \geq c) < 5\%$, so you would reject H_0 if $P(X \geq c) < 0.05$

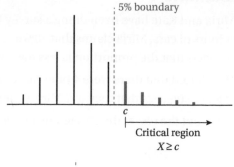

5% boundary

Critical region
$X \geq c$

In a **one-tailed lower tail test**, H_1 is looking for a *decrease* in p

$H_1: p < p_0$

The critical region is in the lower tail and consists of values *less than or equal to c* such that $P(X \leq c) < 5\%$, so you would reject H_0 if $P(X \leq c) < 0.05$

Note that, in both cases, the probability is *less than* 0.05.

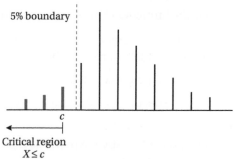

5% boundary

Critical region
$X \leq c$

Two-tailed test

A **two-tailed test** is carried out when the alternative hypothesis looks for a *change* in p, not specifically an increase or a decrease.

The alternative hypothesis is that p does not equal p_0, that is $H_1: p \neq p_0$

The critical region is in two parts, split between the upper tail and the lower tail.

In the *lower tail*, the critical region consists of values *less than or equal to c_1* such that $P(X \leq c_1) < 2.5\%$.

In the *upper tail*, the critical region consists of values *greater than or equal to c_2* such that $P(X \geq c_2) < 2.5\%$.

You would reject H_0
if $P(X \leq c_1) < 0.025$ or $P(X \geq c_2) < 0.025$

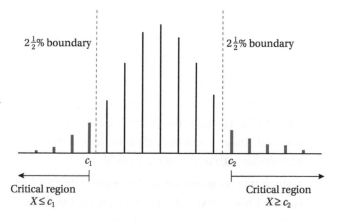

$2\frac{1}{2}\%$ boundary

$2\frac{1}{2}\%$ boundary

Critical region
$X \leq c_1$

Critical region
$X \geq c_2$

Summary of steps in a hypothesis test for a population proportion p

▶ Define the variable, X.
▶ State H_0 and H_1, then give the distribution of X assuming H_0 is true.
▶ State the type and significance level of the test.
▶ State the rejection rule – you could give a probability statement or define the critical region.
▶ Find whether the test value lies in the critical region or not – this usually involves finding a probability and comparing it with the significance level.
▶ Make your conclusion in statistical terms by saying whether H_0 is rejected or not and then *relate it to the situation*.

Example 1

Miria and Kate have been doing a survey for their maths homework on the colours of cars. Miria claims that 30% of cars in the town are blue, but Kate suspects that the proportion is less than 30%.

They found that there were 4 blue cars in a random sample of 25 cars. Their teacher suggests that they test Miria's claim by carrying out a hypothesis test.

Carry out the test at the 5% significance level and state your conclusion.

▶ *Define the variable*

X is the number of blue cars in a random sample of 25 cars, where $X \sim B(25, p)$.

▶ *State H_0 and H_1 and the distribution of X if H_0 is true*

H_0: $p = 0.3$ (30% of cars are blue, as Miria claims)

H_1: $p < 0.3$ (less than 30% of cars are blue, as Kate suspects)

If H_0 is true, then $p = 0.3$ and $X \sim B(25, 0.3)$.

▶ *State the type and level of the test*

Since the alternative hypothesis is $p < 0.3$, the critical region is in the lower tail of the distribution.

Use a one-tailed (lower tail) test at the 5% level of significance.

▶ *State the rejection rule*

Reject H_0 if $P(X \leq x) < 0.05$, where x is the test value.

▶ *Now find whether the test value is in the critical region*

Since there were 4 blue cars in the sample, $x = 4$.

Using Table 1 with $p = 0.30$, $n = 25$:

$P(X \leq 4) = 0.0905 > 0.05$

▶ *Make your conclusion, relating it to the context*

Since $P(X \leq 4) > 0.05$, the test value is not in the critical region, H_0 is not rejected. There is not enough evidence, at the 5% significance level, to reject Miria's claim that 30% of cars are blue.

Note that you cannot say with certainty that 30% of cars *are* blue, as Miria claims, only that you do not have enough evidence to support Kate's claim that less than 30% of cars are blue.

Table 1: $n = 25$

x	$p = 0.30$
0	0.0001
1	0.0016
2	0.0090
3	0.0332
4	0.0905
5	0.1935
etc	....

Example 2

A particular type of tomato seed has a germination rate of 72%. In order to improve this rate, the manufacturer develops a new treatment on the seeds.

Following the introduction of the new treatment, 14 out of the 15 seeds in a pack germinate and the manufacturer claims that the germination rate has increased.

Carry out a hypothesis test at the 10% significance level to test the manufacturer's claim, stating your hypotheses clearly and any necessary assumptions.

X is the number of seeds that germinate in a batch of 15, where $X \sim B(15, p)$.

The null hypothesis is that the germination rate is the same as before. The alternative hypothesis is the manufacturer's claim that the germination rate has increased.

H_0: $p = 0.72$ (the germination rate is unchanged)

H_1: $p > 0.72$ (the germination rate has increased)

Assume that the seeds in the pack constitute a random sample of seeds produced after the introduction of the new treatment.

If H_0 is true, then $p = 0.72$ and $X \sim B(15, 0.72)$

Using a one-tailed upper tail test, at the 10% significance level,

reject H_0 if $P(X \geq x) < 0.10$, where x is the test value.

Since 14 seeds in the sample germinated, $x = 14$.

$P(X \geq 14) = P(X = 14) + P(X = 15)$

$$= \binom{15}{14} \times 0.72^{14} \times 0.28 + 0.72^{15}$$

$$= 0.04950... < 0.10$$

Since $P(X \geq 14) < 0.10$, the test value is in the critical region, so you would reject H_0 in favour of H_1.

There is significant evidence, at the 10% level, that the germination rate has increased, thus supporting the manufacturer's claim.

Notice that, since $0.04950... < 0.05$, you would also reject H_0 at the 5% level.

AS Textbook Chapter 15:

If $X \sim B(n, p)$,

$$P(X = x) = \binom{n}{x} p^x (1-p)^{n-x}, x = 0, 1, ..., n$$

Example 3

A government department claims that 35% of residents in a particular city visit their local library at least once a week. In a random sample of 12 residents in the city, 7 said that they visited their local library at least once a week. Carry out a hypothesis test at the 10% level of significance to test the claim that 35% of residents visit their local library at least once a week.

X is the number of residents in a sample of 12 who visit their local library at least once a week, where $X \sim B(12, p)$.

H_0: $p = 0.35$ (the proportion is 35 per cent)

H_1: $p \neq 0.35$ (the proportion is not 35 per cent)

If H_0 is true, then $p = 0.35$ and $X \sim B(12, 0.35)$

Use a two-tailed test at the 10% level (distributing 5% in each tail).

Reject H_0 if

$\quad P(X \leq x) < 0.05$

or $\quad P(X \geq x) < 0.05$

where x is the test value.

The test value is $x = 7$, so look in the upper tail

$P(X \geq 7) = 1 - P(X \leq 6)$

$\quad = 1 - 0.9154$

$\quad = 0.0846 > 0.05$

Table 1: $n = 12$

x	$p = 0.35$
0	0.0057
1	0.0424
2	0.1513
3	0.3467
4	0.5833
5	0.7873
6	0.9154
7	0.9745
etc	....

(continued)

(continued)

Since $P(X \geq 7) > 0.05$, the test value is not in the critical region. Do not reject H_0.

There is not enough evidence, at the 10% level, to say that the percentage who visit their local library at least once a week is different from 35%.

Exercise 1

1 A random observation x is taken from a binomial distribution $X \sim B(n, p)$.

Test the given hypotheses at the significance level stated.

	Sample value x	n	Hypotheses	Significance level
a	1	10	$H_0: p = 0.45$ $H_1: p < 0.45$	5%
b	6	8	$H_0: p = 0.45$ $H_1: p > 0.45$	5%
c	5	7	$H_0: p = 0.25$ $H_1: p \neq 0.25$	2%
d	2	20	$H_0: p = 0.3$ $H_1: p < 0.3$	1%
e	12	14	$H_0: p = 0.58$ $H_1: p \neq 0.58$	6%
f	8	12	$H_0: p = 0.8$ $H_1: p < 0.8$	10%

2 The records of a particular hospital show that 30% of patients who come to the hospital's casualty department have to wait more than an hour before receiving medical attention. An extra doctor is employed and during the next week 3 out of 20 randomly selected patients have to wait more than an hour to receive medical attention.

Carry out a hypothesis test at the 5% level to test whether the proportion of patients who have to wait more than an hour has been reduced.

3 The random variable X can be modelled by a binomial distribution with $n = 10$.

A random observation, x, is taken from the distribution.

Test, at the 8% level, the hypothesis that $p = 0.45$ against the alternative hypothesis $p \neq 0.45$

a when $x = 7$ **b** when $x = 1$.

4 In a test of 10 true-false questions, Sian got 8 correct. Carry out a hypothesis test at the 5% significance level to test whether she could have obtained this score by guessing all the answers.

5 When Luca used to play darts regularly he scored a bull on 40% of attempts, on average. After a break of three months, he played darts one evening and scored 2 bulls on 12 attempts.

Stating a necessary assumption, carry out a hypothesis test, at the 10% significance level, to test whether the percentage of bulls he scores has decreased.

6 A large college introduced a new procedure to try to ensure that staff arrived on time for the start of lectures. A recent survey by the students had suggested that, in 15% of cases, staff arrived late for the start of the lecture. In the first week following the introduction of the new procedure a random sample of 35 lectures was taken and in only one case did the member of staff arrive late. Stating your hypotheses clearly, test, at the 5% level of significance, whether there is evidence that the new procedure has been successful.

15.2 Testing a Poisson mean, λ

This test is for the mean of the discrete random variable X, where $X \sim \text{Po}(\lambda)$. The steps are the same as those for a binomial proportion described in Section 15.1 on page **321**.

The null hypothesis is

$\quad$ $H_0: \lambda = \lambda_0$ $\quad \leftarrow \lambda_0$ is a specific value

The alternative hypothesis H_1 depends on whether the test is one-tailed or two tailed.

For a 5% significance level:

▶ To test for an *increase* in λ, the alternative hypothesis is $H_1: \lambda > \lambda_0$.
You would reject H_0 if $P(X \geq x) < 5\%$.

▶ To test for a *decrease* in λ, the alternative hypothesis is $H_1: \lambda < \lambda_0$.
You would reject H_0 if $P(X \leq x) < 5\%$.

▶ To test for a *change* in λ, the alternative hypothesis is $H_1: \lambda \neq \lambda_0$.
You would reject H_0 if $P(X \leq x) < 2.5\%$ or $P(X \geq x) < 2.5\%$.

Example 4

The number of minor accidents per week at a particular road junction has a Poisson distribution with a mean of 0.5. Following the introduction of a new road layout, there were 3 minor accidents in a randomly chosen period of 13 weeks.

Test, at the 5% significance level, the claim that the mean number of minor accidents at the road junction has decreased.

▶ *Define the variable*
The sample data is for a period of 13 weeks, so you need to consider this as the unit interval.

X is the number of minor accidents in 13 weeks, where $X \sim \text{Po}(\lambda)$.

▶ *State H_0 and H_1 and the distribution of X if H_0 is true*
λ is the mean number of accidents in 13 weeks, so $\lambda = 13 \times 0.5 = 6.5$

$H_0: \lambda = 6.5$ (the mean number of accidents is unchanged)

$H_1: \lambda < 6.5$ (the mean number of accidents has decreased)

If H_0 is true, then $X \sim \text{Po}(6.5)$

▶ *State the type and level of the test*
Use a one-tailed (lower tail) at the 5% level.

▶ *State the rejection rule*
Since the alternative hypothesis is $\lambda < 6.5$, the critical region is in the *lower tail* of the distribution, so you would reject H_0 if $P(X \leq x) < 5\%$, where x is the test value.

▶ *Calculate the required probability*
The test value is $x = 3$.

Using Table 2: Cumulative Poisson probabilities with $\lambda = 6.5$,

$\quad$ $P(X \leq 3) = 0.1118 > 0.05$ $\qquad$ *(continued)*

Table 2: Cumulative Poisson probabilities

x	$\lambda = 0.65$
0	0.0015
1	0.0113
2	0.0430
3	0.1118
4	0.2237
etc	....

(continued)

▶ *Make your conclusion, relating it to the context*

Since $P(X \leq 3) > 0.05$, the test value is not in the critical region so you would not reject H_0.

There is not enough evidence, at the 5% level, to support the claim that the mean number of accidents per week has decreased.

Example 5

Wendell is the manager of the local football team. During last season the number of goals per match scored by the team followed a Poisson distribution with mean 1.4

Prior to the start of the new season, the team took part in intensive coaching sessions. Wendell decided to use the results of the first 2 matches of the new season to test whether the team now scores more goals on average than they did last season.

In the first 2 matches of the new season, the total number of goals scored by the team was 6.

a Carry out the test at the 7.5% significance level, clearly stating your hypotheses and conclusion.

b State, with a reason, whether the conclusion would be the same at the 5% significance level.

c Comment on the reliability of this test.

a X is the number of goals in 2 *matches*, where $X \sim \text{Po}(\lambda)$.

λ is the mean number of goals in 2 matches so $\lambda = 2 \times 1.4 = 2.8$

$H_0: \lambda = 2.8$ (the mean number of goals is unchanged)

$H_1: \lambda > 2.8$ (the mean number of goals has increased)

If H_0 is true, then $X \sim \text{Po}(2.8)$.

Use a one-tailed upper tail test at the 7.5% level and reject H_0 if $P(X \geq x) < 0.075$, where x is the test value.

Since 6 goals were scored in the 2 matches, $x = 6$.

$$P(X \geq 6) = 1 - P(X \leq 5)$$
$$= 1 - 0.9349$$
$$= 0.0651 < 0.075$$

Since $P(X \geq 6) < 0.075$, the test value is in the critical region, so H_0 is rejected in favour of H_1.

There is significant evidence, at the 7.5% level, that the mean number of goals per match has increased.

b At the 5% significance level, you would reject H_0 if $P(X \geq 6) < 0.05$.

From part **a**, $P(X \geq 6) = 0.06511... > 0.05$

At the 5% level, $x = 6$ is not in the critical region, so you would not reject H_0.

There is not enough evidence, at the 5% level, to support the claim that the mean number of goals per match has increased.

> **Note**
>
> You need to consider 2 matches because the sample data gives the results of 2 matches.

Table 2

x	$\lambda = 2.8$
0	0.0608
1	0.2311
2	0.4695
3	0.6919
4	0.8477
5	0.9349
6	0.9756
etc	....

(continued)

(continued)

c The test is not very reliable as it is based on the results of only two matches. To improve the reliability, consider the results from a greater number of matches.

In Example 6, you are asked to find the critical region.

Example 6

The number of flaws in a certain brand of tape has a Poisson distribution with a mean of 1.40 per 10 metre length. Following an adjustment to the machine producing the tape, an inspection of a 100 metre length revealed 8 flaws.

a Test, at the 10% level of significance, whether there has been a change in the mean number of flaws in the tape.

b Find the critical region for this test.

a The sample value is for a length of 100 m, so you need to use 100 m as the unit interval in your calculations.

X is the number of flaws in 100 m, where $X \sim \text{Po}(\lambda)$.

Since λ is the mean number of flaws in 100 m, $\lambda = 10 \times 1.40 = 14.0$

$H_0: \lambda = 14.0$ (the mean number of flaws is unchanged)

$H_1: \lambda \neq 14.0$ (the mean number of flaws has changed)

If H_0 is true, then $X \sim \text{Po}(14)$.

Use a two-tailed test at the 10% level.

Reject H_0 if $P(X \leq x) < 0.05$ or $P(X \geq x) < 0.05$, where x is the test value.

Since 8 flaws were found in 100 m of material, the test value is 8.

$P(X \leq 8) = 0.0621 > 0.05$

Since $P(X \leq 8) > 0.05$, the test value is not in the critical region, so do not reject H_0.

There is not enough evidence, at the 10% significance level, that the mean number of flaws per 100 m has changed after the new process.

b *Critical region in the lower tail:*

You know from part **a** that $x = 8$ is not in the critical region.

You can see from the cumulative probability table that

$P(X \leq 7) = 0.0316 < 0.05$, so the lower tail critical value is 7 and the lower tail critical region is $X \leq 7$.

Critical region in the upper tail:

You want to find c such that

$$P(X \geq c) < 0.05$$

so $\qquad P(X \leq c - 1) > 0.95$

From Table 2:

$P(X \leq 19) = 0.9235 < 0.95$

$P(X \leq 20) = 0.9521 > 0.95$

so, $\quad c - 1 = 20$

$\qquad c = 21$

The upper tail critical value is 21 and the upper tail critical region is $X \geq 21$.

Therefore, the critical region for the test is $X \leq 7, X \geq 21$.

Table 2

x	$\lambda = 14.0$
0	0.0000
1	0.0000
2	0.0001
3	0.0005
4	0.0018
5	0.0055
6	0.0142
7	0.0316
8	0.0621
....	....
....	....
19	0.9235
20	0.9521
21	0.9712
22	0.9833
....	....

Exercise 2

1 A random observation x is taken from a Poisson distribution X, where $X \sim Po(\lambda)$.

Test the given hypotheses at the significance level stated.

	Sample value x	Hypotheses	Significance level
a	4	$H_0: \lambda = 10.0$ $H_1: \lambda < 10.0$	5%
b	2	$H_0: \lambda = 8.5$ $H_1: \lambda \neq 8.5$	10%
c	3	$H_0: \lambda = 7.0$ $H_1: \lambda \neq 7.0$	8%
d	4	$H_0: \lambda = 1.4$ $H_1: \lambda > 1.4$	3%

2 Over a period of time it is found that the number of reported cases of a particular medical condition follows a Poisson distribution with mean 0.2 per day. However, a consultant thinks that the mean number of cases is increasing.

During the next 7 days, 4 cases are reported. Stating a necessary assumption, perform a hypothesis test at the 5% level to test the consultant's claim.

3 Over a period of time it is found that the number of letters of complaint received by a large store follows a Poisson distribution with mean 1.2 per day. Following the appointment of a new manager, it is believed that the mean number of letters of complaint has changed. In a randomly selected period of 5 days, 2 letters of complaint were received.

Carry out a hypothesis test, at the 10% level, to test whether the mean number of letters of complaint has changed.

4 Over a period of time it was found that the number of telephone calls made to an office may be modelled by a Poisson distribution at an average rate of 1 every 10 minutes.

 a In a particular 30-minute period, 5 calls were received.

 Test, at the 5% level, whether the mean number of calls has increased.

 b On a particular day, there were 3 calls between 11.00 am and 12.30 pm.

 Test, at the 5% level, whether the mean number of calls has decreased.

5 The number of bacterial colonies that develop in dishes of nutrients exposed to an infected environment has a Poisson distribution with mean 1.5.

An experiment was conducted to determine the effectiveness of an antibiotic spray. When 5 dishes were sprayed, the total number of bacterial colonies that developed was 3. Stating suitable null and alternative hypotheses, test, at the 5% significance level, whether the mean number of bacterial colonies has decreased.

6 The number of customers arriving per half-hour at a shop has a Poisson distribution with mean 2.2. An advertising campaign is organised which it is hoped will increase the mean number of customers coming to the store. After the campaign, the manager decides to note how many customers arrive during 5 randomly chosen half-hour periods and to use this figure to carry out a hypothesis test at the 5% level. Calculate the critical region.

15.3 Type I and Type II errors

When you carry out a hypothesis test there are four possible conclusions, two of which lead to a correct decision being made and the other two lead to a wrong decision being made.

The errors associated with making wrong decisions are called **Type I** and **Type II** errors.

The outcomes and errors are summarised as:

▶ H_0 is true and your test leads you to accept H_0 – correct decision.

▶ H_0 is true but your test leads you to reject H_0 – wrong decision – a Type I error is made.

▶ H_0 is false but your test leads you to accept H_0 – wrong decision – a Type II error is made.

▶ H_0 is false and your test leads you to reject H_0 – correct decision.

Type I error

A Type I error **is made when you reject H_0, even though H_0 is true.**

So, **P(Type I error) = P(reject H_0 when H_0 is true)**

Note that you reject H_0 if the test value lies in the critical region. This region is fixed according to the significance level of the text.

Type II error

A Type II error **is made when you accept H_0 but in fact H_0 is false.**

So, **P(Type II error) = P(accept H_0 when H_0 is false)**

To calculate the probability of making a Type II error you would need to be given a specific value for the alternative hypothesis.

Note that:

▶ A Type I error involves rejecting H_0. Therefore, if a test leads to H_0 being accepted, a Type I error cannot be made.

▶ A Type II error involves accepting H_0. Therefore, if a test leads to H_0 being rejected, a Type II error cannot be made.

> **Note**
>
> You will not be expected to find the probability of a Type II error.

Example 7

In a game, Freddie wins points when a coin shows a head when it is tossed. After playing the game for some time, Freddie thinks that the coin is biased in favour of tails. He decides to carry out a hypothesis test at the 8% level to determine whether the coin shows fewer heads than tails.

(continued)

(continued)

He tosses the coin 20 times and counts the number of times that it shows heads.

a **i** Find the critical region for the test.

   ii Explain in the context of this question, the meaning of making a Type I error.

   iii Find the probability of making a Type I error.

b Explain, in the context of the question, the meaning of making a Type II error.

a **i** X is the number of **heads** in 20 tosses of the coin.

   H_0: $p = 0.5$ (the coin is fair)

   H_1: $p < 0.5$ (the coin is biased in favour of tails)

   If H_0 is true, then $p = 0.5$ and $X \sim B(20, 0.5)$.

   Use a one-tailed lower tail test at the 8% level.

   To find the critical value you need the maximum value c such that $P(X \leq c) < 0.08$.

   Using Table 1 with $n = 20$ and $p = 0.5$, you can see that

   $P(X \leq 6) = 0.0577 < 0.08$

   whereas $P(X \leq 7) = 0.1316 > 0.08$

   The critical region is $X \leq 6$, so Freddie will reject H_0 if he obtains at most 6 heads.

   ii A Type I error is made when H_0 is rejected but it is in fact true, that is you conclude that the coin is biased in favour of tails when, in fact, it is a fair coin.

   iii Since Freddie will reject H_0 if $X \leq 6$,

   P(Type I error)

   $= P(X \leq 6$ when $p = 0.5)$

   $= 0.0577$

b A Type II error is made when you accept a false hypothesis, that is when you accept that the coin is fair, when in fact it is biased in favour of tails.

Table 1: $n = 20$

x	$p = 0.5$
0	0.0000
1	0.0000
2	0.0002
3	0.0013
4	0.0059
5	0.0207
6	0.0577
7	0.1316
etc	....

Example 8

A pharmaceutical company produces a new pain-relieving drug for migraine sufferers. In an advertising campaign, it states that the drug has a 90% success rate.

A doctor doubted that the drug was as successful as the company claimed. After prescribing the drug for a randomly selected sample of 15 patients for six months, 11 of these patients said that their migraine symptoms had been relieved by the drug.

a Test the pharmaceutical company's claim that the drug has a 90% success rate against the alternative hypothesis that the success rate is less than 90%. Carry out the test at the 5% level of significance.

b Indicate whether a Type I error or a Type II error might have occurred in carrying out your hypothesis test in part **a**. Give a reason for your answer.

a X is the number of patients in 15 whose symptoms are relieved by the drug.

H_0: $p = 0.9$ (the drug has a 90% success rate)

H_1: $p < 0.9$ (the success rate is less than 90%)

If H_0 is true, then $p = 0.9$ and $X \sim B(15, 0.9)$.

Use a one-tailed lower tail test at the 5% level.

Reject H_0 if $P(X \leq x) < 0.05$, where x is the test value.

The test value is $x = 11$, so you want to find $P(X \leq 11)$.

To do this using Table 1, use the symmetry property of the binomial distribution. AS Textbook Section 15.2

Let Y be the number whose symptoms are *not relieved* by the drug, where $Y \sim B(15, 0.1)$

$$P(X \leq 11) = P(Y \geq 4) \qquad \leftarrow \text{Use Table 1 with } n = 15 \text{ and } p = 0.1$$
$$= 1 - P(Y \leq 3)$$
$$= 1 - 0.9444$$
$$= 0.0556 > 0.05$$

Since $P(X \leq 11) > 0.05$, the test value is not in the critical region, so do not reject H_0.

There is not enough evidence, at the 5% level, that the success rate is less than 90%, so the pharmaceutical company's claim that the success rate is 90% is accepted.

b Since H_0 is accepted, a Type II error might have been made in this test.

Table 1: $n = 15$

y	$p = 0.10$
0	0.2059
1	0.5490
2	0.8159
3	0.9444
4	0.9873
etc	....

Example 9

A single random observation, x, is taken from a Poisson distribution with mean λ and used to test the null hypothesis $\lambda = 8$ against the alternative hypothesis $\lambda < 8$. The significance level of the test is 5%.

a Find the critical region of the test.

b Find the probability of making a Type I error in this test.

a $X \sim Po(\lambda)$

H_0: $\lambda = 8$

H_1: $\lambda < 8$

If H_0 is true, then $X \sim Po(8)$.

To find the critical region, find the greatest value c such that $P(X \leq c) < 0.05$.

From Table 2 you can see that

$$P(X \leq 3) = 0.0424 < 0.05$$
$$P(X \leq 4) = 0.0996 > 0.05$$

so the critical region is $X \leq 3$

b The rejection rule is to reject H_0 if $X \leq 3$.

$$P(\text{Type I error}) = P(\text{reject } H_0 \text{ when } H_0 \text{ is true})$$
$$= P(X \leq 3 \text{ when } \lambda = 8)$$
$$= 0.0424$$

Table 2

x	$\lambda = 8.0$
0	0.0003
1	0.0030
2	0.0138
3	0.0424
4	0.0996
etc	....

Exercise 3

1 The manager at a particular supermarket claims that less than 35% of people pay by cheque. In a random sample of 16 people, 3 people pay by cheque.

    a Test, at the 10% significance level, the null hypothesis $H_0: p = 0.35$ against the alternative hypothesis $H_1: p < 0.35$. State your conclusion clearly.

    b Explain, in the context of the question, the meaning of making

       i a Type I error

      ii a Type II error.

2 The random variable X has distribution $B(10, p)$. A hypothesis test, at the 10% significance level, is to be carried out to test the null hypothesis $p = 0.5$ against the alternative hypothesis $p \neq 0.5$.

    a Find the critical region for the test.

    b Find the probability of making a Type I error in this test.

3 The random variable X is distributed $Po(\lambda)$. A hypothesis test at the 5% level is to be carried out to test the null hypothesis $\lambda = 9$ against the alternative hypothesis $\lambda < 9$. The critical region is $X \leq c$. Find the value of c and hence find the probability of making a Type I error in this test.

4 Rachel claims that 45% of students in her school have watched a particular film. Her friend Lily suspects that less than 45% of students have watched the film. To test Rachel's claim, they decide to carry out a hypothesis test at the 10% significance level, asking 50 randomly selected students whether they have watched the film.

    a State suitable null and alternative hypotheses.

    b **i** Find the critical region for the test.

      ii Find the probability of making a Type I error.

In the random sample of 50 students, 17 had watched the film.

    c State the conclusion of the test.

    d Would your conclusion be different if Lily had suspected that the proportion of students who have watched the film was not 45%?

5 Hester suspected that a die was biased in favour of a six occurring. She decided to carry out a hypothesis test.

    a State suitable null and alternative hypotheses for the test.

    b When she threw the die 10 times, she obtained a six on 4 occasions. Carry out the test, at the 5% level, stating your conclusion clearly.

    c Indicate whether a Type I error or a Type II error might have occurred in carrying out your hypothesis test in part **b**. Give a reason for your answer.

6 The number of breakdowns per day of the lifts in a large block of flats has a Poisson distribution with mean 0.2

When the maintenance contract for the lifts is given to a new company, there are 2 breakdowns over a period of 30 days.

    a Carry out a hypothesis test at the 5% level to decide whether the mean number of breakdowns has decreased.

    b State, in the context of this question, the meaning of a Type I error.

15.4 Introduction to testing a population mean μ

In the preceding sections, hypothesis tests relating to *discrete* random variables were studied, in which exact binomial and Poisson probabilities were calculated.

The tests in the remaining sections of the chapter are for the population mean μ of a *continuous* random variable.

Consider this situation.

Ice packs are produced for use in cool boxes. A machine fills the packs with a liquid and is set so that the volume of liquid dispensed into an ice pack has a normal distribution with a mean of 524 ml and a standard deviation of 3 ml. When the ice packs are frozen, the liquid expands. To allow space for the expansion, it is important that the packs are not over-filled.

The machine breaks down and is repaired. When the next batch of ice packs is produced, the supervisor suspects that the *mean volume of liquid* being dispensed into the packs has increased. This could prove expensive, not only because of the cost of the extra liquid dispensed but also because the packs might crack when frozen.

To investigate her suspicion, the supervisor decides to select a random sample of 50 packs from the production line. She calculates that the mean volume of liquid per pack in the sample is 524.9 ml. Assuming that the standard deviation is still 3 ml, does this provide statistical evidence that the machine is over-filling the packs?

Carrying out the hypothesis test

To carry out the test, first **define the variable** being considered. If X is the volume, in millilitres, of liquid dispensed into a pack after the machine has been repaired, then

$X \sim N(\mu, \sigma^2)$ μ is unknown, $\sigma = 3$

The **null hypothesis**, H_0, is that the mean volume dispensed is the same as it was before the repair.

so $H_0: \mu = 524$

Since it is suspected that the mean volume has *increased*, the **alternative hypothesis**, H_1, is that the mean is *greater than* 524 ml.

so $H_1: \mu > 524$

The **test statistic** is $\bar{X}$, the *mean* volume of a sample of 50 packs.

In Section 14.3 you saw that if $X \sim N(\mu, \sigma^2)$ then, for samples of size n,

$$\bar{X} \sim N\left(\mu, \frac{\sigma^2}{n}\right)$$ ← This is the distribution of the sample mean.

If H_0 is true, then $\mu = 524$.

So, since $\sigma = 3$ and $n = 50$,

$$\bar{X} \sim N\left(524, \frac{3^2}{50}\right)$$

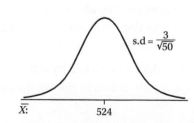

s.d $= \frac{3}{\sqrt{50}}$

$\bar{X}$:   524

The **test value** is 524.9, the mean of the random sample of 50 packs.

The outcome of the test depends on the position of 524.9 in the distribution of $\bar{X}$. If it is *close to* 524, it could have come from the distribution stated by the null hypothesis. However, if it is *far away*, in the extreme upper tail, it is likely to have come from a distribution with mean greater than 524 ml.

A decision needs to be taken about the **critical value**, c, which indicates the boundary of the **critical (rejection) region** in which values of $\bar{X}$ would be unlikely to occur. Values that are not in the critical region are in the **acceptance region**.

As with discrete tests, the critical value and critical region are fixed using probabilities linked to the **significance level** of the test and, again, the most commonly used levels are 10%, 5% and 1%.

Suppose that the supervisor chooses a 5% significance level.

Since the supervisor suspects that there has been an increase in the mean, she will reject H_0 if the test value is in the upper tail 5% of the distribution of sample means.

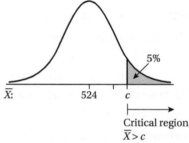

However, since the distribution of $\bar{X}$ is normal, it is usual to work in standardised z-values and the test is often referred to as a **z-test**.

The **standardised test statistic** is given by Z, where

$$Z = \frac{\bar{X} - \mu}{\frac{\sigma}{\sqrt{n}}} \text{ and } Z \sim N(0, 1)$$

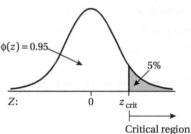

So, instead of finding c (the critical value of $\bar{x}$), you would find z_{crit}, the z-value that gives 5% in the upper tail.

From Table 4, with $p = 0.95$, $z_{crit} = 1.6449$.

The test value, $\bar{x} = 524.9$, is standardised to give z_{test} and if z_{test} is in the critical region, H_0 is rejected.

At the 5% significance level, the **rejection rule** is:

▶ Reject H_0 if $z_{test} > z_{crit}$, that is, if $z_{test} > 1.6449$.

Now calculate z_{test} and compare it with 1.6449.

$$z_{test} = \frac{524.9 - 524}{\frac{3}{\sqrt{50}}} = 2.121\cdots > 1.6449$$

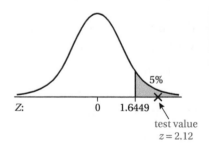

The result of the test is now stated in statistical terms and related to the situation.

Since $z_{test} > z_{crit}$, the test value is in the critical region. The supervisor would reject H_0 in favour of H_1.

It is helpful to put a cross on the diagram to indicate the approximate position of the test value.

The supervisor would conclude that there is evidence, at the 5% significance level, that the *mean volume* of liquid being dispensed by the machine is not 524 ml, but has increased.

With a decision backed by statistical theory, she would then be able to take appropriate action to resolve the problem.

Finding the critical value c

Note that the critical value c can be found as follows:

$$\frac{c - 524}{\frac{3}{\sqrt{50}}} = 1.6449$$

$$c = 524 + 1.6449 \times \frac{3}{\sqrt{50}}$$

$$= 524.69\ldots = 524.7 \text{ (1 dp)}$$

$\bar{X} \sim N\left(524, \frac{3^2}{50}\right)$

5%

$\bar{X}$:   524   524.7

test value
$\bar{x} = 524.9$

The critical value is 524.7 ml and the critical region is $\bar{X} > 524.7$ ml. This means that any test value greater than 524.7 ml is in the critical region.

Since the supervisor's test value of 524.9 ml is greater than 524.7 ml, it is in the critical region, confirming the result obtained above.

One-tailed and two-tailed tests

Suppose that the null hypothesis is

$$H_0: \mu = \mu_0 \qquad \leftarrow \mu_0 \text{ is a specified value.}$$

One-tailed test

You will recall from Sections 15.1 and 15.2 that in a one-tailed test the alternative hypothesis H_1 looks for an increase or a decrease in μ.

If H_1 looks for an increase, then H_1 is $\mu > \mu_0$ and the critical region is in the upper tail.

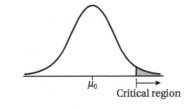

μ_0

Critical region

If H_1 looks for a decrease, then H_1 is $\mu < \mu_0$ and the critical region is in the lower tail.

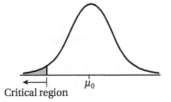

μ_0

Critical region

Two-tailed test

In a two-tailed test, the alternative hypothesis H_1 looks for a change in μ without specifying whether it is an increase or a decrease.

The alternative hypothesis H_1 is $\mu \neq \mu_0$.

The critical region is in two parts, *symmetrically* placed in the lower and upper tails.

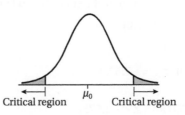

Critical region μ_0 Critical region

Critical *z*-values

Critical values depend on the significance level of the test and whether the test is one-tailed or two-tailed. They can be found using **Table 4: Percentage points of the normal distribution**, described in section 13.4. This is given in the *Formulae and Statistical Tables* booklet and is printed here for reference.

p	0.00	0.01	0.02	0.03	0.04	0.05	0.06	0.07	0.08	0.09	p
0.5	0.0000	0.0251	0.0502	0.0753	0.1004	0.1257	0.1510	0.1764	0.2019	0.2275	0.5
0.6	0.2533	0.2793	0.3055	0.3319	0.3585	0.3853	0.4125	0.4399	0.4677	0.4958	0.6
0.7	0.5244	0.5534	0.5828	0.6128	0.6433	0.6745	0.7063	0.7388	0.7722	0.8064	0.7
0.8	0.8416	0.8779	0.9154	0.9542	0.9945	1.0364	1.0803	1.1264	1.1750	1.2265	0.8
0.9	1.2816	1.3408	1.4051	1.4758	1.5548	1.6449	1.7507	1.8808	2.0537	2.3263	0.9

p	0.000	0.001	0.002	0.003	0.004	0.005	0.006	0.07	0.08	0.09	p
0.95	1.6449	1.6546	1.6646	1.6747	1.6849	1.6954	1.7060	1.7169	1.7279	1.7392	0.95
0.96	1.7507	1.7624	1.7744	1.7866	1.7991	1.8119	1.8250	1.8384	1.8522	1.8663	0.96
0.97	1.8808	1.8957	1.9110	1.9268	1.9431	1.9600	1.9774	1.9954	2.0141	2.0335	0.97
0.98	2.0537	2.0749	2.0969	2.1201	2.1444	2.1701	2.1973	2.2262	2.2571	2.2904	0.98
0.99	2.3263	2.3656	2.4089	2.4573	2.5121	2.5758	2.6521	2.7478	2.8782	3.0902	0.99

For example, when carrying out a one-tailed test at the 1% level, you want to find z such that $\Phi(z) = 0.99$, so look up $p = 0.99$, column 0.000

From Table 4, $z = 2.3263$.

For an upper tail test, $z_{crit} = 2.3263$ and the critical region is $Z > 2.3263$.

For a lower tail test, by symmetry, $z_{crit} = -2.3263$ and the critical region is $Z < -2.3263$.

Upper tail

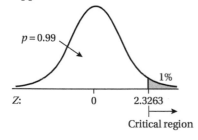

Lower tail

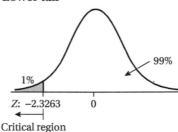

For a two-tailed test at the 1% level, the area of 1% is split equally between the upper and lower tails, with 0.5% in each, that is 0.005. There are two critical values.

To find the upper tail value, you need to find z such that $\Phi(z) = 0.995$

From Table 4, with $p = 0.995$, $z = 2.5758$.

So in the upper tail, $z_{crit} = 2.5758$.

By symmetry, in the lower tail, $z_{crit} = -2.5758$.

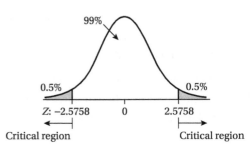

The critical region is $Z > 2.5758$ and $Z < -2.5758$.

This can be written in one statement as $|Z| > 2.5758$.

You need to be able to find the critical values for any given level of significance. For reference, this summary shows the critical z-values and rejection rules for the most commonly used levels of significance of 10%, 5% and 1%.

Significance level	One-tailed (lower tail) $H_0: \mu = \mu_0$ $H_1: \mu < \mu_0$	One-tailed (upper tail) $H_0: \mu = \mu_0$ $H_1: \mu > \mu_0$	Two-tailed $H_0: \mu = \mu_0$ $H_1: \mu \neq \mu_0$
10%	Reject H_0 if $z < -1.2816$	Reject H_0 if $z > 1.2816$	Reject H_0 if $z > 1.6449$ or $z < -1.6449$ that is, if $|z| > 1.6449$
5%	Reject H_0 if $z < -1.6449$	Reject H_0 if $z > 1.6449$	Reject H_0 if $z > 1.96$ or $z < -1.96$ that is, if $|z| > 1.96$
1%	Reject H_0 if $z < -2.3263$	Reject H_0 if $z > 2.3263$	Reject H_0 if $z > 2.5758$ or $z < -2.5758$ that is, if $|z| > 2.5758$

Steps in a *z*-test

The steps follow the same general pattern as the test for a binomial proportion and Poisson mean:

▶ Define the variable.

▶ State H_0 and H_1.

▶ State the test statistic and its distribution when H_0 is true.

▶ State the type and significance level of the test.

▶ State the rejection rule: this is usually given in terms of a critical z-value.

▶ Find whether the test value lies in the critical region: this usually involves finding the standardised test value and comparing it with the critical z-value.

▶ Make your conclusion in statistical terms by saying whether H_0 is rejected or not and then *relate it to the situation*.

There are several cases to consider, depending on whether the population variable is normal or not, on whether the population variance is known and on the size of the sample taken.

15.6 Case 1: Testing the mean of a normal distribution with known variance

This is the case described in the introductory example of the volume of liquid in the ice packs.

Hypothesis test for a population mean μ when

▶ the distribution of X is *normal*,

▶ the population variance σ^2 is *known*.

If $X \sim N(\mu, \sigma^2)$, the test statistic is X, where $\bar{X} \sim N\left(\mu, \dfrac{\sigma^2}{n}\right)$

and the standardised test statistic is

$$Z = \frac{\bar{X} - \mu}{\dfrac{\sigma}{\sqrt{n}}} \quad \text{where} \quad Z \sim N(0, 1)$$

Example 10

Question

Each year a large number of students taking a particular course at a college sit an examination. Over a period of time it is found that the marks of the students at this college follow a normal distribution with mean 70.0 and standard deviation 6.

This year the examination contains questions on a new topic and a lecturer believes that the marks are lower, on average. To test this belief, he calculates the mean mark of a random sample of 25 students and finds it to be 67.3.

a   Carry out a hypothesis test at the 5% level of significance to test whether the mean mark in the examination is lower this year. Assume that the marks are normally distributed and the standard deviation is 6.

b   The critical region is $\bar{X} < c$. Find the value of c.

Answer

a   ▶ *Define the variable*

   X is the examination mark of a student, where $X \sim N(\mu, 6^2)$.

   ▶ *State H_0 and H_1 and the distribution of the test statistic if H_0 is true*

   $H_0: \mu = 70.0$ (the mean mark is unchanged)

   $H_1: \mu < 70.0$ (the mean mark is lower this year)

   The test statistic is $\bar{X}$, where $\bar{X} \sim N\left(\mu, \dfrac{6^2}{25}\right)$.

   If H_0 is true, then $\mu = 70.0$, and the standardised test statistic is

   $$Z = \frac{\bar{X} - 70.0}{\dfrac{6}{\sqrt{25}}}$$

   ▶ *State the type and level of the test*

   Use a one-tailed lower tail test at the 5% level.

   ▶ *State the rejection rule*

   Since you are looking for a *decrease* in μ, you need to find the critical z-value that gives 5% in the lower tail, so z_{crit} will be negative.

   Table 4: $p = 0.95$ gives $z_{crit} = -1.6449$,

   so you will reject H_0 if $z_{test} < -1.6449$.

   ▶ *Calculate the required probability*

   The test value $\bar{x}_{test} = 67.3$

   so $z_{test} = \dfrac{67.3 - 70.0}{\dfrac{6}{\sqrt{25}}}$

   $= -2.25 < -1.6449$

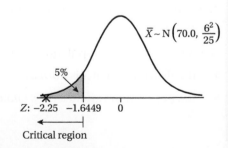

(continued)

(continued)

▶ *Make your conclusion, relating it to the context*

Since $z_{test} < z_{crit}$, the test value is in the critical region and H_0 is rejected in favour of H_1.

There is significant evidence, at the 5% level, to support the lecturer's belief that the mean examination mark is lower this year.

Alternatively:

b The standardised value of c is −1.6449,

$$\frac{c - 70.0}{\frac{6}{\sqrt{25}}} = -1.6449$$

$$c - 70.0 = -1.6449 \times \frac{6}{\sqrt{25}}$$

$$c = 70.0 - 1.6449 \times \frac{6}{\sqrt{25}}$$

$$c = 68.02612 = 68.0 \text{ (1 dp)}$$

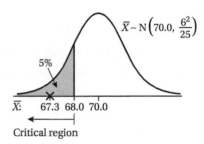

Note that the critical region is $\bar{X} < 68.0$, so any test value less than 68.0 would result in the null hypothesis being rejected.

Since $\bar{x} = 67.3 < 68.0$, H_0 would be rejected, as above.

Calculation note

In Example 10, $\frac{\sigma}{\sqrt{n}} = \frac{6}{\sqrt{25}} = 1.2$. You could calculate this at the outset, but it is often useful to leave it in its uncalculated form as a reminder that the *distribution of the sample mean* is being used.

Example 11

A machine packs flour into bags. A random sample of 11 bags was taken and the bags were weighed. The results, in grams, were as follows:

    1506.8, 1506.4, 1506.3, 1507.2, 1507.5, 1506.8,
    1506.6, 1507.0, 1507.5, 1506.3, 1506.4

Filled bags are supposed to weigh 1506.5 g.

Test whether the sample provides evidence, at the 8% significance level, that the machine is producing overweight bags. Assume that the mass of a bag of flour has a normal distribution with standard deviation 0.4 g.

X is the mass, in grams, of a bag of flour, where $X \sim N(\mu, 0.4^2)$.

$H_0: \mu = 1506.5$ (the mean mass is 1506.5)

$H_1: \mu > 1506.5$ (the mean mass is greater than 1506.5)

The test statistic is $\bar{X}$, where

$$\bar{X} \sim N\left(\mu, \frac{0.4^2}{11}\right)$$

If H_0 is true, then $\mu = 1506.5$, and the standardised test statistic is

$$Z = \frac{\bar{X} - 1506.5}{\frac{0.4}{\sqrt{11}}}$$

> **Note**
>
> $\frac{0.4}{\sqrt{11}}$ is not exact, so leave it in this form.

(continued)

Using a one-tailed (upper tail) test at the 8% level, Table 4: $p = 0.92$ gives

$z_{crit} = 1.4051$

Reject H_0 if $z_{test} > z_{crit}$, that is, if $z_{test} > 1.4051$.

From the sample data, $\bar{x} = 1506.8$

So,

$$z_{test} = \frac{1506.8 - 1506.5}{\frac{0.4}{\sqrt{11}}}$$

$$= 2.48... > 1.4051$$

Since $z_{test} > z_{crit}$, the test value is in the critical region, so reject H_0 in favour of H_1.

There is significant evidence, at the 8% level, that the mean mass has increased and the machine is producing overweight bags.

Example 12

A hospital administrator claims that the mean time spent by patients in the Accident and Emergency department is 170 minutes. The times, in minutes, spent in the Accident and Emergency department by a random sample of 9 patients were:

| 191 | 134 | 218 | 164 | 201 | 196 | 107 | 228 | 294 |

a   Test the hospital administrator's claim, using a 5% level of significance. Assume that the data came from a normal distribution with standard deviation 45 minutes.

b   Further investigation revealed that the times in the sample were not the times from arriving in the department, but the times from arrivals being recorded by a receptionist. The effect of this is that each of the times in the sample increased by 5 minutes.

How, if at all, does this further information affect your conclusion in part **a**?

AQA SS02 January 2006

a X is the time, in minutes, spent by patients in the Accident and Emergency department, where

$X \sim N(\mu, 45^2)$.

$H_0: \mu = 170$ (the mean time is 170 minutes, as the administrator claims)

$H_1: \mu \neq 170$ (the mean time is not 170 minutes)

The test statistic is $\bar{X}$, where $\bar{X} \sim N\left(\mu, \frac{45^2}{9}\right)$, so $\bar{X} \sim N(\mu, 15^2)$.

If H_0 is true, then $\mu = 170$, and the standardised test statistic is

$$Z = \frac{\bar{X} - 170}{15}$$

> **Note**
>
> s.d. $= \frac{45}{\sqrt{9}} = \frac{45}{3} = 15$

(continued)

(continued)

Use a two-tailed test at the 5% level.

Table 4: $p = 0.975$ gives $z_{crit} = 1.96$ in the upper tail.

By symmetry, in the lower tail, $z_{crit} = -1.96$ and the critical values are ±1.96.

Reject H_0 if $|z_{test}| > 1.96$.

From the sample data, $\bar{x} = 192.55...$

So,

$$z_{test} = \frac{192.55... - 170}{15}$$

$$= 1.5037...$$

$$|z_{test}| = 1.5037... < 1.96..$$

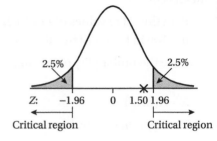

Since $|z_{test}| < 1.96$, the test value is not in the critical region and H_0 is not rejected.

There is not enough evidence, at the 5% level, to doubt the hospital administrator's claim that the mean waiting time is 170 minutes.

b With the further information

$$\bar{x} = 192.55... + 5 = 197.55... \quad \leftarrow \text{Since each value is increased by 5, the mean is increased by 5}$$

so $z_{test} = \dfrac{197.55... - 170}{15}$

$$= 1.8370...$$

$$|z_{test}| = 1.8370... < 1.96$$

Since z_{test} is still not in the critical region, H_0 is not rejected and there is no change to the conclusion made in part **a**.

Type I and Type II errors in a *z*-test

You will recall from section 15.3 that

▶ A **Type I error** is made when you reject H_0, even though H_0 is true.

▶ A **Type II error** is made when you accept H_0 but in fact H_0 is false.

In a *z*-test, there is a relationship between the probability of making a **Type I error** and the significance level of the test.

Suppose you are using a one-tailed upper tail test at the 5% level and that H_0 **is** true. If the test value falls in the critical region you would reject H_0, *even though H_0 is true.*

You would be making a Type I error, with probability 5%. This is equal to the significance level of the test.

So, for a *z*-test,

P(Type I error) = significance level of the test

The higher the significance level, the greater the risk of making a Type I error, that is of rejecting H_0 when it is in fact true. This is illustrated in these diagrams:

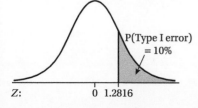
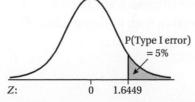
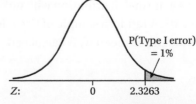

Example 13

A sample of size 16 is taken from the distribution of X, where $X \sim N(\mu, 3^2)$ and a hypothesis test is carried out at the 0.5% level of significance. The sample mean is m.

a Find the set of values of m which result in the rejection of the null hypothesis $\mu = 100$ in favour of the alternative hypothesis $\mu > 100$.

b State the probability of making a Type I error in this test.

a $X \sim N(\mu, 3^2)$

The hypotheses are

$H_0: \mu = 100$

$H_1: \mu > 100$

The test statistic is $\overline{X}$, where $\overline{X} \sim N\left(\mu, \dfrac{3^2}{16}\right)$.

If H_0 is true, then $\mu = 100$, and the standardised test statistic is

$$Z = \frac{\overline{X} - 100}{\dfrac{3}{\sqrt{16}}}$$

so $Z = \dfrac{\overline{X} - 100}{0.75}$

Use a one-tailed, upper tail, test at the 0.5% level.

Since H_0 is rejected, the sample mean m must lie in the critical region.

To define the critical region, find the critical value c by first finding the corresponding critical z-value.

Using Table 4: $p = 0.995$, the critical z-value that gives 0.5% in the upper tail is 2.5758.

$$\frac{c - 100}{0.75} = 2.5758$$

$c - 100 = 2.5758 \times 0.75$

$c = 100 + 2.576 \times 0.75$

$= 101.93.. = 101.9 \ (1 \text{ dp})$

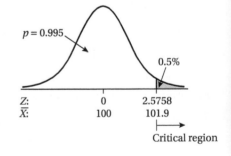

The critical value is 101.9 and the critical region is $\overline{X} > 101.9$.

For H_0 to be rejected, m must lie in the critical region, so $m > 101.9$.

b In a z-test, the probability of making a Type I error is the same as the significance level of the test.

So, $P(\text{Type I error}) = 0.5\% = 0.005$

Exercise 4

1 X follows a normal distribution with unknown mean μ and known variance σ^2. A random sample of size n is taken from the population of X and the sample mean, $\overline{x}$, is calculated.

Test the hypotheses stated, at the significance level indicated.

	Distribution of X	n	$\bar{x}$	Hypotheses	Significance level
a	$X \sim N(\mu, 1.44)$	10	27	$H_0: \mu = 26.3 \quad H_1: \mu > 26.3$	5%
b	$X \sim N(\mu, 17.64)$	49	125	$H_0: \mu = 123.5 \quad H_1: \mu > 123.5$	1%
c	$X \sim N(\mu, 0.18^2)$	100	4.35	$H_0: \mu = 4.40 \quad H_1: \mu < 4.40$	2%
d	$X \sim N(\mu, 3^2)$	30	15.2	$H_0: \mu = 15.8 \quad H_1: \mu \neq 15.8$	5%

2 A machine fills cans with soft drinks so that their contents have a nominal volume of 330 ml. Over a period of time it is found that the volume of liquid in the cans follows a normal distribution with mean 335 ml and standard deviation 3 ml.

A setting on the machine is altered, following which the operator suspects that the mean volume of liquid discharged by the machine into the cans has decreased. He takes a random sample of 50 cans and finds that the mean volume of liquid in these cans is 334.6 ml.

a Carry out a hypothesis test at the 5% significance level to test whether the mean volume of liquid in the cans has decreased. Assume that the standard deviation is 3 ml.

b Explain, in the context of this question, the meaning of making

   i a Type I error

   ii a Type II error.

3 When clients visit a certain firm of solicitors, the duration of consultations, in minutes, may be modelled by a normal distribution with mean 50 minutes and standard deviation 5.3 minutes.

Following the appointment of two new solicitors, the mean duration of a random sample of 15 consultations was 54.2 minutes. Assuming that the standard deviation is 5.3 minutes, test, at the 1% significance level, whether the mean duration of consultations has increased.

4 Czarina carries out a test, at the 5% significance level, using a normal distribution. The null hypothesis is $\mu = 103.5$ and the alternative hypothesis is $\mu < 103.5$. Czarina finds that the value of the test statistic is -1.350. What conclusion should Czarina draw?

5 A marmalade manufacturer produces thousands of jars of marmalade each week. The mass of marmalade in a jar may be modelled by a normal distribution with mean 455 g and standard deviation 0.8 g. Following a slight adjustment to the filling machine, the mass, in grams, of marmalade in a random sample of 10 jars was found to be:

   454.8   453.8   455.0   454.4   455.4

   454.4   454.4   455.0   455.0   453.6

Assume that the standard deviation is unaltered by the adjustment.

a Test at the 5% significance level whether there has been a change in the mean mass of marmalade in jars produced by the manufacturer.

b It is later discovered that the mean mass of marmalade in a jar is 455 grams. Indicate whether a Type I error, a Type II error, or neither has occurred in carrying out the hypothesis test in part **a**. Give a reason for your answer.

6 An athlete finds that her times for running a race are normally distributed with mean 10.75 seconds and standard deviation 0.05 seconds. She trains intensively for a week and then records her time in the next 6 races. Her times, in seconds, are:

   10.70, 10.63, 10.75, 10.81, 10.66, 10.72

   a Is there evidence, at the 5% level, that training intensively has improved her mean time?

   b Explain, in the context of this question, the meaning of making a Type I error.

   c State the probability of making a Type I error in this test.

7 A resident of an urban road in the UK claims that the average speed of vehicles using the road is greater than the 30 mph speed limit. To investigate this claim, a random sample of 25 vehicles is taken and the time each takes to travel along a mile of the road is measured. It is assumed that the speeds, in miles per hour, calculated from these observations may be modelled by a normal distribution with mean μ and standard deviation 12.

   A hypothesis test is carried out to test the null hypothesis $H_0: \mu = 30$ against the alternative hypothesis $H_1: \mu > 30$.

   a The critical region for a hypothesis test, at the 5% level of significance, is $\bar{X} > k$. Find the value of k.

   b State, with a reason, your conclusion for the test when the mean speed calculated from the sample was 35 mph.

15.7 Case 2: Test for the mean of a distribution using a normal approximation

If the population of X is *not normal* then, provided that the sample is large, you can apply the Central Limit Theorem.

See Section 14.3.

By the Central Limit Theorem, if samples of size n are taken from a population X that is not normal, the distribution of $\bar{X}$ is *approximately normal*, provided that n is large ($n \geq 30$, say).

Hypothesis test for a population mean μ when

▶ the distribution of X is *not normal*

▶ the sample is large.

Case 2a: If the population variance σ^2 is _known:_

The test statistic is $\bar{X}$, where $\bar{X} \sim N\left(\mu, \dfrac{\sigma^2}{n}\right)$ approximately

and the standardised test statistic is

$$Z = \frac{\bar{X} - \mu}{\dfrac{\sigma}{\sqrt{n}}}$$

Case 2b: If the population variance σ^2 is <u>unknown</u>:

When σ^2 is unknown, its estimator S^2 may be used where

$$S^2 = \frac{\sum(X_i - \bar{X})^2}{n-1} = \frac{1}{n-1}\left(\sum X_i^2 - \frac{(\sum X_i)^2}{n}\right)$$

See Section 14.1

Provided that the sample size is *large*,

$$\bar{X} \sim N\left(\mu, \frac{S^2}{n}\right) \text{ approximately,}$$

and the standardised test statistic is

$$Z = \frac{\bar{X} - \mu}{\frac{S}{\sqrt{n}}}$$

As these tests use the z-statistic, they are also examples of a z-test.

Example 14

The manager of a large hospital states that the mean age of patients at the hospital is 45.0 years. Records of a random sample of 100 patients at the hospital give a mean age of 41.6 years.

Assuming that the population standard deviation is 18.0 years, test, at the 2% level of significance, whether the manager's statement should be accepted. State clearly the null and alternative hypotheses.

X is the age, in years, of a patient.

The mean of X is μ, the standard deviation $\sigma = 18.0$ and the distribution of X is unknown.

$H_0: \mu = 45.0$ (the mean age is 45.0 years)

$H_1: \mu \neq 45.0$ (the mean is not 45.0 years)

By the Central Limit Theorem, since n is large,

$$\bar{X} \sim N\left(\mu, \frac{18.0^2}{100}\right) \text{ approximately,}$$

If H_0 is true, $\mu = 45.0$ and the standardised test statistic is

$$Z = \frac{\bar{X} - 45.0}{\frac{18.0}{\sqrt{100}}}$$

> **Note**
>
> Use a **z-test (Case 2a)**, since:
> ▶ Distribution of X is unknown
> ▶ Variance is known
> ▶ Sample is large

Using a two-tailed test at the 2% level, Table 4: $p = 0.99$ gives $z_{crit} = \pm 2.3263$, so reject H_0 if $|z_{test}| > 2.3263$.

The test value is $\bar{x} = 41.6$.

$$Z_{test} = \frac{41.6 - 45.}{\frac{18.0}{\sqrt{100}}}$$

$$= -1.88...$$

1%   1%

Z: −2.3263 −1.88... 0   2.3263

Critical region   Critical region

(continued)

Now $|z_{\text{test}}| = 1.88... < 2.3263$

Since $|z_{\text{test}}| < z_{\text{crit}}$, the test value is not in the critical region, so H_0 is not rejected.

There is not enough evidence, at the 2% level, that the mean age has changed. The manager's statement that the mean age is 45.0 years is accepted.

Example 15

The manufacturer of a certain type of light bulb claims that the average lifetime of the firm's light bulbs is 1000 hours. A consumer group thinks that the manufacturer is overstating the lifetime of the light bulbs and tests a random sample of 64 bulbs, recording the life x, in hours, of each bulb.

The results are summarised as:

$$\Sigma x = 63\,910.4, \quad \Sigma x^2 = 63\,824\,061$$

a Calculate $\bar{x}$, the sample mean.

b Calculate s^2, the unbiased estimate of the variance of the lifetime of this type of light bulb.

c Is there evidence, at the 10% level, that the manufacturer is overstating the lifetime of light bulbs produced by the firm?

d State whether it was necessary to use the Central Limit Theorem in your answer in part c and if it was, state at what point it was necessary to use it.

a $\bar{x} = \dfrac{\Sigma x}{n} = \dfrac{63910.4}{64} = 998.6$

b $s^2 = \dfrac{1}{n-1}\left(\Sigma x^2 - \dfrac{(\Sigma x)^2}{n}\right) = \dfrac{1}{63}\left(63824061 - \dfrac{63910.4^2}{64}\right) = 49.77... = 49.8 \text{ (3 sf)}$

c X is the lifetime, in hours, of a light bulb. The mean is μ and the standard deviation is σ (both unknown).

The distribution of X is unknown and the sample size is large.

$H_0: \mu = 1000$ (the mean lifetime is 1000 h)

$H_1: \mu < 1000$ (the mean is less than 1000 h and the manufacturer is overstating the lifetime)

Since n is large (≥ 30), by the Central Limit Theorem

$$\bar{X} \sim N\left(\mu, \dfrac{S^2}{n}\right) \text{ approximately}$$

If H_0 is true, then $\mu = 1000$ and, from part b, $s = \sqrt{49.77...}$, so

$$Z = \dfrac{\bar{X} - 1000}{\dfrac{\sqrt{49.77...}}{\sqrt{64}}}$$

Using a one-tailed lower tail test at the 10% level, Table 4: $p = 0.90$ gives $z_{\text{crit}} = -1.2816$.

Reject H_0 if $z_{\text{test}} < -1.2816$.

Note

Use a **z-test (Case 2b)**, since:
▶ Distribution of X is unknown
▶ Sample is large
▶ Variance is unknown (use s^2)

(continued)

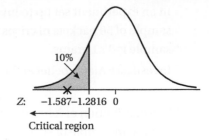

(continued)

The test value is $\bar{x} = 998.6$, so

$$z_{test} = \frac{989.6 - 1000}{\dfrac{\sqrt{49.77...}}{\sqrt{64}}}$$

$$= -1.587 < -1.2816$$

Since $z_{test} < z_{crit}$, reject H_0 in favour of H_1.

There is evidence, at the 10% level, that the mean lifetime is less than 1000 hours and the manufacturer is overstating the length of lifetime of the light bulbs produced by the firm.

d It was necessary to use the Central Limit Theorem because you do not know whether the lifetimes of the bulbs are normally distributed. Since the sample size is large, by the Central Limit Theorem, the distribution of the *mean* lifetimes of the bulbs is approximately normal. This is needed to specify the distribution of the test statistic $\bar{X}$, as stated in part **c**.

Exercise 5

1 The distribution of X has unknown mean μ and known variance σ^2. A random sample of size n is taken from the population of X and the sample mean, $\bar{x}$, is calculated.

Test the hypotheses stated, at the significance level indicated.

	n	$\bar{x}$	σ	Hypotheses	Significance level
a	120	223.3	20.0	$H_0: \mu = 220.0$ $H_1: \mu > 220.0$	5%
b	150	7.9	1.5	$H_0: \mu = 8.2$ $H_1: \mu < 8.2$	1%
c	80	12.0	2.2	$H_0: \mu = 12.5$ $H_1: \mu \neq 12.5$	5%
d	100	0.32	0.13	$H_0: \mu = 0.30$ $H_1: \mu \neq 0.30$	10%

2 An intelligence test is developed in which the mean score is 100 and the standard deviation is 12. When the test was given to a random sample of 50 children from a particular city, the mean score was 105.

a Test, at the 5% significance level, whether the mean score of children in this city is greater than 100. Assume that the standard deviation is 12.

b Explain, in the context of the question, the meaning of making a Type I error.

3 A continuous variable X has mean μ and variance 32. It is thought that $\mu = 55.0$.

The mean of a random sample of 81 observations of X is 56.2.

Does this provide evidence at the 10% level of significance that the mean is not 55.0?

4 Alan's company produces packets of crisps. The standard deviation of the weight of a packet of crisps is known to be 2.5 grams.

Alan believes that, due to the extra demand on the production line at a busy time of the year, the mean weight of packets of crisps is not equal to the target weight of 34.5 grams.

In an experiment set up to investigate Alan's belief, the weights of a random sample of 50 packets of crisps were recorded. The mean weight of this sample is 35.1 grams.

Investigate Alan's belief at the 5% level of significance.

AQA MS2A June 2008

5 The distribution of X has unknown mean μ and unknown variance σ^2. A random sample of size n is taken from the population of X and the values are summarised as shown.

Test the hypotheses stated, at the significance level indicated.

	n	$\sum x$	$\sum(x-\overline{x})^2$	$\sum x^2$	Hypotheses	Significance level
a	65	6500		650842.4	$H_0: \mu = 99.2$ $H_1: \mu \neq 99.2$	5%
b	65	6500		650842.4	$H_0: \mu = 99.2$ $H_1: \mu > 99.2$	5%
c	80	6824	2508.8		$H_0: \mu = 86.2$ $H_1: \mu < 86.2$	10%
d	100	685		4728.25	$H_0: \mu = 7.0$ $H_1: \mu \neq 7.0$	1%

6 A random sample of 75 children performed a simple task and the time taken, t minutes, noted for each. The results were summarised by:

$$\sum t = 1215, \quad \sum(t - \overline{t})^2 = 2026$$

a Calculate an unbiased estimate of the population variance of the time taken to perform the task.

b Test, at the 1% significance level, whether there is evidence that the mean time taken to perform the task is greater than 15 minutes.

c Explain in the context of this question the meaning of making a Type II error.

7 An inspector checks items from a production line. On average, he takes 22.5 s to check each item. After the installation of a new lighting system the times, t seconds, that he takes to check each of 50 randomly chosen items from the production line are summarised by

$$\sum t = 1107, \quad \sum t^2 = 24\,592.35$$

a Calculate an unbiased estimate of the population variance of the time taken to check an item under the new lighting system.

b Test at the 2% significance level whether there is evidence that the mean time to inspect an item has been reduced.

c A technician who carried out the above test concluded with the following incorrect statement *'It is not necessary for the population to be normal since the sample size is large and the Central Limit Theorem states that any sufficiently large sample is normal.'*

Give a corrected version.

8 A sample of 40 observations from a normal distribution gave

$$\sum x = 24, \quad \sum(x - \overline{x})^2 = 581.6$$

Perform a two-tail test, at the 5% significance level, to test whether the mean of the distribution is zero.

9 Cans of lemonade are filled by a machine which is set to dispense a mean amount of 330 ml into each can. The manufacturer suspects that the machine is tending to over-dispense and, in order to test the suspicion, measures the contents, x millilitres, of a random sample of 30 cans. The results are summarised by

$$\sum x = 9925 \text{ and } \sum x^2 = 3\,284\,137$$

a Calculate an unbiased estimate of the population variance of the amount dispensed into each can.

b Test the manufacturer's suspicion at the 10% significance level.

c Indicate where the Central Limit Theorem is used in the test, and state why the use of the Central Limit Theorem is necessary.

10 David is the professional coach at the golf club where Becki is a member. He claims that, after having a series of lessons with him, the mean number of putts that Becki takes per round of golf will reduce from her present mean of 36.

After having the series of lessons with David, Becki decides to investigate his claim.

She therefore records, for each of a random sample of 50 rounds of golf, the number of putts, x, that she takes to complete the round. Her results are summarised below, where $\bar{x}$ denotes the sample mean.

$$\sum x = 1730, \ \sum(x - \bar{x})^2 = 784$$

Using a z-test and the 1% level of significance, investigate David's claim.

AQA MS2A June 2007

15.8 The *t*-distribution

At the beginning of the 20th century a young statistician, William Gosset, discovered that when a random sample of size n is drawn from a *normal* population with *unknown variance*, the test statistic

$$T = \frac{\bar{X} - \mu}{\dfrac{S}{\sqrt{n}}}$$

has a symmetrical, bell-shaped distribution similar to the normal distribution, but with a greater spread. Unlike the normal distribution, there is a different curve for each sample size, so there is a family of distributions known as *t*-distributions.

Each curve is symmetrical about zero and has a single parameter, ν (read as 'new').

ν is the number of **degrees of freedom.** In general, this is the number of independent variables used in calculating a sample statistic, where

$$\nu = \text{number of variables} - \text{number of restrictions}$$

> **Note**
>
> William Gosset wrote under the pen-name of 'Student', so the *t*-distribution is often referred to as the **Student's *t*-distribution**.

For a sample of size n, there are n variables and one restriction (σ has been estimated using S), so there are $(n-1)$ degrees of freedom. Hence $v = n - 1$

$$T = \dfrac{\bar{X} - \mu}{\dfrac{S}{\sqrt{n}}}$$

follows a t-distribution with $(n-1)$ degrees of freedom.

This is written

$T \sim t(n-1)$ or $T \sim t(v)$

For example, for a sample size of 8, T has a t-distribution with 7 degrees of freedom, so $T \sim t(7)$.

The diagram shows two curves, $t(2)$ and $t(10)$.

As v increases, the corresponding $t(v)$ curve gets closer to the curve for Z, the standard normal variable and the difference between the $t(v)$ distribution and the standard normal distribution is negligible.

In fact, when $n \geq 30$ a normal approximation can be used and, in this situation, the z-test described in Case 2 in Section 15.6 is usually carried out.

However, when $n < 30$, the t-distribution *must* be used.

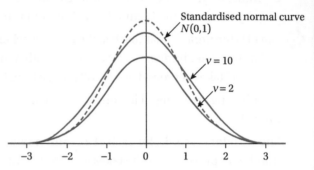

Percentage points for the *t*-distribution

To find critical values of t, **Table 5**, a table of percentage points, is provided in the *Formulae and Statistical Tables* booklet and an extract is reproduced below.

It gives values of t for which $P(T \leq t) = p$, for various values of v.

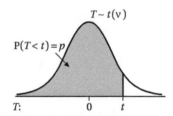

An extract from Table 5: Percentage points of the student's *t*-distribution

p / v	0.9	0.95	0.975	0.99	0.995
1	3.078	6.314	12.706	31.821	63.657
2	1.886	2.920	4.303	6.965	9.925
3	1.638	2.353	3.182	4.541	5.841
4	1.533	2.132	2.776	3.747	4.604
5	1.476	2.015	2.571	3.365	4.032
6	1.440	1.943	2.447	3.143	3.707
7	1.415	1.895	2.365	2.998	3.499
8	1.397	1.860	2.306	2.896	3.355
9	1.383	1.833	2.262	2.821	3.250
10	1.372	1.812	2.228	2.764	3.169
11	1.363	1.796	2.201	2.718	3.106
12	1.356	1.782	2.179	2.681	3.055
13	1.350	1.771	2.160	2.650	3.012
14	1.345	1.761	2.145	2.624	2.977
15	1.341	1.753	2.131	2.602	2.947
16	1.337	1.746	2.121	2.583	2.921

For example, consider a sample with $n = 12$.

Since $v = n - 1$, $v = 11$ and $T \sim t(11)$.

Suppose you want to find the critical value of t for a one-tailed, *upper tail* test at the 5% significance level, that is the value of t such that there is 5% in the upper tail.

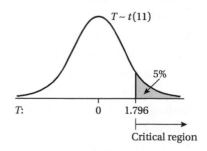

Since $P(T > t) = 0.05$,

$$P(T < t) = 0.95$$

To find t_{crit}

▶ Find row $v = 11$

▶ Go across to the column headed $p = 0.95$

So $t_{crit} = 1.796$

If you want the critical value for a one-tailed *lower tail* test at the 5% significance level then, by symmetry,

$$t_{crit} = -1.796$$

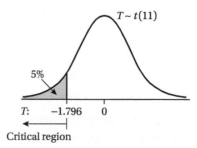

Now consider a sample with $n = 6$.

Since $v = n - 1$, $v = 5$ and $T \sim t(5)$.

Suppose the test is a two-tailed test at the 5% significance level.

Since you need 2.5% in each tail, first find t such that $P(T < t) = 0.975$:

▶ Find row $v = 5$

▶ Go across to the column headed $p = 0.975$

▶ This gives 2.571

The upper tail value of $t_{crit} = 2.571$.

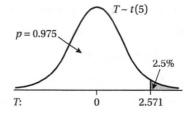

By symmetry, the lower tail value is $t_{crit} = -2.571$

So, $t_{crit} = \pm 2.571$

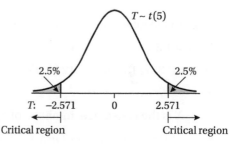

15.9 Case 3: Test for the mean of a normal distribution with unknown variance and small sample size

Hypothesis test for a *population mean* μ when

▶ the distribution of X is *normal*

▶ the population variance σ^2 is *unknown*

▶ the sample size is *small*. See Section 14.1

   The test statistic is

$$T = \frac{\bar{X} - \mu}{\frac{S}{\sqrt{n}}} \quad \text{where} \quad T \sim t(n-1)$$

$$S^2 = \frac{\Sigma(X_i - \bar{X})^2}{n-1} = \frac{1}{n-1}\left(\Sigma X_i^2 - \frac{(\Sigma X_i)^2}{n}\right)$$

As this hypothesis test uses a t-statistic, it is known as a **t-test**.

Steps in a *t*-test

The steps follow the same general pattern as earlier tests:

▶ Define the variable.
▶ State H_0 and H_1.
▶ State the test statistic T and its distribution when H_0 is true.
▶ State the type and significance level of the test.
▶ State the rejection rule: this is usually given in terms of a critical *t*-value.
▶ Find whether the test value lies in the critical region: this usually involves finding the value of the standardised test value and comparing it with the critical *t*-value.
▶ Make your conclusion in statistical terms by saying whether H_0 is rejected or not and then *relate it to the situation*.

Example 16

Question

Five readings of the resistance, X ohms, of a piece of wire gave these results:

   1.51,   1.49,   1.54,   1.52,   1.54

If the wire is pure, the resistance is 1.50 ohms. If the wire is impure, its resistance is higher than 1.50 ohms. Assuming that the resistance can be modelled by a normal variable with mean μ and variance σ^2,

a Calculate the value of

   i $\bar{x}$, the sample mean,

   ii s, where s^2 is the unbiased estimate of σ^2.

b Is there evidence, at the 5% level of significance, that the wire is impure?

Answer

a Using a calculator in statistical mode:

   i $\bar{x} = 1.52$,

   ii $s = 0.021213... = 0.0212$ (3 sf)

b ▶ *Define the variable*

   X is the resistance, in ohms, of a piece of wire, where $X \sim N(\mu, \sigma^2)$.

   ▶ *State H_0 and H_1 and the distribution of the test statistic if H_0 is true*

   $H_0: \mu = 1.50$ (the wire is pure silver)

   $H_1: \mu > 1.50$ (the wire is impure)

   The test statistic is

   $$T = \frac{\bar{X} - \mu}{\frac{S}{\sqrt{n}}} \text{ where } T \sim t(n-1)$$

   $n = 5$, $\nu = n - 1 = 4$, so the distribution $t(4)$ is required.

   If H_0 is true, then $\mu = 1.50$. Since the variance, σ^2, is unknown, use $s = 0.02121...$ found in part **a**.

   $$T = \frac{\bar{X} - 1.50}{\frac{0.02121...}{\sqrt{5}}} \text{ where } T \sim t(4)$$

> **Note**
>
> Use a *t*-test since:
> ▶ X is normal.
> ▶ Variance is unknown.
> ▶ Sample is small.

(continued)

(continued)

► *State the type and level of the test*

Use a one-tailed upper tail test at the 5% level.

► *State the rejection rule*

Find the critical *t*-value that gives 5% in the upper tail.

Table 5 with $v = 4$, $p = 0.95$ gives $t_{crit} = 2.132$,

so reject H_0 if $t_{test} > 2.132$.

► *Calculate the required probability*

The test value is $\bar{x} = 1.52$.

$$t_{test} = \frac{1.52 - 1.50}{\frac{0.02121...}{\sqrt{5}}}$$

$$= 2.108... < 2.132$$

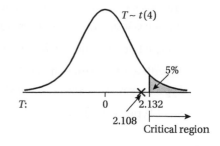

$T \sim t(4)$

5%

$T:$ 0   2.132

2.108

Critical region

► *Make your conclusion, relating it to the context*

Since $t_{test} < t_{crit}$, the test value is not in the critical region. Do not reject H_0.

There is not enough evidence, at the 5% level, that the wire is impure.

Example 17

A mechanic claims that, on average, he spends 45 minutes on an oil change for a car. A random sample of 15 cars being serviced is selected and for this sample the mean time, $\bar{x}$, is 43.6 minutes and the standard deviation, s, is 2.3 minutes.

a Test whether the mechanic is over-estimating the mean time for the oil change. Use the 10% significance level and assume that the distribution is normal.

b State the probability of making a Type I error in this test.

a Let X be the time, in minutes, the mechanic spends on an oil change, where $X \sim N(\mu, \sigma^2)$.

σ is estimated, using $s = 2.3$

$H_0: \mu = 45$ (the mean time is 45 minutes)

$H_1: \mu < 45$ (the mechanic is over-estimating the time)

The test statistic is

$$T = \frac{\bar{X} - \mu}{\frac{S}{\sqrt{n}}} \quad \text{where} \quad T \sim t(n-1)$$

Since $n = 15$, $T \sim t(14)$.

If H_0 is true, then $\mu = 45$

$$T = \frac{\bar{X} - 45}{\frac{2.3}{\sqrt{15}}} \sim t(14)$$

Use a one-tailed, lower tail test at the 10% level.

Table 5 with $v = 14$, $p = 0.9$ gives $t = 1.345$

The critical value of t is in the lower tail so, by symmetry,

$t_{crit} = -1.345$

Reject H_0 if $t_{test} < t_{crit}$, that is, if $t_{test} < -1.345$

(continued)

The test value, $\bar{x} = 43.6$

$$t_{\text{test}} = \frac{43.6 - 45}{\frac{2.3}{\sqrt{15}}}$$

$$= -2.357... < -1.345$$

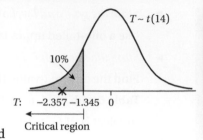

$T \sim t(14)$

T: -2.357 -1.345 0

Critical region

Since $t_{\text{test}} < t_{\text{crit}}$, the test value is in the critical region, so reject H_0 in favour of H_1.

There is significant evidence, at the 10% level that the mean time for a standard service is less than 45 minutes, so the mechanic is over-estimating the time.

b A Type I error is made when you reject H_0 but it is in fact true.

As with a z-test, since the variable is continuous, the probability of making a Type I error is the same as the significance level of the test.

So P(Type I error) = 0.10

Example 18

Lorraine bought a new golf club. She then practised with this club by using it to hit golf balls on a golf range.

After several such practice sessions, she believed that there had been no change from 190 metres in the mean distance that she had achieved when using her old club.

To investigate this belief, she measured, at her next practice, the distance, x metres, of each of a random sample of 10 shots with her new club. Her results gave

$$\Sigma x = 1840 \quad \text{and} \quad \Sigma(x - \bar{x})^2 = 1240$$

Investigate Lorraine's belief at the 2% level of significance, stating any assumptions that you make.

AQA MS2B January 2010

Let X be the distance, in metres, of a shot.

Assume that X is normally distributed, where $X \sim N(\mu, \sigma^2)$.

H_0: $\mu = 190$ (there has not been a change in the distance)

H_1: $\mu \neq 190$ (there has been a change in the distance)

The test statistic is

$$T = \frac{\bar{X} - \mu}{\frac{S}{\sqrt{n}}} \quad \text{where } T \sim t(n-1)$$

Since $n = 10$, $T \sim t(9)$.

If H_0 is true, then $\mu = 190$.

Since σ is unknown, use s, calculated from the sample data.

From the sample

$$s^2 = \frac{\Sigma(x_i - \bar{x})^2}{n-1} = \frac{1240}{9}$$

$$s = \sqrt{\frac{1240}{9}} = 11.737...$$

> **Note**
>
> Since the sample is small and the population variance is unknown, a t-test is needed, for which the distribution of X **must** be normal.

(continued)

(continued)

$$T = \frac{\bar{X}-190}{\frac{\sqrt{11.737...}}{\sqrt{10}}} \quad \text{where} \quad T \sim t(9)$$

Use a two-tailed test at the 2% level.

Find the critical *t*-value that gives 1% in the upper tail.

Table 5 with $v = 9$, $p = 0.99$ gives $t = 2.821$.

So, by symmetry, $t_{\text{crit}} = \pm 2.821$

Reject H_0 if $|t_{\text{test}}| > 2.821$.

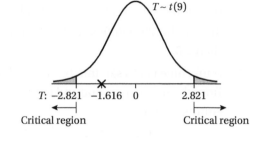

From the sample data, $\bar{x} = \dfrac{\sum x}{n} = \dfrac{1840}{10} = 184$

The test value is $\bar{x} = 184$.

$$t_{\text{test}} = \frac{184-190}{\frac{\sqrt{11.737...}}{\sqrt{10}}}$$

$$= -1.616...$$

$$|t_{\text{test}}| = 1.616... < 2.821$$

Since $|t_{\text{test}}| < 2.821$, the test value is not in the critical region.

Do not reject H_0.

There is not enough evidence, at the 2% level, that the length of shot has changed. Therefore, Lorraine's belief that it has not changed after using the new club is supported.

Exercise 6 (*t*-tests)

1 *X* follows a normal distribution with unknown mean μ and unknown variance σ^2. A random sample of size *n* is taken from the population of *X* and the sample mean, $\bar{x}$, is calculated.

Test the hypotheses stated, at the significance level indicated.

	n	$\sum x$	$\sum(x-\bar{x})^2$	$\sum x^2$	Hypotheses	Level
a	12	298.8		7542.42	$H_0: \mu = 24.1$ $H_1: \mu > 24.1$	5%
b	17	605.2		23016.92	$H_0: \mu = 40.0$ $H_1: \mu \neq 40.0$	5%
c	6	9034.8	50.8		$H_0: \mu = 1503$ $H_1: \mu \neq 1503$	10%
d	10	1298	97.6		$H_0: \mu = 133.0$ $H_1: \mu < 133.0$	1%

2 An athlete finds that her times for running a race are normally distributed with a mean of 10.6 seconds. She trains intensively for a week and then records her times for the next 5 races, with these results:

  10.70 10.65 10.75 10.80 10.60

Test, at the 5% significance level, whether, following the intensive training, there has been an improvement in her times.

3 Family packs of cheese are sold in 1.5 kg packs. A sample of 12 packs was selected at random and their weights, in kg, noted with these results:

  $\sum x = 17.81$ $\sum x^2 = 26.4357$

Assuming that the weights are normally distributed, test at the 10% level whether the packs are underweight if

a σ^2 is unknown

b $\sigma^2 = 0.0003$

4 It is thought that a normal population has a mean of 1.6. A random sample of 10 observations gives a mean of 1.49 and variance 0.09.

Test, at the 5% significance level, the claim that the population mean is less than 1.6.

5 A machine packs salt into bags. Four bags of salt were selected at random from the production line and their mass, x grams, noted. The results are summarised by

$$\Sigma x = 2992.8 \qquad \Sigma(x - \bar{x})^2 = 90$$

It is claimed that the mean mass of salt in the packs is less than the stated amount of 755 grams. Test this claim at the 5% significance level. You may assume that the mass, in grams, of salt in bags packed by the machine is normally distributed.

6 A random sample of 8 observations of a normal variable gave

$$\Sigma x = 36.5 \qquad \Sigma(x - \bar{x})^2 = 0.74$$

Test, at the 5% significance level, the null hypothesis that the mean of the distribution is 4.3 against the alternative hypothesis that the mean is greater than 4.3.

7 The cholesterol levels of 8 women were measured, with these results:

3.1	2.8	1.5	1.7	2.4	1.9	3.3	1.6

a   Stating any necessary assumptions, test at the 5% level of significance whether the sample has been selected from a distribution with mean cholesterol level of 3.1.

b   The test carried out in part **a** may have resulted in an error. State, with a reason, whether this is a Type I or a Type II error.

8 The lengths of components produced by a machine are known to be normally distributed. The machine has been set to produce components with a mean length of 13.65 mm, but the operator suspects that it may need further adjustment.

A random sample of 10 components has a mean length of 13.79 mm and standard deviation of 0.24 mm. Use the 10% significance level to test the hypothesis that the mean length is equal to 13.65 mm.

9 A gas company claimed that the length of time between a gas leak being reported and an engineer arriving to investigate the report has a mean of 115 minutes. A random sample of these times, in minutes, was

141	85	152	114	150	103	66	234

Using a 5% significance level, test whether there is evidence that the mean time exceeds 115 minutes. Assume that the distribution of times is normal.

AQA SS04 January 2012

10. Jasmine's French teacher states that a homework assignment should take, on average, 30 minutes to complete.

Jasmine believes that he is understating the mean time that the assignment takes to complete and so decides to investigate. She records the times, in minutes, that it takes for a random sample of 10 students to complete the French assignment, with the following results:

29   33   36   42   30   28   31   34   37   35

a   Test, at the 1% level of significance, Jasmine's belief that her French teacher has understated the mean time that it should take to complete the homework assignment.

b   State an assumption that you must make in order for the test used in part **a** to be valid.

AQA MS2B January 2007

11. Giles persuaded 11 students to allow him to measure the lengths of their bare left feet. The results, in centimetres, were as follows.

28.3   32.9   26.6   25.4   26.2   26.8   29.3   30.7

30.2   31.6   28.7

a   Using the 5% significance level, test the hypothesis that the mean length of a male student's left foot is 30.5 cm. The data may be assumed to be a random sample from a normal distribution.

b   Use the data and the result of your test in part **a** to make **two** comments on the claim that, on average, a male student's foot is one foot in length. (One foot in length is equal to 30.5 cm, correct to one decimal place.)

AQA SS04 June 2012

15.10 Further applications

Example 19

A mill produces cloth in 100-metre lengths. It is common for lengths to contain faults that have to be treated before the cloth is sold. These faults are distributed over the cloth independently, at random and at a constant average rate.

a   Name a distribution that will provide a suitable model for the number of faults in a 100-metre length of cloth.

b   A new manager states that it is unacceptable for the mean number of faults per 100-metre length of cloth to exceed 2. The next 100-metre length of cloth produced contains 3 faults.

Carry out a hypothesis test, using the 10% significance level, to test the hypothesis that the mean number of faults per 100-metre length of cloth does not exceed 2.

c   The results of the tests are presented to the new manager who states that a significance level of 10% is too high and that a level of 0.1% should be used in order to reduce the possibility of error.

Without carrying out further calculations, comment on the new manager's statement and point out one disadvantage of reducing the significance level.

AQA SS04 January 2006

a A Poisson distribution would provide a suitable model.

b X is the number of faults in a 100 m length of cloth, where $X \sim Po(\lambda)$.

H_0: $\lambda = 2$ (the mean number of faults is 2)

H_1: $\lambda > 2$ (the mean number of faults exceeds 2)

If H_0 is true, then $X \sim Po(2)$.

Use a one-tailed upper tail test at the 10% level and reject H_0 if $P(X \geq x) < 0.10$, where x is the test value.

There were 3 faults in the next 100 m length of cloth, so the test value is 3.

$$P(X \geq 3) = 1 - P(X \leq 2)$$

$$= 1 - 0.6767$$

$$= 0.3233 > 0.10$$

Since $P(X \geq 3) > 0.10$, the test value is not in the critical region, so H_0 is not rejected.

There is not enough evidence, at the 10% level, that the mean number of faults exceeds 2.

c If a significance level of 0.1% is used, instead of 10%, the critical region is much smaller and the risk of making a Type I error, that is of wrongly claiming that the mean is greater than 2, can be reduced.

A disadvantage, however, is that this increases the risk of accepting that the mean is 2 when in fact it is greater than 2, that is of making a Type II error.

Table 2: Cumulative Poisson probabilities

x	$\lambda = 2.0$
0	0.1353
1	0.4060
2	0.6767
3	0.8571
etc	......

Example 20

A market trader sells bags of cherries. A sign on his stall says '1 lb bags of cherries'. The unit of weight 1 lb is equal to 453.6 grams. Sophie, the market inspector, suspects that the bags may, on average, contain less than 453.6 grams.

Sophie asks her assistant, Kevin, to investigate her suspicion. She tells Kevin that, from previous measurements, the weights of bags of cherries from this trader may be assumed to be a normal distribution with standard deviation 10 grams. Kevin weighs 6 bags of cherries and obtains the following weights in grams.

448.2   461.9   455.8   437.0   442.5   441.4

a Assuming that the distribution of weights of bags of cherries is still normal with standard deviation 10 grams, investigate Sophie's suspicion at the 10% significance level.

b Sophie discovers that Kevin has included the weight of the bag in the data, as well as the weight of the cherries. Each bag weighs 5 grams. With this information, show that, on the basis of the above data, Sophie can now confirm her suspicion at the 1% significance level.

c Kevin suggests that, in rejecting the trader's claim, Sophie may have made a Type I error or a Type II error.

State, giving a reason, which type of error (Type I or Type II) Sophie may have made in this case.

AQA SS02 June 2013

a X is the weight, in grams, of a bag of cherries from the market stall, where $X \sim N(\mu, 10^2)$.

$H_0: \mu = 453.6$ (the mean weight is 1 lb as advertised)

$H_1: \mu < 453.6$ (the mean weight is less than 1 lb)

The test statistic is $\bar{X} \sim N\left(\mu, \dfrac{102}{6}\right)$.

If H_0 is true, then $\mu = 453.6$, and the standardised test statistic is

$$Z = \frac{\bar{X} - 453.6}{\dfrac{10}{\sqrt{6}}}$$

Using a one-tailed lower tail test at the 10% level, Table 4: $p = 0.90$ gives $z_{crit} = -1.2816$,

so reject H_0 if $z_{test} < -1.2816$.

From the sample data: $\bar{x} = 447.8$.

$$z_{test} = \frac{447.8 - 453.6}{\dfrac{10}{\sqrt{6}}} = -1.42... < -1.2816$$

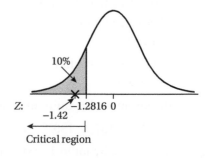

Since $z_{test} < z_{crit}$, the test value is in the critical region, so H_0 is rejected in favour of H_1.

There is significant evidence, at the 10% level, that bags of cherries, on average, weigh less than 453.6 grams, (1 lb) thus confirming Sophie's suspicion.

b Let Y be the weight, in grams, of the cherries in a bag, where $Y \sim N(\mu, 10^2)$. Using a one-tailed lower tail test at the 1% level, Table 4 with $p = 0.99$ gives $z_{crit} = -2.3263$.

So this time she will reject H_0 if $z_{test} < -2.3263$.

To find the sample value, you need to use the fact that each bag weighs 5 grams, so $\bar{y} = 447.8 - 5 = 442.8$ ← Since each value is reduced by 5, the mean is reduced by 5

So this time

$$z_{test} = \frac{442.8 - 453.6}{\dfrac{10}{\sqrt{6}}} = -2.645... < -2.3263$$

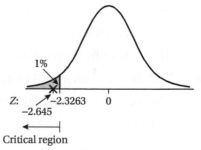

Since $z_{test} < z_{crit}$, H_0 is again rejected, and Sophie can confirm her suspicion at the 1% significance level.

c A Type I error is made when H_0 is rejected when in fact H_0 is true.

Since H_0 is rejected in both tests, a Type I error may have been made in both cases.

Answer

Example 21

Question

A catering company provides prepacked meals for consumption by airline passengers. Gamma Airlines buys large quantities of these meals and finds that 40 per cent of passengers, when surveyed, rate the meals as poor. Passengers also complain about the small quantity of meat in non-vegetarian meals.

The catering company undertakes to improve the meals and to ensure that there will be a mean of more than 140 grams of meat in non-vegetarian meals.

a   Following this undertaking, a random sample of 20 passengers was asked to rate the meals, and 6 of these passengers rated them as poor.

Carry out a hypothesis test, using the 5% significance level, to examine whether the proportion of passengers rating the meals as poor has been reduced.

b   Following the undertaking, the meat content of 10 non-vegetarian meals was measured.

The amounts, in grams, of meat which they contained were:

145   157   139   162   133   156   148   131   164   159

Carry out a hypothesis test, using the 5% significance level, to examine whether the mean meat content of the meals was more than 140 grams. You may treat the data as a random sample from a normal distribution.

c   Summarise the evidence as to whether or not the catering company has complied with its undertakings.

AQA SS04 January 2006

Answer

a X is the number of passengers in a random sample of 20 who rate the meals as poor, where $X \sim B(20, p)$.

H_0: $p = 0.4$ (the proportion is unchanged)

H_1: $p < 0.4$ (the proportion has reduced)

If H_0 is true, then $p = 0.4$ and $X \sim B(20, 0.4)$.

Use a one-tailed lower tail test at the 5% level.

Reject H_0 if $P(X \le x) < 0.05$, where x is the test value.

The test value is $x = 6$.

Using Table 1 with $n = 20$, $p = 0.4$:

$P(X \le 6) = 0.2500 > 0.05$

Since $P(X \le 6) > 0.05$, the test value is not in the critical region, so do not reject H_0.

There is not enough evidence, at the 5% level, that the proportion of passengers rating the meals as poor has been reduced.

(continued)

Table 1: $n = 20$

x	$p = 0.40$
0	0.0000
1	0.0005
2	0.0036
3	0.0160
4	0.0510
5	0.1256
6	0.2500
etc	......

(continued)

b Let X be the meat content, in grams, of a meal, where $X \sim N(\mu, \sigma^2)$.

$H_0: \mu = 140$ (the mean meat content of the meals is 140 grams)

$H_1: \mu > 140$ (the mean meat content of the meals is greater than 140 grams)

The test statistic is

$$T = \frac{\bar{X} - \mu}{\frac{S}{\sqrt{n}}} \quad \text{where} \quad T \sim t(n-1)$$

Since $n = 10$, $T \sim t(9)$.

If H_0 is true, then $\mu = 140$.

Since σ is unknown, s is used instead.

Using a calculator in statistical mode, $s = 12.029...$

$$T = \frac{\bar{X} - 140}{\frac{12.029...}{\sqrt{10}}} \quad \text{where} \quad T \quad t(9)$$

Using a one-tailed upper tail test at the 5% level, Table 5 with $v = 9$, $p = 0.95$ gives $t_{crit} = 1.833$.

Reject H_0 if $t_{test} > t_{crit}$, that is, if $t_{test} > 1.833$.

From the sample data, $\bar{x} = 149.4$.

$$t_{test} = \frac{149.4 - 140}{\frac{12.029...}{\sqrt{10}}}$$

$$= 2.4710... > 1.833$$

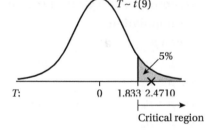

Since $t_{test} > t_{crit}$, the test value is in the critical region.

Reject H_0 in favour of H_1.

There is significant evidence, at the 5% level, that the mean meat content of the meals is greater than 140 grams.

c Although less than 40% in the sample rated the meals as poor, the evidence was not significant, so the first undertaking has not been met.

There was significant evidence that the mean meat content of the meals has increased, so the second undertaking has been met.

Summary

Steps in a hypothesis test

▶ Define the variable.

▶ State H_0 and H_1 and the distribution of the test statistic, assuming H_0 is true.

▶ State the type and level of significance of the test.

▶ State the rejection rule.

▶ Find whether the test value lies in the critical region.

▶ Make your conclusion in statistical terms by saying whether H_0 is rejected or not and then relate it to the situation.

Rejection rule

▶ If the test value is in the critical region, reject H_0.

▶ If the test value is not in the critical region, do not reject H_0.

Tests

Testing a population proportion or Poisson mean at α % level of significance

Hypotheses for proportion p	Hypotheses for Poisson λ	Critical region	Rejection rule (x is the test value)
$H_0: p = p_0$ $H_1: p > p_0$	$H_0: \lambda = \lambda_0$ $H_1: \lambda > \lambda_0$	Upper tail α %	Reject H_0 if $P(X \geq x) < \alpha$ %
$H_0: p = p_0$ $H_1: p < p_0$	$H_0: \lambda = \lambda_0$ $H_1: \lambda < \lambda_0$	Lower tail α %	Reject H_0 if $P(X \leq x) < \alpha$ %
$H_0: p = p_0$ $H_1: p \neq p_0$	$H_0: \lambda = \lambda_0$ $H_1: \lambda \neq \lambda_0$	Upper tail $\frac{\alpha}{2}$ % and lower tail $\frac{\alpha}{2}$ %	Reject H_0 if $P(X \geq x) < \frac{\alpha}{2}$ % or $P(X \leq x) < \frac{\alpha}{2}$ %

Testing a population mean (z-tests and t-tests) at α% level of significance

Hypotheses for population mean μ	Critical region	z-test rejection rule (z_{test} is the test value)	t-test rejection rule (t_{test} is the test value)
$H_0: \mu = \mu_0$ $H_1: \mu > \mu_0$	Upper tail α%	Reject H_0 if $z_{test} > z_{crit}$	Reject H_0 if $t_{test} > t_{crit}$
$H_0: \mu = \mu_0$ $H_1: \mu < \mu_0$	Lower tail α%	Reject H_0 if $z_{test} < z_{crit}$	Reject H_0 if $t_{test} < t_{crit}$
$H_0: \mu = \mu_0$ $H_1: \mu \neq \mu_0$	Upper tail $\frac{\alpha}{2}$ % and lower tail $\frac{\alpha}{2}$ %	Reject H_0 if $|z_{test}| > z_{crit}$	Reject H_0 if $|t_{test}| > t_{crit}$

Testing a population mean μ

	Type of test	Test statistic
Case 1 ▶ X is normal ▶ σ^2 is known ▶ sample can be any size	z-test	$\bar{X} \sim N\left(\ , \dfrac{\sigma^2}{n}\right)$, so $Z = \dfrac{\bar{X} - }{\dfrac{\sigma}{\sqrt{n}}}$
Case 2a ▶ X is not normal ▶ σ^2 is known ▶ sample must be large ($n \geq 30$)	z-test (normal approximation)	$\bar{X} \sim N\left(\ , \dfrac{\sigma^2}{n}\right)$ approximately, so $Z = \dfrac{\bar{X} - }{\dfrac{\sigma}{\sqrt{n}}}$
Case 2b ▶ X is not normal ▶ σ^2 is unknown ▶ sample must be large ($n \geq 30$)	z-test (normal approximation)	$\bar{X} \sim N\left(\ , \dfrac{S^2}{n}\right)$ approximately, so $Z = \dfrac{\bar{X} - }{\dfrac{S}{\sqrt{n}}}$
Case 3 ▶ X is normal ▶ σ^2 is unknown ▶ sample is small ($n < 30$)	t-test	$T = \dfrac{\bar{X} - }{\dfrac{S}{\sqrt{n}}}$ where $T \sim t(n-1)$

Note that

$$S^2 = \frac{\sum\left(X - \bar{X}\right)^2}{n-1} = \frac{1}{n-1}\left[\sum X^2 - \frac{\left(\sum X\right)^2}{n}\right]$$

Type I and Type II errors

A Type I error is made when the null hypothesis is rejected when it is true.

P(Type I error) = P(reject H_0 when H_0 is true)

For continuous variables (z-test or a t-test)

P(Type I error) = significance level

A Type II error is made when the null hypothesis is accepted when it is false.

P(Type II error) = P(accept H_0 when H_0 is false)

You will not be asked to find the probability of a Type II error.

Review

1 A survey is being carried out in a particular city on the methods of transport used by commuters to travel to work. The local authority claims that 40 per cent of commuters cycle to work, but the Cycling Association believes that more than 40 per cent of commuters cycle to work. In a random sample of 12 commuters, 8 cycle to work.

Is there evidence, at the 5% significance level, to justify the Cycling Association's belief that more than 40 per cent of commuters cycle to work?

2 The number of breakdowns per day of an office photocopier can be modelled by a Poisson distribution with mean 0.9. The photocopier is serviced and during the following working week of 5 days it broke down twice. Is there evidence, at the 10% significance level, of an improvement in the reliability of the photocopier?

3 A company manufactures components for fridge motors. The components are designed to have a length of 135.0 millimetres.

The lengths, in millimetres, of a random sample of components manufactured on a given Monday were

135.2   135.7   134.8   135.1   136.2   135.7   136.0   135.8   135.5

a Examine whether the mean length of components manufactured on that Monday was 135.0 millimetres. Use the 5% significance level and assume that the lengths of the components are normally distributed with a standard deviation of 0.45 millimetres.

b Explain, in the context of this question, the meaning of a Type I error.

AQA SS02 June 2010

4 An ambulance service serves a large rural area. A sample of the times taken, in minutes, from receipt of an emergency call to the arrival of an ambulance at the location of the emergency was recorded as follows.

23   14   27   15   12   26   20   18   22

a Assume that these times may be regarded as a random sample from a normal distribution with standard deviation 3.5 minutes. Carry out a hypothesis test, at the 5% significance level, to examine whether the mean time could be equal to 19 minutes.

b It is now decided to examine, at the 5% significance level, whether the mean time is less than 19 minutes. State the:

i hypotheses; **ii** critical value(s); **iii** conclusion.

c There is a target of at most 19 minutes for the mean time from receipt of an emergency call to the arrival of an ambulance at the location of the emergency. The director of the ambulance service states that the data provide significant evidence that this target has been achieved. Comment on this statement.

AQA SS02 June 2009

5 In previous years, the marks obtained in a French test by students attending Topnotch College have been modelled satisfactorily by a normal distribution with a mean of 65 and a standard deviation of 9.

Teachers in the French department at Topnotch College suspect that this year their students are, on average, underachieving.

In order to investigate this suspicion, the teachers selected a random sample of 35 students to take the French test and found that their mean score was 61.5.

a Investigate, at the 5% level of significance, the teachers' suspicions.

b Explain, in the context of this question, the meaning of a Type I error.

AQA MS2B January 2006

6 A company developed a new vaccine for malaria and claimed that it is 80% effective. A professor of tropical medicine doubted that the vaccine would be as successful as this and decided to administer the vaccine to 20 people who were travelling to high-risk areas. On their return, 13 out of the 20 were symptom-free.

a Stating a necessary assumption, test, at the 10% level of significance, whether the company is overstating the effectiveness of the new vaccine.

b Would the conclusion be different if the test is carried out at the 5% level of significance?

7 The number of minor accidents at a particular factory has a Poisson distribution with mean 2.5 per week. Health and safety advice is given to all employees, following which the number of accidents in a randomly selected period of 4 weeks was 5.

a Does this provide evidence, at the 5% significance level, that the mean number of accidents per week has decreased?

b State, in the context of this question, the meaning of making a Type I error.

8 The mean age of people attending a large concert is claimed to be 35 years.

A random sample of 100 people attending the concert was taken and their mean age was found to be 37.9 years.

a Given that the standard deviation of the ages of people attending the concert is 12 years, test, at the 1% level of significance, the claim that the mean age is 35 years.

b Explain, in the context of the question, the meaning of a Type II error.

AQA MS2B June 2005

9 The management of the Wellfit gym claims that the mean cholesterol level of those members who have held membership of the gym for more than one year is 3.8.

A local doctor believes that the management's claim is too low and investigates by measuring the cholesterol levels of a random sample of 7 such members of the Wellfit gym, with the following results:

4.2   4.3   3.9   3.8   3.6   4.8   4.1

Is there evidence, at the 5% level of significance, to justify the doctor's belief that the mean cholesterol level is greater than the management's claim? State any assumption that you make.

AQA MS2A June 2008

10 The lifetime, X hours, of Everwhite camera batteries is normally distributed. The manufacturer claims that the mean lifetime of these batteries is 100 hours.

   a The members of a photography club suspect that the batteries do not last as long as it is claimed by the manufacturer. In order to investigate their suspicion, the members test a random sample of five of these batteries and find the lifetimes, in hours, to be as follows:

   85   92   100   95   99

   Test the members' suspicion at the 5% level of significance.

   b The manufacturer, believing the mean lifetime of these batteries has not changed from 100 hours, decides to determine the lifetime, x hours, of each of a random sample of 80 Everwhite camera batteries. The manufacturer obtains the following results, where $\bar{x}$ is the sample mean:

   $$\sum x = 8080 \quad \text{and} \quad \sum(x - \bar{x})^2 = 6399$$

   Test the manufacturer's belief at the 5% level of significance.

AQA MS2B June 2006

11 A consumer organisation is investigating the service offered by companies supplying household gas.

   a The waiting times, in seconds, between a telephone call connecting to a gas company, Northgas, and the caller actually speaking to one of its employees, were recorded for nine telephone calls as follows:

   76   157   62   56   193   34   89   185   134

   Test, using the 5% significance level, whether the mean waiting time for calls made to Northgas exceeds 90 seconds. Assume that this sample is a random sample from a normal distribution with standard deviation 55 seconds.

   b Another gas company, Southgas, claims that the waiting time for incoming telephone calls to its offices does not, on average, exceed 90 seconds.

   A random sample of 85 telephone calls made to Southgas had a mean waiting time of 94 seconds and a standard deviation of 12 seconds. Test the claim made by Southgas using the 5% significance level.

c Angus, a member of the consumer organisation, compared the sample means from parts **a** and **b** and also the conclusions reached. He expressed surprise.

   i Give a possible reason why Angus expressed surprise.

   ii Identify one feature of the samples which explains why, despite his surprise, the conclusions are plausible.

AQA SS02 January 2010

12 During weekday mornings, customers arrive at Hanif's newsagents at random at a constant average rate of 1 customer every 3 minutes. The amount spent per visit by each customer can be modelled by a normal distribution with mean £4.89 and standard deviation £0.92 . The amounts spent are independent from customer to customer.

a **i** Find the probability that, during a 30-minute period on a weekday morning, exactly 10 customers will arrive at Hanif's newsagents.

   ii Find the probability that the **total** amount spent by 10 customers will be less than £45.

   iii What is the probability that, between 10.45 am and 11.15 am on a given Wednesday, exactly 10 customers will arrive **and** a total of at **least** £45 will be spent by these 10 customers?

b After a sales promotion, Hanif believes that the mean amount spent is more than £5. To investigate this, he records the amount spent by a random sample of 7 customers. These amounts (in £) are as follows.

   5.42   2.50   4.98   6.72   5.09   7.15   6.14

   Investigate Hanif's belief at the 5% level of significance.

   Assume that the amount spent per visit by each customer can be modelled by a normal distribution with unknown standard deviation.

AQA SS04 June 2014

Assessment

1 National records show that 35 per cent of train passengers buy their tickets in advance. A random sample of 25 passengers using a particular railway station is selected, and it is found that 13 of them bought their tickets in advance.

a Investigate, at the 10% level of significance, whether the data support the view that the percentage of passengers from this station who buy their tickets in advance is different from the national figure of 35 per cent.   [6]

b It was suggested that, for a follow-up survey, it would be easier to collect data from all the passengers in a particular railway carriage. Explain in context why it would **not** be appropriate to then apply the test that you have used in part **a**.   [2]

AQA SS04 June 2014

2 At the Berbekan Bakery a machine is used to measure out quantities of dough for making loaves. For each batch, the mean amount of dough per loaf is set using a control on the machine. The weights of the resulting loaves are normally distributed with standard deviation 24 grams.

The control is set to produce a batch of loaves with a mean weight of 1000 grams. The weights, in grams, of a random sample of nine loaves from this batch were:

| 998 | 996 | 936 | 1002 | 957 | 968 | 920 | 943 | 1011 |

Using a 5% significance level, examine whether the mean weight of this batch of loaves is 1000 grams. [8]

AQA SS02 January 2008

3 The number of complaints received by a restaurant about the quality of service follows a Poisson distribution.

Over recent months, the mean number of complaints had been 2.2 per day. As a result, the restaurant manager organised a training session for all serving staff. During a five-day period after this training session, a total of 7 complaints were received about the quality of service.

Carry out a hypothesis test to investigate whether there has been a reduction in the mean number of complaints per day. Use an exact Poisson distribution and the 10% significance level. [6]

AQA SS04 January 2009

4 Twelve pears, selected at random from Aaron's market stall, are weighed. For this sample of pears, the mean weight, $\bar{x}$, is 137.24 grams and the standard deviation, s, is 22.79 grams.

Test Aaron's claim that the mean weight of his pears is 140 grams. Use the 5% significance level and assume that the distribution is normal. [8]

AQA SS04 June 2011

5 A company produces low-energy light bulbs. The bulbs are described as using 9 watts of power. Amir, the production manager, asked Jenny to measure the power used by each of a sample of 120 bulbs from the latest batch produced and to test the hypothesis that the mean for the batch is 9.0 watts. Jenny is to carry out the test at the 5% significance level.

a What assumption must be made about the sample of 120 bulbs if the result of this test is to be valid? [1]

b Jenny found that the mean for the sample was 9.2 watts and that the standard deviation was 1.3 watts. Carry out the test asked for by Amir. [8]

c When Jenny reported the conclusion of the test to Amir, he said that he had intended to ask Jenny to test whether the bulbs in the batch use more than 9.0 watts on average.

For the test in part **b**, state the effect that this new information would have on:

i the alternative hypothesis; **ii** the critical value(s);

iii the conclusion. [4]

AQA SS02 June 2014

6 The handicap committee of a golf club has indicated that the mean score achieved by the club's members in the past was 85.9.

A group of members believes that recent changes in the golf course have led to a change in the mean score achieved by the club's members and decides to investigate this belief.

A random sample of the scores, x, of 100 club members was taken and is summarised by

$$\sum x = 8350 \text{ and } \sum(x-\bar{x})^2 = 15321$$

where $\bar{x}$ denotes the sample mean.

Test, at the 5% level of significance, the group's belief that the mean score of 85.9 has changed.

AQA MS2A January 2007

7 A jam producer claims that the mean weight of jam in a jar is 230 grams.

  a A random sample of 8 jars is selected and the weight of jam in each jar determined. The results, in grams, are

    220   228   232   219   221   223   230   229

    Assuming that the weight of jam in a jar is normally distributed, test, at the 5% level of significance, the jam producer's claim. [9]

  b It is later discovered that the mean weight of jam in a jar is indeed 230 grams.

    Indicate whether a Type I error, a Type II error, or neither has occurred in carrying out the hypothesis test in part **a**. Give a reason for your answer. [2]

AQA MS2B June 2007

8 A machine fills paper bags with flour. Before maintenance on the machine, the weight of the flour in a bag could be modelled by a normal distribution with mean 1005 grams and standard deviation 2.1 grams. Following this maintenance, the flour in each of a random sample of 8 bags was weighed. The weights, in grams, were as follows.

    1006.1   1004.9   1005.8   1007.9   1004.7   1006.3   1007.4   1007.2

  a Carry out a test, at the 10% significance level, to decide whether the mean weight of flour in a bag filled by the machine had **changed**. Assume that the distribution of weights was still normal with standard deviation 2.1 grams. [7]

  b The flour in each of a random sample of 90 bags was then weighed. For this sample, the mean weight of flour in a bag was 1005.48 grams and the standard deviation was 2.41 grams.

    Carry out a test, at the 2% significance level, to decide whether the mean weight of flour in a bag filled by the machine had **increased** from the value of 1005 grams. [5]

  c Explain why you did not have to assume that the weight of flour in a bag was normally distributed in order to carry out the test in part **b**. [2]

  d State, with a reason, which of the tests carried out in parts **a** and **b** might have resulted in a Type II error.

AQA MS2A January 2007

Glossary

A

acceptance region values of the test statistic for which the null hypothesis is not rejected

alternative hypothesis H_1: a hypothesis which looks for an increase, decrease or change in the value of the parameter in the null hypothesis

B

binomial distribution the distribution of the number of successes in n independent trials

binomial series the series formed when $(1+x)^n$ is expanded where n is any real number

C

census a survey of every member of a population

Central Limit Theorem when large samples ($n \geq 30$ say) are taken from a non-normal population, the distribution of $\bar{X}$ is approximately normal

chain rule the formula $\dfrac{dy}{dx} = \dfrac{dy}{du} \times \dfrac{du}{dx}$ for differentiating a composite function

common logarithms logarithms to the base 10

component one of a set of vectors that are equivalent to a single (resultant) vector

composite function a function of the form $fg(x)$ where f and g are both functions of x

continuous random variable a random variable that cannot be given exactly, such as height, weight, time

cosecant the reciprocal of the sine of an angle

cotangent the reciprocal of tangent of an angle

critical region values of the test statistic for which the null hypothesis is rejected

critical value boundary of the critical region and acceptance region

cumulative distribution function a function F where $F(x) = P(X \leq x)$

D

degrees of freedom $v =$ number of variables – number of restrictions, where $v = n - 1$ in t-distribution

differential equation an equation relating a function with its derivative

discrete random variable a random variable whose value is determined by chance, taking individual values with a given probability

displacement the distance and direction of one point from another

distribution function cumulative distribution function

distribution of the sample mean distribution of the random variable $\bar{X}$, where $\bar{X} = \dfrac{1}{n} \sum X_i$, $i = 1, 2, ..., n$

domain the values for which a function is defined, i.e. the input values

E

estimate the value from the sample data that an estimator gives for a population parameter

estimator random variable used to estimate a population parameter

expectation of a function of X, $E(g(X))$ the mean or expected value of $g(X)$

expectation of X, $E(X)$ the mean μ (expected value) of the random variable X

expected value expectation, average value, mean

exponential decay the rate of decrease of a quantity x that is proportional to x

exponential distribution continuous distribution giving the time between successive events in a Poisson distribution

exponential function a function of the form a^x where a is a positive constant

exponential growth the rate of increase of a quantity x that is proportional to x

F

first order differential equation a differential equation containing the first derivative to the power 1, for example $xy + \dfrac{dy}{dx} = 0$

free vector a vector that can be in any position

function a function is an expression containing one variable where every value of the variable in the domain gives only one value of the function

function of a function a composite function

G

growth factor the constant by which a quantity grows in equal intervals of time

H

half-life the time it takes for an initial mass to decay to half its original mass

hypothesis test procedure to decide between two hypotheses (theories or claims)

I

implicit function an equation containing two variables where one variable is not isolated on one side of the equation, for example $x^2 + y^2 = 1$

improper (algebraic fraction) a fraction where the variable in the numerator has the same or higher power than the variable in the denominator

independent random variables observations of one variable are not affected by observations of any other variable

integrating by parts using the formula $\int v\dfrac{du}{dx}dx = uv - \int u\dfrac{dv}{dx}dx$ to integrate a product of functions

interquartile range difference between the quartiles, upper quartile – lower quartile

inverse function the function that maps the range of a function to its domain

iteration a process where one value is put into a function to get another value, and that value is put into the same function to get a third value, and this is repeated

L

linear combination of X and Y combination of the form $aX + bY$, where a and b are constants

lower quartile value of q_1 such that $F(q_1) = 0.25$

M

magnitude the size of a quantity, the magnitude of a vector is its length

mean of a random variable X μ, the expectation or expected value of X

median value of m such that $F(m) = 0.5$

mid-ordinate rule a formula for finding an approximate value of the area under a curve

modulus the modulus of a vector is its magnitude, the modulus of a function is its size irrespective of its sign, for example, the modulus of -2 is 2

N

Naperian logarithms logarithms to the base e

natural logarithms logarithms to the base e

no-memory rule property of an exponential distribution where there is no memory of previous events

normal curve graph of $y = f(x)$ for the normal variable

normal distribution distribution of $X \sim N(\mu, \sigma^2)$

normal distribution table Table 3 giving $P(Z < z)$, cumulative probabilities of the standard normal variable

normal variable a special continuous random variable whose distribution is specified completely by its mean μ and standard deviation σ

null hypothesis H_0, an assertion that a population parameter takes a particular value, for example $H_0: \mu = \mu_0$, where μ_0 is specified

O

one-tailed lower tail test the alternative hypothesis looks for a decrease ($H_1: \mu < \mu_0$)

one-tailed test the alternative hypothesis looks for either an increase ($H_1: \mu > \mu_0$) or a decrease ($H_1: \mu < \mu_0$)

one-tailed upper tail test the alternative hypothesis looks for an increase ($H_1: \mu > \mu_0$)

P

parameter a variable that other variables can be expressed in terms of

parametric equation an equation expressing one variable in terms of a parameter

partial fractions expressing a fraction as the sum or difference of two or more simpler fractions

percentage points of the normal distribution Table 4 giving z for particular values of p, where $P(Z < z) = p$

percentage points of the Student's t-distribution Table 5 giving t for particular values of p, where $P(T < t) = p$

Poisson distribution the distribution of the number of occurrences of an event in a given interval of space or time

population parameter defines the population for example mean μ, standard deviation σ, proportion p

position vector a vector from a fixed point (the origin) to another point

probability density function a function f where $f(x)$ defines a continuous random variable

proper (algebraic fraction) a fraction where the variable in the numerator has a lower power than the variable in the denominator

R

random sample every member of the population has an equal chance of being selected and all possible samples of size n have an equal chance of being selected

range the range of a function is the set of values it can take

rational function a fraction whose numerator and denominator are polynomials

real numbers numbers that can be represented as points on a number line

rejection region critical region

resultant a single vector equivalent to a set of component vectors

S

sample a selection taken from a population

sample mean $\bar{X}$ the random variable $\bar{X} = \dfrac{1}{n}\sum X_i$, where $i = 1, 2, ..., n$;

sample mean $\bar{x}$ the mean of a set of sample data

sample statistic random variable formed from independent observations of X

scalar a quantity that is fully defined by size

secant the reciprocal of the cosine of an angle

significance level probability used to define the critical region in a hypothesis test

Simpson's rule a formula for finding an approximate area under a curve

skew (lines) lines that do not intersect and are not parallel

standard deviation of X σ, a measure of the spread of the random variable X about μ

standard error of the mean standard deviation of the distribution of the sample mean

standardise linearly scale a normal variable by subtracting the mean then dividing by the standard deviation

standard normal variable normal variable Z with mean 0 and variance 1 where $Z \sim N(0, 1)$

statistical mode calculator mode giving statistical values such as mean and standard deviation directly

Student's t-distribution another name for the t-distribution

summary data a summarised form of data, such as Σx or Σx^2, rather than individual observations

T

t-distribution a family of distributions similar to the standard normal distribution but with greater variability

t-test hypothesis test involving a t-statistic

test statistic random variable used in a hypothesis test, such as $\bar{X}$ or T

test value the value of the test statistic obtained from the sample

triangle law the law that states when two vectors are represented by two sides of a triangle in the same sense, their resultant is represented by the third side of the triangle in the opposite sense

two tailed test looks for a change in the null hypothesis where the alternative hypothesis is $H_1 : \mu \neq \mu_0$

Type I error error made when a true hypothesis is rejected

Type II error error made when a false hypothesis is accepted

U

unbiased estimate of population mean μ the value of the mean calculated from the sample where $\bar{x} = \dfrac{1}{n} \Sigma X_i$

unbiased estimate of population variance σ the value of s^2 calculated from sample values, where

$$s^2 = \frac{\Sigma(x_i - \bar{x})^2}{n-1} = \frac{1}{n-1}\left(\Sigma x_i^2 - \frac{\left(\Sigma x_i\right)^2}{n} \right)$$

unbiased estimator of a population parameter the expected value of the estimator is equal to the true value of the parameter

unit interval the interval defined in a Poisson distribution

unit vector a vector whose magnitude is one unit

upper quartile value of q_3 such that $F(q_3) = 0.75$

V

variance of X $\sigma^2 = \text{Var}(X)$, the expectation of $(X - \mu)^2$

vector a quantity that is defined by its magnitude and direction

vertical line diagram diagram illustrating the probabilities of a discrete random variable

volume of revolution the volume formed when the area between a curve and one axis is rotated about that axis

Z

z-test hypothesis test involving a z-statistic

Exercise 8

1 $4\cos 4x$

2 $2\sin(\pi - 2x)$

3 $\dfrac{1}{2}\cos\left(\dfrac{1}{2}x + \pi\right)$

4 $-2\cos x \sin x$

5 $(\cos x)e^{\sin x}$

6 $-\dfrac{\sin x}{\cos x} = -\tan x$

7 $2x\cos x^2$

8 $-(\sin x)e^{\cos x}$

9 $\dfrac{\cos x}{\sin x} = \cot x$

10 $-4\sin x\cos^3 x$

11 $(2x-2)e^{(x^2-2x)}$

12 $6\sec^2 x$

13 $\dfrac{4x-3}{2x^2-3x}$

14 $5\cos(5x-8)$

Exercise 9

1 $2x + 2y\dfrac{dy}{dx} = 0$

2 $2x + y + (x+2y)\dfrac{dy}{dx} = 0$

3 $2x + x\dfrac{dy}{dx} + y = 2y\dfrac{dy}{dx}$

4 $-\dfrac{1}{x^2} - \dfrac{1}{y^2}\dfrac{dy}{dx} = e^y\dfrac{dy}{dx}$

5 $-\dfrac{2}{x^3} - \dfrac{2}{y^3}\dfrac{dy}{dx} = 0$

6 $\dfrac{x}{2} - \dfrac{2y}{9}\dfrac{dy}{dx} = 0$

7 $\cos x + \cos y\dfrac{dy}{dx} = 0$

8 $\cos x\cos y - \sin x\sin y\dfrac{dy}{dx} = 0$

9 $e^y + xe^y\dfrac{dy}{dx} = 1$

10 $\dfrac{dy}{dx} = \pm\dfrac{1}{\sqrt{2x+1}}$

11 $\pm\dfrac{1}{4}\sqrt{2}$

12 $\dfrac{dy}{dx} = \dfrac{1}{1+x^2}$

13 a $8y + 4x\sqrt{5} + 4 = 0, 8y = 4x\sqrt{5} - 4$

b $y(2 + 3y_1) = xx_1 + 3y_1^2 + 2y_1 - x_1^2$

14 $y = 2 - x$ and $y = -2$

15 $3x + 12y - 7 = 0$

Exercise 10

1 a $x = 2y^2$

b $x^2 + y^2 = 1$

c $xy = 4$

2 a

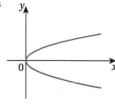

b

c

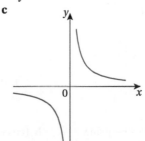

3 $y = \dfrac{x^2}{1+x}$

5 $y = x(x^2 - 1)$

6 $\dfrac{x^2}{4} + \dfrac{y^2}{9} = 1$

7 b centre $(0, 0)$, radius 4

Exercise 11

1 a $\dfrac{1}{4t}$

b $-\cot\theta$

c $-\dfrac{4}{t^2}$

2 a $\dfrac{dy}{dx} = 2t - t^2$

b $\dfrac{3}{4}$

3 $\dfrac{3t}{2}$

4 $\left(-\dfrac{1}{3}\sqrt{3}, \dfrac{2}{9}\sqrt{3}\right)$, max; $\left(\dfrac{1}{3}\sqrt{3}, \dfrac{-2}{9}\sqrt{3}\right)$, min

5 $\dfrac{1}{2}\pi$

6 $2x + y + 2 = 0$

7 a $6y = 4x + 5\sqrt{2}$

b $\left(-\dfrac{137\sqrt{2}}{97}, -\dfrac{21\sqrt{2}}{194}\right)$

8 a $y - t = \dfrac{1}{4t}(x - 2t^2),\ y - \dfrac{4}{t} = -\dfrac{4}{t^2}(x - t)$

b $y - t = -4t(x - 2t^2);\ y - \dfrac{4}{t} = \dfrac{t^2}{4}(x - t)$

9 a $y - \dfrac{2}{s} = s^2(x - 2s)$

b $\left(-\dfrac{2}{s^3}, -2s^3\right)$

10 a $t^2 y + x = 2t$

b $\left(0, \dfrac{2}{t}\right); (2t, 0)$

11 a $2y + tx = 8t + t$

b $\left(0, \dfrac{t}{2}(8 + t^2)\right); (8 + t^{2,0})$

Review

1 a $-4\cos 4\theta$ **b** $1+\sin\theta$ **c** $3\sin^2\theta\cos\theta+3\cos 3\theta$

2 a $3x^2+e^x$ **b** $2e^{(2x+3)}$ **c** $e^x(\sin x+\cos x)$

3 a $-\dfrac{3}{x}$ **b** $-\dfrac{2}{x}$ **c** $\dfrac{1}{2x}$

4 a $3\cos x+e^{-x}$ **b** $\dfrac{1}{2x}+\dfrac{1}{2}\sin x$ **c** $4x^3+4e^x-\dfrac{1}{x}$ **d** $-\dfrac{1}{2}\left(e^{-x}+x^{\frac{-3}{2}}\right)-\dfrac{1}{x}$

5 $\ln x+\dfrac{x+1}{x}$ **6** $6\sin 3x\cos 3x$ **7** $\dfrac{8}{3}(4x-1)^{\frac{-1}{3}}$ **8** $\left(3\sqrt{x}-2x\right)\left(\dfrac{3}{\sqrt{x}}-4\right)$

9 $\dfrac{(x^4+4x^3+3)}{(x+1)^4}$ **10** $\dfrac{(x-1)\ln(x-1)-x\ln x}{x(x-1)\{\ln(x-1)\}^2}$ **11** $\dfrac{-1}{\sin x\cos x}$ **12** $2x\tan x+x^2\sec^2 x$

13 $e^x\dfrac{(x-2)}{(x-1)^2}$ **14** $\dfrac{2\cos x}{(1-\sin x)^2}$ **15** $\dfrac{x(5x-4)}{2\sqrt{x-1}}$ **16** $-2(1-x)^2(2x+1)$

17 $\dfrac{3}{2(x+3)}-\dfrac{x}{(x^2+2)}$ **18** $\cos^2 x(4\cos^2 x-3)$ **19** $-e^{\cos^2 x}\sin 2x$

20 a $x=\ln 3$ **b** $x=1\,(not-1)$

21 a 1 **b** $y-x=1-\dfrac{1}{2}\pi$ **c** $y+x=1+\dfrac{1}{2}\pi$

22 a $1+e$ **b** $y=x(1+e)$ **c** $y(1+e)+x=(1+e)^2+1$

23 a 2 **b** $y=2x+1$ **c** $2y+x=2$

24 a -1 **b** $y=\dfrac{\pi}{2}-x$ **c** $y=\dfrac{\pi}{2}+x$

25 a for example $\left(\dfrac{1}{6}\pi,\left(\dfrac{1}{2}\pi-\sqrt{3}\right)\right)$ **b** $(1,-1)$

26 a $4y^3\dfrac{dy}{dx}$ **b** $y^2+2xy\dfrac{dy}{dx}$ **c** $-\dfrac{1}{y^2}\dfrac{dy}{dx}$ **d** $\ln y+\dfrac{x}{y}\dfrac{dy}{dx}$ **e** $\cos y\dfrac{dy}{dx}$

f $e^y\dfrac{dy}{dx}$ **g** $\dfrac{dy}{dx}\cos x-y\sin x$ **h** $(\cos y-y\sin y)\dfrac{dy}{dx}$ **27** $\dfrac{x}{2y}$

28 $-\dfrac{y^2}{x^2}$ **29** $-\dfrac{2y}{3x}$ **30** $-\dfrac{y(y+1)(3x+2)}{x(x+1)(y+2)}$ **31** $\dfrac{3t}{2}$

32 $\dfrac{t}{t+1}$ **33** $-\dfrac{3}{2}\cos\theta$ **34** $-\dfrac{1}{t^2}$ **35** $-\dfrac{1}{e^t}$

36 $2t-t^2$ **37** $y=2x+2\sqrt{2}$

Assessment

1 a $\dfrac{dy}{dx}=\dfrac{1}{x}-2x$ **b** $x+y=3$ **c** $y=x+1$

2 a $\dfrac{dy}{dx}=\dfrac{1}{6y-2}$ **b** $x-4y+3=0$ **c** $4x+y=5$

3 a $\dfrac{dy}{dx}=-4x$ **b** $(0,1)$

5 a $2y\dfrac{dy}{dx}-2x\dfrac{dy}{dx}-2y+3\dfrac{dy}{dx}=2$ **b** $(1,1),(1,-2)$ **c** $3y-4x+1=0$ and $3y+2-8=0$

6 a $y=a^2x-a^3+\dfrac{1}{a}$ **b** $\dfrac{1}{b}=a^2b-a^3+\dfrac{1}{a}\Rightarrow b=-\dfrac{1}{a^3}$

7 $2y\sin\theta+3x\cos\theta=6$ **8** $y\cos\theta-x\sin\theta=\cos\theta$ **9** $\left(\dfrac{1}{4},1+\ln 4\right)$ **10 a** $(0,27)$

12 a $3e^x + \dfrac{1}{x}$ **13 a** $-\dfrac{1}{2}$ **b** $y = 2x - \dfrac{1}{2}$ **c** $-\dfrac{1}{8}$

14 $\left(\dfrac{1}{3}, 1\right)$ and $\left(-\dfrac{1}{3}, -1\right)$

6 Integration

Exercise 1

1 $\dfrac{1}{4}e^{4x} + k$ **2** $-4e^{-x} + k$ **3** $\dfrac{1}{3}e^{(3x-2)} + k$ **4** $-\dfrac{2}{5}e^{(1-5x)} + k$ **5** $-3e^{-2x} + k$ **6** $5e^{(x-3)} + k$

7 $2e^{\left(\frac{x}{2}+2\right)} + k$ **8** $e^{2+x} + k$ **9** $\dfrac{1}{2}e^{2x} - \dfrac{1}{2e^{2x}} + k$ **10** $\dfrac{1}{2}\{e^4 - 1\}$ **11** $2\{e^2 - 1\}$ **12** $1 - \dfrac{1}{e}$

13 $-1 - e^{-2}$

Exercise 2

1 $2\ln x + k$ **2** $\dfrac{1}{4}\ln x + k$ **3** $\left(\dfrac{3}{2}\right)\ln x + k$ **4** $x + \ln x + k$ **5** $\dfrac{1}{2}x^2 + x - \ln x + k$

6 $e^x + 2\ln|x| + k$ **7** $\dfrac{1}{3}\ln 2$ **8** $2 - \ln 3$ **9** $\ln 2 - 1$ **10** $e^3 - e^2 + \ln\left(\dfrac{2}{3}\right)$

11 $\dfrac{1}{3}\left(2\ln\left(\dfrac{5}{4}\right) - 1\right)$ **12** $\ln\dfrac{3}{2} + \dfrac{38}{3}$

Exercise 3

1 $\dfrac{1}{8}(2x-3)^4 + k$ **2** $\dfrac{1}{15}(3x+1)^5 + k$ **3** $\dfrac{1}{25}(5x-2)^5 + k$ **4** $(2-x)^{-1} + k$ **5** $-(x+3)^{-1} + k$

6 $\dfrac{2}{3}(1+x)^{\frac{3}{2}} + k$ **7** $\dfrac{1}{18}(3x+1)^6 + k$ **8** $\dfrac{-1}{25}(2-5x)^5 + k$

Exercise 4

1 $-\dfrac{1}{2}\cos 2x + k$ **2** $\dfrac{1}{7}\sin 7x + k$ **3** $\dfrac{1}{4}\tan 4x + k$ **4** $-\cos\left(\dfrac{1}{4}\pi + x\right) + k$

5 $\dfrac{3}{4}\sin\left(4x - \dfrac{1}{2}\pi\right) + k$ **6** $\dfrac{1}{2}\tan\left(\dfrac{1}{3}\pi + 2x\right) + k$ **7** $-\dfrac{2}{3}\cos(3x - \alpha) + k$ **8** $-10\sin\left(\alpha - \dfrac{1}{2}x\right) + k$

9 $\dfrac{1}{3}\sin 3x - \sin x + k$ **10** $\dfrac{1}{2}\tan 2x + k$ **11** $\dfrac{1}{3}$ **12** $-\dfrac{1}{4}$

13 0 **14** $\dfrac{1}{2}$ **15** $\dfrac{1}{2}\cos\left(\dfrac{1}{2}\pi - 2x\right) + k$ **16** $\dfrac{1}{4}e^{(4x-1)} + k$

17 $\dfrac{1}{7}\tan 7x + k$ **18** $\dfrac{1}{2}\ln|2x-3| + k$ **19** $\sqrt{2x-3} + k$ **20** $\dfrac{-1}{3(3x-2)} + k$

21 $\dfrac{1}{5}e^{5x} + k$ **22** $1 - \ln|x| + k$ **23** $\dfrac{1}{9}(3x-5)^3 + k$ **24** $\dfrac{1}{4}e^{(4x-5)} + k$

25 $\dfrac{1}{6}(4x-5)^{\frac{3}{2}} + k$ **26** $-\dfrac{1}{5}\cos\left(5x - \dfrac{1}{3}\pi\right) + k$ **27** $-\dfrac{3}{2}\ln|1-x| + k$ **28** $-\dfrac{4}{3}(x-6)^{-3} + k$

29 $\dfrac{1}{3}\sin\left(3x - \dfrac{1}{3}\pi\right) + k$ **30** $\dfrac{2}{3}x^3 - 8x + k$ **31** $\dfrac{1}{4}x^4 - 2x^3 + \dfrac{9}{2}x^2 + k$ **32** $3x - x^2 + 2x^3 + k$

33 $-\dfrac{1}{3x^2} + k$ **34** $\dfrac{3}{2}\ln|x| + k$ **35** $\dfrac{2}{9}(x-3)^3 + k$

Exercise 5

1 $e^{x+\sin x}+k$

2 $-e^{\cos x}+k$

3 $e^{\tan x}+k$

4 $e^{x^2+x}+k$

5 $e^{(1-\cot x)}+k$

6 $e^{(x+\sin x)}+k$

7 $e^{(1+x^2)}+k$

8 $e^{(x^3-2x)}+k$

9 $\dfrac{1}{10}(x^2-3)^5+k$

10 $-\dfrac{1}{3}(1-x^2)^{\frac{3}{2}}+k$

11 $\dfrac{1}{6}(\sin 2x+3)^3+k$

12 $-\dfrac{1}{6}(1-x^3)^2+k$

13 $\dfrac{2}{3}(1+e^x)^{\frac{3}{2}}+k$

14 $\dfrac{1}{5}\sin^5 x+k$

15 $\dfrac{1}{4}\tan^4 x+k$

16 $\dfrac{1}{3(n+1)}(1+x^{n+1})^3+k$

17 $\dfrac{3}{5}(1-\cos x)^5+k$

18 $\dfrac{4}{9}\left(1+x^{\frac{3}{2}}\right)^{\frac{3}{2}}+k$

19 $\dfrac{1}{12}(x^4+4)^3+k$

20 $-\dfrac{1}{4}(1-e^x)^4+k$

21 $\dfrac{2}{3}(1-\cos\theta)^{\frac{3}{2}}+k$

22 $\dfrac{1}{3}(x^2+2x+3)^{\frac{3}{2}}+k$

23 $\dfrac{1}{2}e^{(x^2+1)}+k$

24 $\dfrac{1}{2}(1+\tan x)^2+k$

Exercise 6

1 $x\sin x+\cos x+k$

2 xe^x-e^x+k

3 $x^2\ln\dfrac{(3x)}{2}-\dfrac{x^2}{4}+k$

4 $-e^{-x}(x+1)+k$

5 $3(\sin x-x\cos x)+k$

6 $\dfrac{1}{4}\sin 2x-\dfrac{1}{2}x\cos 2x+k$

7 $\dfrac{1}{2}xe^{2x}-\dfrac{1}{4}e^{2x}+k$

8 $\dfrac{1}{32}e^{4x}(8x^2-4x+1)+k$

9 $\sin x-x\cos x+k$

10 $x(\ln|2x|-1)+k$

11 xe^x+k

12 $\dfrac{1}{72}(8x-1)(x+1)^8+k$

13 $\sin\left(x+\dfrac{1}{6}\pi\right)-x\cos\left(x+\dfrac{1}{6}\pi\right)+k$

14 $\dfrac{1}{n^2}(\cos nx+nx\sin nx)+k$

15 $\dfrac{x^2}{4}(2\ln x-1)+k$

16 $\dfrac{3}{4}(2x\sin 2x+\cos 2x)+k$

17 $\sin x-\dfrac{1}{3}\sin^3 x+k$

18 $\dfrac{1}{2}e^{x^2-2x+4}+k$

19 $(x^2+1)e^x+k$

20 $-\dfrac{1}{4}(4+\cos x)^4+k$

21 $e^{\sin x}+k$

22 $\dfrac{2}{15}\sqrt{(1+x^5)^3}+k$

23 $\dfrac{1}{5}(e^x+2)^5+k$

24 $\dfrac{1}{4}e^{2x-1}(2x-3)+k$

25 $-\dfrac{1}{20}(1-x^2)^{10}+k$

26 $\dfrac{1}{6}\sin^6 x+k$

Exercise 7

1 $\ln(4+\sin x)+k$

2 $\dfrac{1}{3}\ln|3e^x-1|+k$

3 $\dfrac{1}{4(1-x^2)^2}+k$

4 $\dfrac{1}{2\cos^2 x}+k$

5 $\dfrac{1}{4}\ln(1+x^4)+k$

6 $\ln|x^2+3x-4|+k$

7 $\dfrac{2}{3}\sqrt{2+x^3}+k$

8 $\dfrac{1}{2-\sin x}+k$

9 $\ln|\ln x|+k$

10 $\dfrac{-1}{5\sin^5 x}+k$

11 $-\ln|1-x^2|+k$

12 $-2\sqrt{1-e^x}+k$

13 $\dfrac{1}{6}\ln|3x^2-6x+1|+k$

14 $\dfrac{-1}{(n-1)\sin^{n-1} x}+k\ (n\neq 1)$

15 $\dfrac{1}{(n-1)\cos^{(n-1)} x}+k\ (n\neq 1)$

16 $-\ln(4+\cos x)+k$

17 $\dfrac{1}{2}\ln|x(x-2)|+k$

18 $-\dfrac{1}{e^x-x}+k$

19 $\ln 3$

20 $\ln\sqrt{2}$

21 $\dfrac{55}{1152}$

22 $\dfrac{e-1}{2(e+1)}$

23 0

24 $\ln 2$

Exercise 8

1 $2\ln\left|\dfrac{x}{x+1}\right|+k$

2 $\ln\left|\dfrac{x-2}{x+2}\right|+k$

3 $\dfrac{1}{2}\ln|x^2-1|+k$

4 $\dfrac{1}{2}\ln\left|\dfrac{(x+2)^3}{x}\right|+k$

5 $\ln\dfrac{(x-3)^2}{|x-2|}+k$

6 $\dfrac{1}{2}\ln\dfrac{|x^2-1|}{x^2}+k$

7 $x-\ln|x+1|+k$

8 $x+4\ln|x|+k$

9 $x-4\ln|x+4|+k$

10 $\ln\dfrac{|1-x|}{x^4}+k$

11 $x-\dfrac{1}{2}\ln\dfrac{|x+1|}{|x-1|}+k$

12 $x+\ln\dfrac{|x+1|}{(x+2)^4}+k$

13 $\dfrac{1}{2}\ln|x^2-1|+k$

14 $\dfrac{-1}{x^2-1}+k$

15 $\ln\left|\dfrac{x-1}{x+1}\right|+k$

16 $\ln|x^2-5x+6|+k$

17 $\ln\dfrac{(x-3)^6}{(x-2)^4}+k$

18 $\ln\left|\dfrac{(x-3)^3}{x-2}\right|+k$

19 $4+\ln 5$

20 $\ln\dfrac{1}{6}$

21 $\dfrac{1}{2}\ln\dfrac{12}{5}$

22 $\ln\dfrac{5}{3}$

23 $\dfrac{5}{36}$

24 $1-\dfrac{3}{2}\ln\dfrac{7}{5}$

Exercise 9

1 $\dfrac{1}{4}(2x+\sin 2x)+k$

2 $\sin x-\dfrac{1}{3}\sin^3 x+k$

3 $-\dfrac{1}{15}\cos x\,(15-10\cos^2 x+3\cos^4 x)+k$

4 $\tan x-x+k$

5 $\dfrac{1}{32}\{12x-8\sin 2x+\sin 4x\}+k$

6 $\dfrac{1}{2}\tan^2 x-\ln|\sec x|+k$

7 $\dfrac{1}{32}\{12x+8\sin 2x+\sin 4x\}+k$

8 $\dfrac{1}{3}\cos x\,(\cos^2 x-3)+k$

Exercise 10

1 $\dfrac{512}{15}\pi$

2 $\dfrac{1}{2}\pi\,(e^6-1)$

3 $\dfrac{1}{2}\pi$

4 $\dfrac{64}{5}\pi$

5 $\left(\dfrac{\pi}{2}\right)^2$

6 8π

7 8π

8 $\dfrac{3}{5}\pi\left(\sqrt[3]{32}-1\right)$

9 $\dfrac{1}{2}\pi\,(e^2-1)$

Review

1 $-3e^{-x}+k$

2 $e^{x^2}+k$

3 $\dfrac{1}{6}(3\tan x-4)^2+k$

4 $\dfrac{4}{3}\ln|x|+k$

5 $\dfrac{1}{4}\tan^4 x+k$

6 $\dfrac{1}{18}(3x+4)^6+k$

7 $-e^{\cos x}+k$

8 $\dfrac{1}{4}\sin\left(4x+\dfrac{\pi}{7}\right)+k$

9 $\dfrac{1}{2}e^{\left(x^2-2x+3\right)}+k$

10 $\dfrac{-1}{30}(1-x^3)^{10}+k$

11 $\dfrac{1}{3}(2x+1)^{\frac{3}{2}}+k$

12 $\dfrac{1}{3}(x^2+1)^{\frac{3}{2}}+k$

13 $\dfrac{-1}{3}(2-\sin x)^3+k$

14 $\dfrac{1}{2}e^{2x}\left(x^2-x+1\right)+k$

15 $\dfrac{1}{2}(x+1)^2\ln(x+1)\dfrac{-1}{4}(x+1)^2+k$

16 $x^2\sin x+2x\cos x-2\sin x+k$

17 $\dfrac{1}{2}e^{2x+3}+k$

18 $\dfrac{1}{6}(2x^2-5)^{\frac{3}{2}}+k$

19 xe^x-e^x+k

20 $x\ln x-x+k$

21 $\dfrac{1}{2}x-\dfrac{1}{12}\sin 6x+k$

22 $-\dfrac{1}{2}e-x^2+k$

23 $\dfrac{1}{3}\sin^3 x+k$

24 $\dfrac{10u-7}{110}(u+7)^{10}+k$

25 $-\dfrac{1}{12}(x^3+9)^{-4}+k$

26 $\dfrac{1}{2}\ln(1-\cos 2y)+k$

27 $\dfrac{1}{2}\ln|2x+7|+k$ **28** $-\dfrac{2}{9}(1+\cos 3x)^{\frac{3}{2}}+k$ **29** $\dfrac{1}{2}\ln|x^2+4x-5|+k$ **30** $\dfrac{2}{3}\ln|x+5|+\dfrac{1}{3}\ln|x-1|+k$

31 $\dfrac{1}{3}x\sin 3x+\dfrac{1}{9}\cos 3x+k$ **32** $x\ln|5x|-x+k$ **33** $\dfrac{3}{8}\pi^2$

Assessment

1 a $-\dfrac{1}{2}e^{(1-x)^2}+k$ **b** $\ln(x^2+1)+k$ **c** $\dfrac{1}{5}\sin^5 x+k$

2 a $-\dfrac{1}{4}(4+\cos x)^4+k$ **b** $\dfrac{1}{16}(2x+3)^8\left(\dfrac{8}{9}x-\dfrac{1}{6}\right)+k$ **c** $-\dfrac{1}{9}(3x+1)e^{(2-3x)}+k$ **3** $3(x^2-3)^{\frac{1}{2}}+k$

4 a $\dfrac{1}{2(x-3)}-\dfrac{1}{2(x+1)}$ **b** $\dfrac{1}{2}\ln\left(\dfrac{x-3}{x+1}\right)+k$ **5 a** $-(9-y^2)^{\frac{3}{2}}+k$ **b** $-\dfrac{\pi}{8}$

6 $\dfrac{33\pi}{5}$ **7** $\left(\dfrac{\pi}{2}\right)(e^2-4e+5)$

8 a $A=2,\ B=3$ **b** $y=x^2+3\ln(2x^2-x+2)+1-3\ln 5$

9 a $-\dfrac{1}{2}x^2\cos 2x+\dfrac{1}{2}x\sin 2x+\dfrac{1}{4}\cos 2x+c$ **b** $\pi\left(\dfrac{\pi^2}{8}-\dfrac{1}{2}\right)$

7 Differential Equations

Exercise 1

1 $y^2=A-2\cos x$ **2** $\dfrac{1}{y}-\dfrac{1}{x}=A$ **3** $2y^3=3(x^2+4y+A)$ **4** $x=A\sec y$

5 $(A-x)y=1$ **6** $y=\ln\dfrac{A}{\sqrt{1-x^2}}$ **7** $y=A(x-3)$ **8** $x+A=4\ln|\sin y|$

9 $u^2=v^2+4v+A$ **10** $16y^3=12x^4\ln|x|-3x^4+A$ **11** $y^2+2(x+1)e^{-x}=A$ **12** $\sin x=A-e^{-y}$

13 $2r^2=2\theta-\sin 2\theta+A$ **14** $u+2=A(v+1)$ **15** $y^2=A+(\ln|x|)^2$ **16** $y^2=Ax(x+2)$

17 $4v^3=3(2+t)^4+A$ **18** $1+y^2=Ax^2$ **19** $Ar=e^{\tan\theta}$ **20** $y^2=A-\csc^2 x$

21 $v^2+A=2u-2\ln|u|$ **22** $e^{-x}=e^{1-y}+A$ **23** $A-\dfrac{1}{y}=2\ln|\tan x|$ **24** $y-1=A(y+1)(x^2+1)$

Exercise 2

1 $y^3=x^3+3x-13$ **2** $e^t(5-2\sqrt{s})=1$ **3** $3(y^2-1)=8(x^2-1)$ **4** $y=e^x-2$

5 $y^2=-e^{-3x}+e^{-3}+4$ **6** $(y+1)^2(x+1)=2(x-1)$ **7** $4y^2=(y+1)^2(x^2+1)$

Exercise 3

1 $s\dfrac{ds}{dt}=k$ **2** $\dfrac{dh}{dt}=k\ln(H-h)$ **3** $\dfrac{dh}{dt}=kV$ **4** $\dfrac{dm}{dt}=-km$

5 $\dfrac{dh}{dt}=-k(H-h)$ **6** $\dfrac{dd}{dt}=\dfrac{k}{d}$ **7 a** $\dfrac{dn}{dt}=0$ **b** $\dfrac{dn}{dt}=k\sqrt{n}$

8 a $\dfrac{dn}{dt}=k_1 n$ **b** $\dfrac{dn}{dt}=\dfrac{k_2}{n}$ **c** $\dfrac{dn}{dt}=-k_3$

Exercise 4

1 a $\dfrac{dh}{dt}=\dfrac{k}{h^3}$ **b** $h^4=5t+1$ **c** 16 minutes **2 a** $\dfrac{dy}{dx}=k\sqrt{x}$

b $y=0.4x^{\frac{3}{2}}+1.6$ **c** 1.2 **3 a** $\dfrac{dn}{dt}=kn$ **b** 17.1 hours (3 sf)

4 a $\dfrac{dm}{dt}=-km$ **b** $m=1000e^{-\frac{t\ln 2}{500}}$ **c** 1000 years

5 a $-\dfrac{dm}{dt}=km;\ m=50e^{-kt}$ where $k=0.002\,554\ldots$ **b** 26.8 g (3 sf)

Review

1 $\dfrac{dr}{dt} = kr^2$

2 $x^3 y = y - 1$

3 $y = \tan\left\{\dfrac{1}{2}\left(x^2 - 4\right)\right\}$

4 a $y^2 = 2x$

5 a $\dfrac{dN}{dt} = kN$

b 33

c E.g. there were not 200 rabbits on the island

6 a $y = 100e^{-0.105x}$

b E.g. the rate of inflation is very unlikely to remain constant.

Assessment

1 $y^2 = 4x^2 + 5$

2 $\sin y = -1 - \cos x$

3 a $\dfrac{dy}{dx} = \dfrac{k}{xy}$

b $y^2 = 100 - \dfrac{48\ln|x|}{\ln 2}$

4 c $k = 0.0357$ (3.s.f.)

d 44°C (nearest degree)

5 About 4 to $4\dfrac{1}{2}$ hours before discovery assuming Newton's Law of Cooling

6 $y = \dfrac{9}{3x\cos x - \sin 3x + 10}$

7 a $x = \dfrac{1}{20}\left(\dfrac{4+5t}{1+t}\right)^2 - \dfrac{4}{5}$

b i $\dfrac{dr}{dt} = \dfrac{k}{r^2}$

ii 3 m

8 Numerical Methods

Exercise 1

9 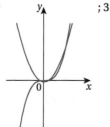 ; 3

10  ; none

11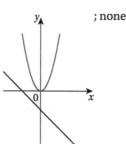

13 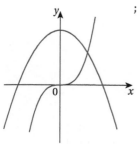 ; 1 and 2

15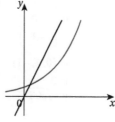

Exercise 2

1 b $x = \dfrac{1}{10}(2 + x^2 - x^3)$ **c** yes

2 b $x = \dfrac{1}{9}(3x^3 - 2x^2 + 2)$ **c** yes

3 b $x = \dfrac{1}{6}(1 - 2x^3 - x^2)$ **c** yes

4 a $x = \dfrac{1}{8}(8 - x^2)$ **c** yes

5 b 0.71, 0.61

6 b 0.97, 0.92

7 b 0.198, 0.194

c 2.6786, 2.608..., 2.389..., 2.389..., 1.787... The values are moving away from the larger root. (They converge to the smaller root.)

Exercise 3

1 a 21   **b** 21.3   **2 a** 0.648   **b** 0.671   **3 a** 1.79   **b** 1.62

4 a 1.32   **b** 1.29   **5 a** 5.58   **b** 5.54

2 c 2.20 (3 sf)   **3** 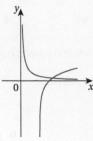   ; 3 and 4   **4** 9360 (3 sf)   **5** −2.60 (3 sf)   **6** 1.81 (3 sf)

Assessment

1 a 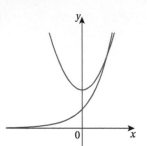   **d** 1.24, 1.26   **3 a** 79.8 (3 sf)   **b** 80.5 (3 sf)

4 a 2.00   **b** $\dfrac{1}{2}(x^2+1)\ln(x^2+1)-1+k$ **c** $\dfrac{1}{2}(5\ln 5-4)=2.02$

5 2.449   **6** 2.541   **7 c** $x_2=3.578,\ x_3=3.568$

9 Vectors

Exercise 1

1 a $\dfrac{1}{2}(\mathbf{a}+\mathbf{c})$ **b** $\dfrac{1}{2}(\mathbf{b}+\mathbf{d})$ **2 a** $\dfrac{1}{4}(\mathbf{a}+2\mathbf{b}+\mathbf{c})$ **b** $\dfrac{1}{4}(\mathbf{b}+2\mathbf{c}+\mathbf{d})$

Exercise 2

1 a $\begin{bmatrix} 3 \\ 6 \\ 4 \end{bmatrix}$ **b** $\begin{bmatrix} 1 \\ -2 \\ -7 \end{bmatrix}$ **c** $\begin{bmatrix} 1 \\ 0 \\ -3 \end{bmatrix}$

2 a $(5,-7,2)$ **b** $(1,4,0)$ **c** $(0,1,-1)$
3 a $\sqrt{21}$ **b** 5   **c** 3
4 a 6   **b** 7   **c** $\sqrt{206}$

5 a $\begin{bmatrix} 3 \\ 0 \\ 4 \end{bmatrix}$ **b** $\begin{bmatrix} 2 \\ -2 \\ 2 \end{bmatrix}$ **c** $\begin{bmatrix} 2 \\ 3 \\ 3 \end{bmatrix}$ **d** $\begin{bmatrix} -6 \\ 12 \\ -8 \end{bmatrix}$

6 $\lambda=\dfrac{1}{2}$ **7 b, e** and **f**

8 a neither   **b** parallel   **c** equal

9 a $\begin{bmatrix} -2 \\ -3 \\ 7 \end{bmatrix}$ **b** $\begin{bmatrix} 0 \\ -1 \\ 3 \end{bmatrix}$ **c** $\begin{bmatrix} -2 \\ -2 \\ 4 \end{bmatrix}$

10 $\sqrt{62}, \sqrt{10}, 2\sqrt{6}$ **11** $\sqrt{5}$ **12** $\overrightarrow{AB} = \begin{bmatrix} -1 \\ 3 \\ 0 \end{bmatrix}, \overrightarrow{BD} = \begin{bmatrix} -0 \\ -1 \\ 4 \end{bmatrix} \overrightarrow{CD} = \begin{bmatrix} -2 \\ 2 \\ -6 \end{bmatrix} \overrightarrow{AD} = \begin{bmatrix} -1 \\ 2 \\ -4 \end{bmatrix}$

Exercise 3

1 $\sqrt{13}; 3$ **2 a** no **b** no **c** no

Exercise 4

1 a $\begin{bmatrix} 2 \\ -1 \\ -5 \end{bmatrix}$ **b** $\begin{bmatrix} 0 \\ 3 \\ -5 \end{bmatrix}$ **c** $\begin{bmatrix} -2 \\ 4 \\ -1 \end{bmatrix}$

2 a $\mathbf{r} = \begin{bmatrix} 1 \\ -3 \\ 2 \end{bmatrix} + t \begin{bmatrix} 5 \\ 4 \\ -1 \end{bmatrix}$ **b** $\mathbf{r} = \begin{bmatrix} 2 \\ 1 \\ 0 \end{bmatrix} + t \begin{bmatrix} 0 \\ 3 \\ -1 \end{bmatrix}$ **c** $\mathbf{r} = t \begin{bmatrix} 1 \\ -1 \\ -1 \end{bmatrix}$

3 $\mathbf{r} = \begin{bmatrix} 4 \\ 5 \\ 10 \end{bmatrix} + s \begin{bmatrix} 1 \\ 1 \\ 3 \end{bmatrix}$ $\mathbf{r} = \begin{bmatrix} 2 \\ 3 \\ 4 \end{bmatrix} + s \begin{bmatrix} 1 \\ 1 \\ 5 \end{bmatrix}$ $\mathbf{r} = \begin{bmatrix} 4 \\ 5 \\ 10 \end{bmatrix} + s \begin{bmatrix} 1 \\ 1 \\ 5 \end{bmatrix}$

4 a $\mathbf{r} = \begin{bmatrix} 1 \\ 7 \\ 8 \end{bmatrix} + s \begin{bmatrix} 3 \\ 1 \\ 4 \end{bmatrix}; (7, 9, 0), \left(0, \frac{20}{3}, \frac{28}{3}\right), (-20, 0, 36)$ **b** $\mathbf{r} = \begin{bmatrix} 1 \\ 1 \\ 7 \end{bmatrix} + s \begin{bmatrix} 2 \\ 3 \\ -6 \end{bmatrix}; \left(\frac{10}{3}, \frac{9}{2}, 0\right), \left(0, -\frac{1}{2}, 10\right), \left(\frac{1}{3}, 0, 9\right)$

Exercise 5

1 a parallel **b** intersect at $\begin{bmatrix} 1 \\ 2 \\ 0 \end{bmatrix}$ **c** skew **2** $-3, \begin{bmatrix} -1 \\ 3 \\ 4 \end{bmatrix}$

Exercise 6

1 a 30 **b** 0 **c** −1

2 a $7, \frac{1}{3}\sqrt{7}$ **b** $14, \sqrt{\frac{7}{19}}$ **c** $3, \frac{3}{58}\sqrt{58}$ **d** $1, \frac{1}{5}$

3 4 **4 a** $-\sqrt{\frac{7}{34}}$ **b** 33.2°

7 a $10\sqrt{3}$ **b** $41 - 20\sqrt{3}$ **8** 112.2°

9 a i $\begin{bmatrix} 0 \\ 4 \\ -4 \end{bmatrix}$ **ii** $\begin{bmatrix} 4 \\ 4 \\ 0 \end{bmatrix}$ **b** $\hat{A} = 90°, \hat{E} = 19° \hat{H} = 71°$

10 no; **a** and **b** may be perpendicular **11 b** 122° **c** 17.3 sq units
12 a 0° **b** 27.9° **c** 90°

13 b $\left(\frac{8}{3}, \frac{5}{6}, \frac{23}{6}\right)$ **c** $\frac{\sqrt{66}}{6} = 1.35 \,(3\,\text{sf})$

14 b $\left(\frac{1}{13}, \frac{3}{13}, \frac{9}{13}\right)$ **c** $\frac{\sqrt{(3861)}}{13} = 4.78$

Review

1 a $\overrightarrow{AC}$ **b** $\overrightarrow{AD}$
2 a $\mathbf{b} - \mathbf{a}$ **b** $\mathbf{c} - \mathbf{b}$ **c** $\mathbf{c} - \mathbf{a}$
3 $(2, 1, 0)$

4 a 3   **b** $\begin{bmatrix} -2 \\ 2 \\ -1 \end{bmatrix}$ **c** $\begin{bmatrix} 4 \\ 0 \\ \frac{-1}{2} \end{bmatrix}$

5 $\mathbf{r} = \begin{bmatrix} 2 \\ 1 \\ 0 \end{bmatrix} + t \begin{bmatrix} -1 \\ 1 \\ -2 \end{bmatrix}$ **6 a** $\mathbf{r} = \begin{bmatrix} 3 \\ 2 \\ -1 \end{bmatrix} + t \begin{bmatrix} 3 \\ 4 \\ -5 \end{bmatrix}$ **b** $\left(\frac{3}{2}, 0, \frac{3}{2} \right)$

7 b $(11, 5, -12)$ **8** 0   **9** 86.5°   **10** 27.2°

11 a $(0, 0, 3)$ **b** $\sqrt{3}$

Assessment

1 a i $\begin{bmatrix} 2 \\ 2 \\ -2 \end{bmatrix}$ **ii** $\begin{bmatrix} -2 \\ 2 \\ 0 \end{bmatrix}$ **b** 90°

2 a $\mathbf{r} = \begin{bmatrix} 2 \\ 1 \\ 0 \end{bmatrix} + t \begin{bmatrix} 2 \\ -1 \\ 3 \end{bmatrix}$ **b** $\mathbf{r} = \begin{bmatrix} 2 \\ 1 \\ 0 \end{bmatrix} + s \begin{bmatrix} 1 \\ 0 \\ -5 \end{bmatrix}$ **c** 47.0°   **3 a** $\mathbf{r} = \begin{bmatrix} 13 \\ -4 \\ 2 \end{bmatrix} + t \begin{bmatrix} 5 \\ 0 \\ 1 \end{bmatrix}$

4 $\dfrac{\sqrt{1386}}{14} = 2.66$ **5** $a + b - 3c = 6$

6 a 60°   **b** $(15, 6, 2)$ **c** $(11, 0, 0)$ and $(23, 4, -8)$

7 a $\mathbf{r} = \begin{bmatrix} 5 \\ 1 \\ -2 \end{bmatrix} + \lambda \begin{bmatrix} -1 \\ -2 \\ 5 \end{bmatrix}$ or $\mathbf{r} = \begin{bmatrix} 4 \\ -1 \\ 3 \end{bmatrix} + \lambda \begin{bmatrix} -1 \\ -2 \\ 5 \end{bmatrix}$ **b i** $(7, 5, -12)$ **ii** $(-23.5, 0)$

10 The Poisson Distribution

Exercise 1

1 a 0.230   **b** 0.00843   **c** 0.857   **d** 0.143
2 a 0.0302   **b** 0.970   **c** 0.321   **d** 0.463
3 a 0.233   **b** 0.264   **c** 0.494   **d** 0.0342
4 a 0.125   **b** 0.197   **c** 0.928
5 a 0.268   **b** 0.269   **c** 0.964
6 a 0.603   **b** 0.178   **c** 0.620
7 a independently, randomly, constant average rate   **b** 0.155   **c** 0.209
8 a i 0.268   **ii** 0.377   **iii** 0.181   **b** 0.494
9 a 0.0758   **b** 0.223   **c** 0.109
10 a 0.0902   **b** 0.0613   **c** approx. 7
11 a 0.0821   **b** 0.560   **c** 0.0631
12 a 4   **b** 0.433   **c** 0.140
13 a 0.983   **b** 0.184   **c** 0.199
14 a 0.5   **b** 6
15 a 4   **b** 4   **c** 14   **d** 100   **e** 6

Exercise 2

1 a 0.269   **b** 0.527   **c** 0.162   **d** 0.976
2 a 0.333   **b** 0.391   **c** 0.0072   **d** 0.390
3 a 0.378   **b** 0.224   **c** 0.579
4 a randomly, independently
   b i 0.826   **ii** 0.703   **iii** 0.989   **c** 0.658
5 a 0.242   **b** 0.156

6 a 0.244   b 0.464   c 0.815

7 Emails arrive independently.   a 0.158   b 0.463   c 0.835

8 a 0.423   b 0.594   c 0.472

9 a 0.0335   b 0.220   c 0.825

10 a 0.224   b 0.868   c no, visitors don't arrive independently if they are in groups

11 a 0.215   b 0.404   12 a 0.532   b 2.12

13 a i 0.368   ii 0.920

    b i 0.908   ii 0.195   iii 0.785

14 a 0.156   b 0.0244   c 0.430

15 a 15   b 14

16 a i 0.189   ii 0.0902   iii 0.748

    b GRBs random or independent

Exercise 3

1 a i 0.0476   ii 0.0498   b i 0.225   ii 0.224

    c i 0.171   ii 0.168   2 a 0.0537   b 0.0486

3 a i 0.184   ii 0.019   b 0.135   c 0.0498

4 a 0.287   b 0.251   5 a $\frac{1}{36}$ b 0.713

6 0.185

7 a i 0.407   ii 0.0629

    b Binomial: p may not be constant, with more care taken; Poisson suitable as n large and p small

8 a 0.353   b 0.903   9 a 0.468   b 0.703

10 a 0.180   b 0.538   c 0.282

Exercise 4

1 a 0.157   b 0.0203   c 0.854   d 0.0177

2 a 0.0100   b 0.126   3 a 0.329   b 0.0231

4 a 6, 6   b 0.285   5 a 0.744   b 0.191

6 0.282   7 a 0.857   b 0.0248

8 a 0.181   b i 0.917   ii X and Y are independent variables

9 a i 0.910   ii 0.257   b i $W = X + Y$, $W \sim \text{Po}(3)$, $P(W = 2) = 0.224$ ii 0.00252

    c i 12.0   ii 0.0630

10 a 18, 18   b 18, 48

    c i yes; sum of Poisson variables is Poisson   ii no; mean ≠ variance

Review

1 a 0.285   b 0.310   c 0.0688   d 2.45

2 a 0.336   b 0.732

3 a 0.251   b i $Y \sim \text{Po}(13)$ ii 0.0111

4 Fault occurs at a constant average rate, independently, randomly   b i 0.183   ii 0.723

5 Breakdowns occur independently, randomly.   a 0.0959   b 6   6 a 3   b 2.45

7 a 0.174   b 0.468   8 a i 0.251   ii 0.0158

    b i $Y \sim \text{Po}(9)$ ii 0.197   c attacks patients randomly, independently

9 a i 0.216   ii 0.715

    b i $T \sim \text{Po}(9.5)$ ii 0.480   iii 0.111

10 a i 0.126   ii 0.264   iii 0.653

    b 16   c average rate of failure unlikely to be constant, very little use of lights over this period

Assessment

1 a i 0.549   ii 0.0231   b i 0.259   ii 0.0333

2 a $X \sim \text{B}(300, 0.015)$ b i 0.169   ii 0.468

3 a i 0.966   ii 0.230   b 0.143

4 a 0.369   b 0.777   c i 0.122   ii 0.777

5 a 0.103   b i 9   ii 0.413

    c i $T \sim \text{Po}(15)$ ii 0.917   iii 0.000599

11 Continuous Random Variables

Exercise 1

1 a $\frac{3}{8}$ **b** $\frac{7}{8}$ **c** $\frac{13}{32}$ **d** 0

2 a i 0.2 **ii** 0.74 **iii** 0.9
  b i 0.4 **ii** 0.25

3 a $\frac{1}{8}$ **b** $\frac{1}{2}$ **c** 1

4 a $\frac{1}{50}$ **b i** 0.64 **ii** 0.48 **iii** 0

5 a $\frac{3}{56}$ **b i** $\frac{19}{56}$ **ii** $\frac{37}{56}$

6 a $\frac{3}{52}$ **b** $\frac{19}{26}$ **7 a** $\frac{3}{2}$ **b** 0.75

8 a $\frac{1}{4}$ **b** $\frac{1}{4}$ **c** $\frac{5}{16}$ **d** 0.3475

9 a 0.6 **b** 0.1372 **c** 0 **d** 0.85
  e 1

10 a no, area $\neq$ 1 **b** no, f(x) < 0 for some x **c** yes, area = 1, f(x) $\geq$ 0

Exercise 2

1 a $\frac{64}{125}$ **b** $\frac{61}{125}$ **c** $\frac{26}{125}$ **d** 3.97

  e f(x) = $\frac{3}{125}$ x^2, $0 \leq x \leq 5$; f(x) = 0 otherwise

2 a $\frac{4}{9}$ **b** $\frac{5}{9}$ **c** 0

3 a i 0.36 **ii** 0.36 **iii** 0.77 **iv** 1

  b ii 0.5 **c** f(t) = 2 − 2t, $0 \leq t \leq 1$; f(t) = 0 otherwise

4 a 1 **b** $\frac{1}{3}$ **c** $\frac{3}{4}, \frac{7}{8}$

  d f(x) = $\frac{2}{3}$, $0 \leq x \leq 1$; f(x) = $\frac{1}{3}$, $1 \leq x \leq 2$; f(x) = 0 otherwise

5 a F(x) = 0, x < 0; F(x) = $\frac{1}{8}x^3$, $0 \leq x \leq 2$; F(x) = 1, x > 2

  b i $\frac{1}{8}$ **ii** 0.150 **iii** 1.78

6 a f(x)

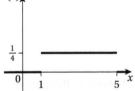

  b F(x) = 0, x < 1; F(x) = $\frac{1}{4}(x-1)$, $1 \leq x \leq 5$; F(x) = 1, x > 5

  c F(x) **d** 0.5

7 a F(t) = 0, t < 1; F(t) = $t - \frac{1}{8}t^2 - \frac{7}{8}$, $1 \leq t \leq 3$; F(t) = 1, t > 3 **b** 0.281

  c 1.76 **d** 0.25

8 a $F(y) = 0, y < 2$; $F(y) = \frac{4}{3}\left(1 - \frac{4}{y^2}\right), 2 \leq y \leq 4$; $F(y) = 1, y > 4$ **b** 2.53

9 a $F(t) = 0, t < 0$; $F(t) = \frac{1}{6}t^5 + \frac{5}{6}t, 0 \leq t \leq 1$; $F(t) = 1, t > 1$ **b** 0.445

10 a $F(x) = 0, x < 0$; $F(x) = \frac{1}{6}x^2, 0 \leq x \leq 2$; $F(x) = 2x - \frac{1}{3}x^2 - 2, 2 \leq x \leq 3$; $F(x) = 1, x > 3$

b 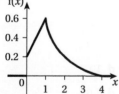   **c i** $\frac{1}{6}$ **ii** 0.12

iii 0.12   **iv** 0.713   **v** 1.73

11 a $F(x) = 0, x < 0$; $F(x) = \frac{1}{4}x, 0 \leq x \leq 2$; $F(x) = 1 + \frac{1}{4}x^2 - \frac{3}{4}x, 2 \leq x \leq 3$; $F(x) = 1, x > 3$

b i $\frac{1}{4}$ **ii** 0.5475   **iii** 2   **iv** 0

12 a i $\frac{1}{4}$ **ii**

b i $f(x) = 1 - \frac{1}{2}x, 0 \leq x \leq 2$; $f(x) = 0$ otherwise   **ii**

13 a 1   **b** $f(x) = x, 0 \leq x \leq 1$; $f(x) = 2 - x, 1 \leq x \leq 2$; $f(x) = 0$, otherwise   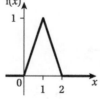   **c** 1

14 a   **b ii** 0.4   **iii** 0.5   **iv** $\frac{1}{2}(\sqrt{6} - 1)$

Exercise 3

1 a $4\frac{1}{3}$ **b** 11

2 a $2\frac{4}{15}$ **b** 6.4   **c** $9\frac{13}{15}$ **3** 0.5625

4 a

b 0

c 5

5 $2\frac{23}{35}$

6 a f(x)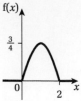

b 1

7 a 1.6   **b** 0.4096   **c** 0.4096 < 0.5 so μ < median

8 b 2.5   **c** 0.5   **d** 0.25

9 a 0.625   **b** 5

10 a $\frac{1}{8}$ **b** 5   **c** $\frac{1}{9}$

11 a $\frac{49}{34}$ **b** 9   **12** 2.875

13 b i $3p - 4q = 1$ **ii** $p = 1, q = 0.5$

Exercise 4

1 a $2\frac{8}{15}$ **b i** 6.5   **ii** 0.0822   **iii** 0.74

2 b 0.186

3 a i $\frac{45}{28}$ **ii** $\frac{93}{35}$ **iii** 0.272

  b i $\frac{9}{14}$ **ii** $\frac{3}{7}$ **iii** 0.124

4 a $\frac{1}{5}$ **b** 0.5   **c** $2\frac{1}{12}$ **d** 1.44

5 a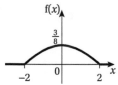

b 0   **c** 0.8   **d** 0, 2.4

6 a $\frac{1}{2}$ **b** $2\frac{1}{3}$ **c** 0.869

7 a $\frac{41}{12}$ **b** 1.02   **c** 25.38

8 a 16   **b** 416   **9** $\frac{17}{32}$, 0.324

10 a 24   **b** 22   **c** $66\frac{1}{4}$ **d** 19.7

11 a 5   **b** 34   **c** 8   **d** 7.68

12 a 8, 0.949   **b** 12, 0.6   **c** 22, 1.41

13 a $\frac{4}{3}, \frac{8}{9}$ **b** $\frac{8}{3}, \frac{8}{9}$ **c i** 4   **ii** $\frac{16}{9}$

  d i $-\frac{4}{3}$ **ii** $\frac{272}{9}$

Review

1 a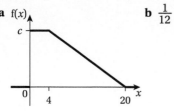
f(x), c, 0, 4, 20, x

b $\frac{1}{12}$

c $\frac{71}{96}$

2 b 0.4   **c** 3.2, 1.6

3 a $\frac{1}{2}c$ **b** $\frac{1}{12}c^2$ **c** 1.73

4 a $f(x) = \frac{1}{9}, -4 \le x \le 5; f(x) = 0,$ otherwise

b

f(x), $\frac{1}{9}$, −4, 0, 5, x

c $\frac{1}{3}$ **d** 0.5, 6.75

5 a

f(x), $\frac{3}{32}$, 0, $\frac{1}{2}$, 11, x

c i $5\frac{1}{3}$ **ii** $5\frac{2}{3}$ **d** $\frac{1}{3}$

6 a

f(x), $\frac{1}{2}$, $\frac{3}{8}$, 0, 1, 2, 3, x

b $\frac{7}{8}$ **c ii** $4\sqrt{3} - 5 = 1.93$

7 a 1, 0.5   **c** $\frac{7}{54}$ **d ii** 1.619

8 a 5   **b** 3.25   **c** 9

Assessment

1 b $\frac{4}{3}$, 0.471   **c** 0.25

2 a

f(x), k, 0, 40, 60, x

c $\frac{23}{32}$

3 a ii $F(x) = 0, x < 0; F(x) = \frac{1}{12}\left(\frac{x^3}{3} + x\right), 0 \le x \le 3; F(x) = 1, x > 3$

iii

F(x), 1, 0, 3, x

iv $\frac{11}{18}$ **b** 2.0625

4 a 0.794   **b** $f(x) = 3x^2, 0 \le x \le 1; f(x) = 0,$ otherwise

c i 0.75   **ii** 0.0781   **d i** 0.0375   **ii** 0.668

5 a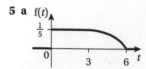

b 0   **c** 0.4   **e** 0.01

6 a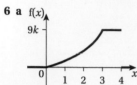

c i 3   **ii** 2.38

7 a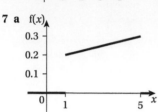

b $3\frac{2}{15}$ **d i** $\frac{21}{40}$ **e** 3.198

8 a 34   **b** 5   **c** 200   **d** 7.68

12 The Exponential Distribution

Exercise 1

1 a 0.368   **b** 0.503   **c** 0.492   **d** 0
 e 2   **f** 4
2 a 0.183   **b** 0.549   **c** 0.632   **d** 1
3 a $\frac{1}{8}$ **b** 0.0866   **c** 0.132

4 a $\frac{3}{2}$ **b** 0.00248   **c** 0.632   **d** 0.462
 e 2.00
5 a 0.177   **b** 0.0719   **c** $f(t) = 4e^{-4t}, t \geq 0$ **d** $\frac{1}{16}$

6 a i 0.699   **ii** 0.111   **iii** 0.449   **iv** 1.73

  v 0.713, 3.466, 2.747   **b i** $f(x) = 0.4\,e^{-0.4x}, x \geq 0$ **ii** 2.5

Exercise 2

1 a 0.0631   **b** Exp(10)   **c i** 0.0357   **ii** 0.107
2 a i $\frac{5}{8}$ **ii** 0.105   **b i** 1.6 hrs = 1 hr 36 mins   **ii** $f(t) = \frac{5}{8}e^{-\frac{5}{8}t}, t \geq 0$
 c i 0   **ii** 0.445
3 a 0.189   **b** 0.487   **c** 0.189   **d** 0.487
4 a 1000   **b** 0.741   **c** 0.632
5 a i 0.472   **ii** 0.632   **b** 0.0933
6 b i 0.181   **ii** 0.819

Review

1 a i 0.528   **ii** 0.32   **iii** 0.417
 b i 0.368   **ii** 64
2 a $\frac{1}{4}$ **b** 4   **c** $F(y) = 0, y < 0; F(y) = 1 - e^{-\frac{1}{4}y}, y \geq 0$

 d i 7.59   **ii** 0.650
3 a $f(x) = e^{-x}, x \geq 0, \lambda = 1$ **b** 0.0855   **c** 0.693   **d** 0.288
4 a 0.551   **b** 0.359
5 a i 1.25   **ii** 1250 hours   **iii** 0.551
 b i 0.670   **ii** 0.699
6 a 8   **b** 0.382   **c** 0.0564

7 a i 0.195   **ii** 0.424   **b i** $T \sim \text{Exp}(0.4)$ **ii** 0.798

   iii 0.0724   **iv** 0.135   **8 b** $\dfrac{\ln\left(\frac{4}{3}\right)}{\lambda}, \dfrac{\ln 4}{\lambda}$

Assessment

1 a i 0.551   **ii** 0.314   **2 a** 0.393   **b** 0.246

3 a $F(t) = 1 - e^{-5t}, \; t \geq 0$ **b** 0.368   **c** $\frac{1}{5}\ln 20 = 0.599$

4 a 50 hours   **b** 0.148   **c** 0.449
   d makes no difference (no memory)   **e i** 0.8   **ii** 0.449

5 a 16   **b i** 0.287   **ii** 0.133   **c** 0.287

   d $\text{Po}\left(\dfrac{5}{16}\right)$

6 a 1   **b** 0.135   **c** 0.632
   d $p = 0.000045$, extremely rare event, casts doubt on model

13 The Normal Distribution

Some answers have been given to more than 3 significant figures as a check that you are reading the tables correctly.

Exercise 1

1 a 0.80785   **b** 0.80785   **c** 0.19215   **d** 0.19215
2 a 0.03593   **b** 0.25785   **c** 0.99305   **d** 0.91309
   e 0.00494   **f** 0.99111   **g** 0.96856   **h** 0.23576
3 a 0.04406   **b** 0.81859   **c** 0.30233   **d** 0.13362
4 a 0.16965   **b** 0.54649   **c** 0.36209   **d** 0.45818
   e 0.47982   **f** 0.96248   **g** 0.03412   **h** 0.90781
   i 0.27290   **j** 0.03063   **5 a** 0.92492   **b** 0.90106
6 a 0.9   **b** 0.7   **7 a** 0.55   **b** 0.15
8 a 0.9   **b** 0.1

Exercise 2

1 a 0.05480   **b** 0.14476   **c** 0.9545
2 0.06681   **b** 0.59871   **c** 0.17466   **d** 0
   e 0.15731
3 a 0.71904   **b** 0.06178   **c** 0   **d** 0.46214   **e** 0.0455
4 a 0.01072   **b** 0.98574   **c** 1
5 a 0.30153   **b** 0.52392   **c** 0.37877
6 a i 0.841   **ii** 0.926   **b** 206   **7** 0.0000386
8 a 0.655   **b** 0.117   **c** $8.2 \approx 8$
9 a 0.106   **b** 0.773   **c** 0.668
10 a 0.00379   **b** 0.908   **c** 0.249   **d** 0.0928
11 0.159, 0.775, 0.067   **12 a** 0.866   **b** 0.988   **c** 0.0027

Exercise 3

1 a 0.0251   **b** 0.8064   **c** −1.8957   **d** −0.4677
   e −0.5534   **f** 1.0364
2 a 1.9431   **b** −0.7063   **c** −0.9154   **d** 0.7063
   e 1.0364

Exercise 4

1 a 73.739   **b** 4.6504   **c** 190.6712   **d** 1.4776
2 a 176.6   **b** 166.7   **3 a** 625.6   **b** 567.1
4 a (384.32, 415.68)   **b** 10.8   **5 a** 0.383   **b** 96.7
6 (128.9, 191.1)

Exercise 5

1 a 54.39   b 129.04   2 35.45   b 0.087
3 a 1.53   b 0.00054   4 1.75   5 2.97
6 a 8.32   b 35.9%   7 3040
8 a 0.507   b 0.00258   9 a 16.2, 1.64   b 0.279
10 5.03, 0.565   11 a 6.99, 0.105   b 0.0105   12 39.5, 5.32
13 0.203   14 535.1, 12.2   15 4.46

Exercise 6

1 a 50, 4.61   b $X + Y \sim N(50, 21.25)$ c i 0.985   ii 0.135
2 a $T \sim N(530, 13.25)$ b i 0.915   ii 0.277
3 a 243, 122.5   b 0.0618   c 261.2
4 a 0.630   b 0.106
5 a 0.0207   b i 0.0287   ii 0.0197   iii 0.624
6 a 0.0274   b 0.00193
7 0.0314   8 a 0.5   b 0.885
9 a 0.111   b 0.0418

Review

1 a 0.977   b 0.749   c 0.0228   d 0.00043
   e 0
2 a 0.0926   b 0.344   c 0.387
3 a i 0.977   ii 0.886   b 0.561
4 a i 0.00357   ii 1   b 0.110
5 a i 107.0   ii 110.3   iii 95.85   b 5.396
6 a 70.24   b 0.434
7 a 0.927   b 1.32   c 2% approx
8 a 828.45   b 0.615
9 a i 0.841   ii 0.0668   b 108.2
10 a 109.0   b 0.429
11 a i 0.0591   ii 0.0144   iii 0.927   iv 0
   b ii 3.83   12 100, 5   13 0.933
14 0.0239   b 0.560   c 0.286   d 0
15 a 0.174   b 0.613

Assessment

1 a 0.945   b 0.992   c 0.576   d 1
2 a i 0.734   ii 0.234   b 63.7
3 a i 0.970   ii 0.894   iii 0.864   b 97.5
4 a i 0.788   ii 0.885   b 0.0574   c 1.94
5 a i 0.894   ii 0.970   iii 0.864   b 3.42
6 a i 0.0764   ii 0.437   b ii 304.4, 2.84   7 b 36.7, 4.1
8 0.206   9 0.928
10 a 9.6, 0.522   b i 0.0174   ii 0.221

14 Estimation

Exercise 1

1 a 10, $34\frac{2}{9}$ b 15, 43.1
2 Assumption: 48.875, 6.98   b 1.69, 0.000008   c 22.8, 1.81
3 0.5, 1.43   4 Sample is random; 69, 16.2   5 76, 12.1
6 1122, 19.6   7 175.25, 10.2   8 a 2.543, 2.58   b yes; mean ≈ variance ≈ 2.5

Exercise 2

1 a 0.0174   b 0.645
2 Assumption: Random sample of all carrots sold by wholesaler;
   a 0.908   b 0.962   c 0.800

3 a 0.0823   **b** 0.0836   **4 a** 0.391   **b** 6 approx
5 a i 0.239   **ii** 0.0793   **iii** 0.0126   **b** 109
6 251   **7** 12.2   **8** 17.2
9 a i 0.894   **ii** 0.788   **b i** 0.168   **ii** 0.159
10 a i 0.977   **ii** 0.576   **b** 137.41   **c** 0.788

Exercise 3

1 a 0.106   **b** 0.311   **2 a** 0.722   **b** 0.117

3 a 6, 4.2   **b** $\bar{X} \sim N\left(6, \dfrac{4.2}{50}\right)$ approx.   **c i** 0.992   **ii** 0.981

4 0.00368   **5 a i** 0   **ii** 0.4

   b i 0, 0.008   **ii** $\bar{X} \sim N(0, 0.008)$ approx.   **iii** 0.487

6 a $\bar{X} \sim N\left(5.8, \dfrac{5.8}{65}\right)$ approx.   **b** 0.943   **7** 0.949

8 a $\bar{Y} \sim N\left(20, \dfrac{3^2}{80}\right)$ approx.   **b i** 0.309   **ii** 0

Review

1 9.49, 0.0129   **2** 375.31, 8.29   **3 a** 0.0197   **b** 0.348
4 a 0.922   **b** yes; distribution of masses is unknown

5 a $\bar{W} \sim N\left(240, \dfrac{25^2}{150}\right)$ approx.

   b i 0.993   **ii** 0.312   **iii** 0.724   **iv** 0.673
6 a 0.919   **b** yes; distribution of X not normal   **7** 0.953
8 a $\dfrac{3}{32}$ **b** 2, 0.8   **c i** 0   **ii** 0.922

9 a i 0.933   **ii** 0.841   **iii** 0.775   **b** 0.159
10 a i 0.549   **ii** 0.0231   **b** 0.606, 1.104
   c i The observed proportion of 3 or more **P** symbols is much larger than the probability predicted by the Poisson model
    ii In a Poisson model, mean = variance, whereas the observed mean is much less than the observed variance
    iii The distribution of car parks not random; they are more likely to be near shopping centres and attractions
11 a i 0.841   **ii** 0.0668   **b** 108.2
   c i 108.5, 1.6   **ii** 0.118
12 a i 0.970   **ii** 0.939   **iii** 1
   b 514.5   **c i** 0.953   **ii** 0.937

Assessment

1 205.16, 9.22   **2** 10.5, 1.52
3 a i 0.818   **ii** 0.637   **iii** 0
   b 68.8   **c i** 0.018   **ii** 0.979
4 a i 0.894   **ii** 0.599
   b i 0   **ii** 0.00418   **iii** 0.271
5 a i 0.726   **ii** 0.542   **iii** 73.6
   b i $\mu - 2\sigma$ gives negative time   **ii** Use CLT since n large   **iii** 0.239
6 a i 0.631   **ii** 0.952   **b ii** 0.156   **c** 0.110
7 a i 0.189   **ii** 0.762   **iii** 0.653
   b 0.434   **c i** 9.2, 9.29   **ii** mean $\approx$ variance, Poisson may be appropriate
8 a 4.55, 4.55   **b** 0.333   **c** 0.483   **d** 0.984
9 a $\dfrac{1}{18}$ **b** 1.25, 0.829   **c** 0.0268

15 Hypothesis Tests

Exercise 1

1 a $P(X \le 1) < 5\%$; reject H_0 **b** $P(X \ge 6) > 5\%$; do not reject H_0 **c** $P(X \ge 5) > 1\%$; do not reject H_0
   d $P(X \le 2) > 1\%$, do not reject H_0 **e** $P(X \ge 12) < 3\%$; reject H_0 **f** $P(X \le 8) > 10\%$; do not reject H_0

2 0.1071 > 5%, not enough evidence of reduction

3 a Do not reject H_0, since 0.102 > 0.04

   b Reject H_0, since 0.0233 < 0.04

4 $P(X \geq 8) > 5\%$, not enough evidence $p > 0.5$, she could have obtained score by guessing

5 Assume p constant for each attempt; evidence that percentage decreased, since $P(X \leq 2) < 10\%$

6 $H_0: p = 0.15$, $H_1: p < 0.15$, reject H_0, since $P(X \leq 1) < 5\%$, evidence of success

Exercise 2

1 a $P(X \leq 4) < 5\%$, reject H_0 **b** $P(X \leq 2) < 5\%$, reject H_0

   c $P(X \leq 3) > 4\%$, do not reject H_0 **d** $P(X \geq 4) > 3\%$, do not reject H_0

2 Assume cases of medical condition occur independently; 0.054 > 5%, not enough evidence of change

3 0.062 > 5%, not enough evidence of change

4 a $X \sim Po(3)$; $P(X \geq 5) > 5\%$, not enough evidence of increase

   b $X \sim Po(9)$; $P(X \leq 3) < 5\%$, evidence of decrease

5 $H_0: \lambda = 7.5$, $H_1: \lambda < 7.5$, $P(X \leq 3) > 5\%$, not enough evidence of reduction

6 $X \geq 18$, reject H_0 if, at least 18 customers arrive

Exercise 3

1 a Do not reject H_0; not enough evidence to support manager's claim

   b i Upholding manager's claim that proportion is less than 0.35 when in fact it is 0.35

     ii Accepting proportion is 0.35 when in fact it is less than 0.35

2 a $X \leq 1$, $X \geq 9$ **b** 0.0214

3 3; 0.0212

4 a $H_0: p = 0.45$; $H_1: p < 0.45$

   b i $X \leq 17$ **ii** 0.0765

   c Reject Rachel's claim in favour of Lily's suspicion

   d Yes, not enough evidence to reject Rachel's claim

5 a $H_0: p = \frac{1}{6}$, $H_1: p > \frac{1}{6}$ **b** Not enough evidence of bias, since 0.0697 > 0.05   **c** Accepted H_0, Type II

6 a $P(X \leq 2) = 0.0620$; not enough evidence that mean has decreased

   b Concluding that the mean number of breakdowns has decreased when in fact it has not

Exercise 4

1 a $z = 1.845$, reject H_0 **b** $z = 2.5$, reject H_0 **c** $z = -2.778$, reject H_0 **d** $z = -1.095$, do not reject H_0

2 a $z = -0.943$, not enough evidence of decrease

   b i Concluding that the mean volume has decreased when in fact it has not

     ii Concluding that the mean volume is 335 ml when in fact it has decreased

3 $z = 3.069$, evidence that mean duration has increased

4 Not enough evidence that $\mu < 103.5$

5 a $z = -1.66$, not enough evidence of change in mean   **b** Since H_0 is accepted and H_0 is true, neither error has been made

6 a $z = -1.878$, evidence of improvement in times

   b Concluding mean running times have decreased when they have not

7 a 33.95   **b** Evidence that mean speed is greater than 30 mph, since sample mean is in critical region

Exercise 5

1 a $z = 1.81$, reject H_0 **b** $z = -2.45$, reject H_0 **c** $z = -2.03$, reject H_0

   d $z = 1.54$, do not reject H_0

2 a $z = 2.946$, evidence mean greater   **b** Concluding mean greater than 100 when it is not

3 $z = 1.909$, yes

4 $z = 1.70$, do not reject H_0, not enough evidence that mean has changed, so Alan's belief is not supported

5 a $z = 1.778$, do not reject H_0 **b** $z = 1.778$, reject H_0 **c** $z = -1.428$, reject H_0

   d $z = -2.487$, do not reject H_0 **6 a** 27.4   **b** $z = 1.99$, not enough evidence

   c Concluding mean is 15 minutes when it is not

7 a 1.70   **b** $z = -1.95$, evidence that mean time has decreased

   c ...and the Central Limit Theorem states that the distribution of the sample mean is approximately normal when the sample size is large.

8 $z = 0.983$, accept mean is zero

9 a 21.25
  b $z = 0.99$, not enough evidence to support manufacturer's suspicion
  c CLT needed for distribution of sample mean, necessary as distribution of X is unknown
10 $z = -2.47$, reject H_0; evidence to support David's claim

Exercise 6

1 a $t = 0.909$, do not reject H_0 **b** $t = -1.89$, do not reject H_0 **c** $t = 2.15$, reject H_0
  d $t = -3.07$, do not reject H_0 **2** $t = 2.828$, evidence of improved times
3 a $t = -3.54$, underweight   **b** $t = -3.2$, underweight
4 $t = -1.16$, not enough evidence mean < 1.6
5 $t = -2.48$, evidence mean less than stated amount
6 $t = 2.284$, evidence mean greater than 4.3
7 a Random sample, normal variable, $t = -3.23$, reject that mean is 3.1   **b** Type I, since H_0 is rejected
8 $t = 1.845$, mean not 13.65 (evidence that greater)
9 $t = 0.85$, do not reject H_0; not enough evidence that mean time exceeds 115 minutes
10 a $t = 2.60$, do not reject H_0; not enough evidence that teacher is understating time to complete homework assignments
  b Assume times are normally distributed
11 a $t = 2.36$, reject H_0; evidence that mean length of male students' feet is not equal to 30.5 cm, in fact that it is less than 30.5
  b Assuming sample is representative, evidence that the mean length a male student's foot is less than a foot, However, the sample shows that some male students' feet are longer than 30.5 cm

Review

1 $P(X \geq 8) > 5\%$, not enough evidence to support C.A's belief
2 $P(X \leq 2) > 10\%$, not enough evidence of improvement
3 a $z = 3.70$, reject H_0, evidence that mean length of components on a Monday was not equal to 135 cm
  b Rejecting H_0 when H_0 is true: concluding that the mean length of components was not 135 cm when in fact it was 135 cm
4 a $z = 0.571$, do not reject H_0, not enough evidence that mean time for ambulance to arrive is not 19 minutes
  b i $H_0: \mu = 19$, H1: $\mu < 19$ **ii** -1.6449
  iii Not enough evidence that the mean time for ambulance to arrive is less than 19 minutes
  c No significant evidence that target has been achieved; indeed, since sample mean is 19.66, no evidence at all; however not significance evidence that target has not been achieved
5 a $z = -2.30$, reject H_0; evidence to suggest students may be underachieving
  b Rejecting H_0 when H_0 is true; concluding students are underachieving when in fact they are not
6 a Random sample, $P(X \leq 13) < 10\%$, evidence that the effectiveness has been overstated
  b Yes: since $P(X \leq 13) > 5\%$, there is not enough evidence of overstatement
7 a No: since $P(X \leq 5) > 5\%$, there is not enough evidence of decrease in the mean number of accidents in a week
  b Rejecting H_0 when H_0 is true; concluding that the mean number of accidents per week has decreased when it has not
8 a $z = 2.42$, do not reject H_0; not enough evidence to doubt claim that mean age is 35 years
  b Accepting H_0 when H_0 is false: accepting mean age is 35 years when it is not 35 years
9 $t = 2.03$, reject H_0, evidence to support doctor's belief that cholesterol level is higher than the management's claim of 3.8. Assume cholesterol levels are normally distributed.
10 a $t = -2.14$, reject H_0; evidence to support members' belief that the batteries last less than 100 hrs
  b $z = 0.99$, do not reject H_0; there is enough evidence to support the manufacturer's belief
11 a $z = 1.07$, do not reject H_0, there is not enough evidence that mean waiting time for calls made to Northgas exceeds 90 minutes
  b $z = 3.07$, reject H_0, there is evidence that mean waiting time for calls made to Southgas exceeds 90 minutes
  c i Southgas' sample mean was closer to the null hypothesis, yet it was the one that was rejected
  ii larger sample in part b makes any difference from 90 more likely to be detected
12 a i 0.125   **ii** 0.09012   **iii** 0.11384
  b $t = 0.74$, do not reject H_0, not enough evidence that mean amount spent is over £5

Assessment

1 a $P(X \geq 13) > 5\%$, do not reject H_0, not enough evidence of difference from 35 per cent
  b Not a random sample; buying tickets may not be independent as people travelling together may have bought tickets in the same way
2 $z = -3.74$, reject H_0; evidence mean weight of loaves is not equal to (less than) 1000 g
3 $P(X \leq 7) > 10\%$, do not reject H_0, not enough evidence that there has been a reduction in the mean number of complaints per day

4 $t = -0.420$, do not reject H_0; accept Aaron's claim, not enough evidence that mean weight of pears is not 140 g

5 a Sample must be random

b $z = 1.6853$, do not reject H_0, not enough evidence that mean power of batch is different from 9.0 watts

c i $H_1: \mu > 9.0$ **ii** 1.6449

   iii reject H_0, evidence that mean power of batch is more than 9.0 watts

6 $z = -1.929$, do reject H_0; reject the claim, not enough evidence that mean has changed from 85.9

7 a $t = -2.65$, reject H_0; no evidence to support producer's claim

b Have rejected H_0 when H_0 is true: Type I

8 a $z = 1.734$, reject H_0, evidence mean has changed

b $z = 1.89$, do not reject H_0, not enough evidence that mean has increased

c Since sample size is large, CLT can be used so sample mean is approximately normally distributed

d The test must be the that accepts H_0, so test in part b

9 a $P(X \geq 6) = > 1\%$

b i $P(X \geq 18) > 1\%$, do not reject H_0, not enough evidence to support her claim

  ii Random sample, independent trials, p constant for each shot

Index